动物微量元素硒营养

弓剑 著

中国原子能出版社

图书在版编目(CIP)数据

动物微量元素硒营养/弓剑著.--北京:中国原子能出版社,2020.7

ISBN 978-7-5221-0712-7

Ⅰ.①动… Ⅱ.①弓… Ⅲ.①动物营养—微量元素营养 Ⅳ.①S816

中国版本图书馆 CIP 数据核字(2020)第 134690 号

内 容 简 介

硒是人和动物的必需微量元素,在生产、生殖、抗氧化、免疫及疾病预防等方面具有重要的生物学功能。本书收集了国内外的有关最新研究成果,也包括作者自己的相关系列研究工作,系统地研究了硒在动物营养科学中的应用,主要内容包括:微量元素硒概述、硒蛋白的生物合成及其生物学功能、硒在家禽生产中的应用、硒在猪生产中的应用、硒与反刍动物营养等。本书论述严谨,结构合理,条理清晰,内容丰富新颖,是一本值得学习研究的著作。

动物微量元素硒营养

出版发行	中国原子能出版社(北京市海淀区阜成路 43 号 100048)
责任编辑	张 琳
责任校对	冯莲凤
印 刷	北京亚吉飞数码科技有限公司
经 销	全国新华书店
开 本	787mm×1092mm 1/16
印 张	16
字 数	287 千字
版 次	2021 年 3 月第 1 版 2021 年 3 月第 1 次印刷
书 号	ISBN 978-7-5221-0712-7 定 价 78.00 元

网址:http://www.aep.com.cn E-mail:atomep123@126.com

发行电话:010－68452845

前　　言

硒(Selenium)是人和动物的必需微量元素，在生产、生殖、抗氧化、免疫及疾病预防等方面具有重要的生物学功能。1817 年，瑞典化学家 Berzelius 在焙烧黄铁矿制硫酸时发现了这种元素。在之后长达 100 年的时间里，硒一直被认为是一种对人和动物体有毒的元素。1957 年，Schwarz 和 Folz 两位科学家以小鼠为研究对象首次证实缺硒会影响其健康状况，之后在其他动物以及人上也相继证实了硒的营养重要性。1973 年，世界卫生组织宣布硒是人和动物的必需微量元素。在人和动物体内，硒主要以硒蛋白的形式发挥其广泛的生物学功能。时至今日，在人和动物体内已发现 25 种具有生物活性的硒蛋白(selenoprotein)，这些蛋白在动物体内也几乎全部以硒蛋白的形式存在。人们对这些硒蛋白生物学功能以及作用机制的揭示也取得了显著的成就，对硒与动物生产、生殖、健康、抗氧化、免疫等方面的研究也积累了丰硕的成果。

为了系统地介绍硒在动物营养科学中的应用，以推动这个学科更广泛深入地发展，作者编写了这部著作。本书收集了国内外的有关最新研究成果，也包括作者自己的相关系列研究工作。全书共五章。第一章为微量元素硒概述，阐述了硒的发现与存在、硒的营养及毒性作用、动物对硒的摄取、硒的添加形式与剂量以及硒在动物体内的消化、吸收和代谢。第二章为硒蛋白的生物合成及其生物学功能，阐述了硒蛋白的生物合成过程以及已知在哺乳动物体内发现的硒蛋白的主要生物学功能。第三章为硒与家禽营养，介绍了硒对种公鸡精子质量的影响、种母鸡硒营养及其向鸡胚的转移以及对新生雏鸡的影响、硒在商品蛋鸡以及肉鸡生产中的应用、有机硒在家禽营养中的应用优势以及硒对家禽健康和免疫的影响。第四章为硒在猪生产中的应用，阐述了硒对种公猪精子质量的影响、母猪硒营养及其向胚胎和猪乳的转移以及对新生仔猪的影响、硒在生长肥育猪生产中的应用、有机硒在猪营养中的应用优势以及硒对猪健康的影响。第五章为硒在反刍动物生产中的应用，阐述了牛对硒的需要量、硒缺乏和添加对牛抗氧化功能、免疫功能、生产性能、繁殖性能以及健康的影响，重点阐述了围产期奶牛硒营养。

本书可供高等院校、科研院所相关专业的教师、科研人员、研究生以及相关工作者参考使用。

本书的出版得到国家自然科学基金项目(项目号:31560644,31960670)的资金资助,特此致谢!

由于作者水平有限,书中缺点和不足之处在所难免,敬请读者批评指正。

弓 剑

2020年5月于呼和浩特

目　　录

第一章　微量元素硒概述 …… 1

第一节　硒的发现 …… 1
第二节　硒的理化性质 …… 1
第三节　环境中的硒 …… 4
第四节　硒的必要性和对动物的毒性 …… 14
第五节　人和动物对硒的需要量 …… 19
第六节　硒在动物体内的吸收、代谢和分布 …… 20
第七节　动物硒营养状况评价 …… 25
参考文献 …… 32

第二章　硒蛋白的生物合成及其生物学功能 …… 45

第一节　硒蛋白的生物合成 …… 45
第二节　硒蛋白及其生物学功能 …… 57
参考文献 …… 94

第三章　硒在家禽生产中的应用 …… 122

第一节　硒与种公鸡精子质量 …… 123
第二节　硒与种母鸡 …… 125
第三节　硒在商品蛋鸡生产中的应用 …… 129
第四节　硒在肉鸡生产中的应用 …… 131
第五节　富硒鸡蛋和肉 …… 135
第六节　硒与家禽免疫 …… 137
第七节　硒与家禽健康 …… 144
参考文献 …… 149

第四章　硒在猪生产中的应用 …… 168

第一节　硒需要量 …… 169
第二节　公猪硒营养与精子质量 …… 170
第三节　母猪硒营养对胎儿和新生仔猪的影响 …… 174

第四节　断奶仔猪硒营养 …… 176
第五节　硒与猪的健康 …… 177
第六节　有机硒在猪营养中的应用优势 …… 179
参考文献 …… 181

第五章　硒与反刍动物营养 …… 189

第一节　硒需要量 …… 189
第二节　硒缺乏 …… 190
第三节　硒添加 …… 194
第四节　硒与围产期奶牛 …… 202
参考文献 …… 221

第一章　微量元素硒概述

硒(Selenium,Se)是人和动物的必需微量元素,在生产、生殖、抗氧化、免疫及疾病预防等方面具有重要的生物学功能。硒缺乏不仅会影响动物的生产性能和生殖机能,严重硒缺乏还会引起疾病的发生,然而过量的硒摄入也会导致中毒,而且从缺乏到中毒的剂量范围相对较窄。本章主要阐述了硒的发现和理化性质、硒在环境中的分布、硒的毒性和必要性、硒在动物体内的吸收、代谢、组织分布和排泄以及机体硒营养状况的评价。

第一节　硒的发现

1817 年,瑞典化学家 Berzelius 在焙烧黄铁矿制硫酸时,发现在铅室的底部以及壁上附着有红色的残泥,残泥接触皮肤会导致皮肤起泡,对残泥进行加热会发出一股类似萝卜腐烂时的臭味,这种味道与碲(Tellurium,Te)化合物燃烧时发出的味道相似,进一步分析发现残泥中并不包含碲元素,由此推测应该含有一种与碲元素性质非常相似的新的元素。1818 年,Berzelius 在他的正式出版物中详细描述了这种新的元素的发现过程,并参照碲(拉丁语中指地球)的命名,将该元素取名为硒(拉丁语中指月亮)。

第二节　硒的理化性质

硒在元素周期表中位于第 4 周期、第 6 主族,是介于金属与非金属之间的半金属或准金属元素,其很多理化性质既有金属的一面,也有非金属的一面。

一、硒的物理性质

硒的原子和物理性质见表 1-1。

表 1-1　硒的原子及物理性质

原子序数	34
原子质量	78.96
电子结构	$[Ar]3d^{10}4s^2 4p^4$
原子半径(pm)	103
共价半径(pm)	116
范德华半径(pm)	190
离子半径(pm)	198(−2)、50(+4)、42(+6)
原子体积(cm^3/mol)	无定形硒:18.55、单斜晶体硒:17.72、六方晶体硒:16.30～16.50
M-M 结合能(kcal/mol)	44
M-H 结合能(kcal/mol)	67
电离能(kJ/mol)	第一:941.0、第二:2 045.0、第三:2 973.7
电子亲和能(eV)	−4.21
电负性	2.55
熔点	494 K,221 ℃
沸点	958 K,685 ℃

单质硒呈红色、黑色或灰色,粉末状,有金属光泽。与硫和碲相似,硒存在多种同素异形体,包括无定形硒以及三种结晶形硒。无定形硒常被描述为两种形式,一种为玻璃状硒,一种为砖红色硒,其微观结构相同,只是外观不同。无定形硒没有固定熔点,当环境温度高于 230 ℃时变为液态,当温度降低到 80 ℃时黏度增加,温度继续降低时黏度又降低。元素硒在 31～230 ℃时呈玻璃状,低于 31 ℃时是一种硬而脆的玻璃。结晶形硒主要为 α 单斜晶体硒、β 单斜晶体硒和六方晶体硒。单斜晶体硒的单晶含有 Se_8 环,α 单斜晶体硒由平面六边形和多边形构成,称为红硒,β 单斜晶体硒由针形或棱柱形构成,称为暗红硒。六方晶体硒亦称为金属样灰硒、黑硒、γ 晶体硒或三角晶体硒,由螺旋状 Se_n 链构成,是热力学上最稳定的硒的同素异形体。无定形硒在 70～210 ℃范围内可转变为六方晶体硒,温度高于 110 ℃时,两种单斜晶体硒可转变为六方晶体硒。

硒有 6 种天然形成的稳定同位素:^{74}Se(0.89%)、^{76}Se(9.37%)、^{77}Se(7.63%)、^{78}Se(23.77%)、^{80}Se(49.61%)和^{82}Se(8.73%),稳定同位素可用

于研究硒的生物学作用和生物利用率。硒还存在不稳定同位素，大约有 24 种，半衰期从 20 毫秒到 295 000 年不等，硒放射性同位素主要用于医药及工业溴放射性同位素的生产。

二、硒的化学性质

硒元素的很多化学性质与硫元素相似，可与高电负性的氧、氯、氟等元素以及高电正性的碱金属元素形成稳定的化合物，在生物体内，硒还可以合成多种有机硒化合物[1,2]。硒在环境介质中存在 5 种氧化态（−2、0、+2、+4 和+6）。

以−2 价存在的硒化合物包括硒化氢（H_2Se）和金属硒化物。H_2Se 是一种具恶臭气味的无色气体，金属硒化物水解或元素硒在空气中加热（400 ℃）可生成 H_2Se，H_2Se 的毒性较硫化氢（H_2S）强，是已知毒性最强的化合物之一。H_2Se 易溶于水，尽管其溶解度比 H_2S 低，但其水溶液的酸性比 H_2S 强，呈中强酸性。

以 Se^{4+} 形式存在的硒化合物主要为二氧化硒（SeO_2）、亚硒酸（H_2SO_3）和亚硒酸盐（SO_3^{2-}）。元素硒在空气中燃烧可以生成 SeO_2，发出蓝色火焰，散发出烂萝卜的臭味。元素硒被浓硝酸氧化也可生成 SeO_2。富硒化石燃料的燃烧是 SeO_2 的重要来源。在水溶液中，SeO_2 可被转变生成 H_2SO_3，H_2SO_3 是一种二元弱酸，经常被用作氧化剂。在 pH 3.5～9 范围内，水中溶解的亚硒酸盐以 HSO_3^- 形式存在。在低 pH 下，SO_3^{2-} 可被一些温和的还原剂如抗坏血酸和 SO_2 还原为元素硒。

以 Se^{6+} 形式存在的硒化合物主要为三氧化硒（SeO_3）、硒酸（H_2SO_4）和硒酸盐（SO_4^{2-}）。SeO_3 易吸潮，化学性质与 SO_2 相似，但较 SO_2 活泼，是强氧化剂。H_2SeO_4 是一种强酸，$NaHCO_3$ 存在时，元素硒或 H_2SeO_3 可被强氧化剂 $NaBrO_3$ 氧化生成 H_2SeO_4。在水中，Br_2、Cl_2 或 H_2O_2 可将元素硒或 H_2SeO_3 氧化为 H_2SeO_4。与亚硒酸盐相比，硒酸盐更易溶于水，在碱性环境中其溶解度大且稳定，不易转变为亚硒酸盐和元素硒或转变缓慢。

硒与卤素反应可形成多种硒的卤化物，主要以+2、+4 和+6 价形式存在。硒的卤化物熔点低、易挥发、易水解，与对应的硫的卤化物相比热稳定性高。

硒在生物体内可合成多种有机硒化合物，主要以−2 价形式存在，包括硒醇（RSeH）及其衍生物、硒醚（RSeR）及其衍生物、硒杂环化合物和含硒氨基酸。

第三节 环境中的硒

硒在自然界中几乎无处不在，岩石、土壤、空气、水、植物、动物以及人体中均含有一定量的硒。环境中的硒主要来自于火山喷发以及岩石的自然侵蚀和风化等自然活动，人类活动也是环境中硒的重要来源[3]。每年通过自然活动进入环境中的硒大约为 4 500 t，其中，因火山喷发进入环境中的硒约为 400～1 200 t[4]。在夏威夷火山群周围的土壤中，硒含量高达 6～15 mg/kg，大约是世界土壤平均含硒量的 15～40 倍[5]。全球范围内每年因人类活动释放进入环境中的硒约为 76 000～88 000 t。一方面来自硒的工业生产，美国、日本和加拿大是硒的主要生产大国，其次是中国、俄罗斯、比利时、芬兰、澳大利亚和德国；另一方面来自工业生产副产品或工业废物的处理以及化石燃料如煤和石油的燃烧；第三方面主要来自农牧业活动，包括土壤灌溉和施肥、含硒杀虫剂和杀菌剂的使用以及动物饲粮中硒添加剂的广泛使用。

一、地壳中的硒

硒在地壳中的含量很低，平均约为 90 g/kg，将地壳中存在的 88 种元素按含量由高到低排序，硒排在第 77 位[6]。硒在地壳中的分布极不平衡，在岩石中的总量大约占到在地壳中总量的 40%，因岩石类型的不同硒含量差别极大。通常情况下，岩浆岩中硒含量较低，很少超过 0.05 mg/kg，沉积岩中硒含量相对较高，但很少超过 0.1 mg/kg，受黏土中硒含量的影响，沉积岩中页岩的硒含量较高，可达 0.6 mg/kg，磷灰石、煤炭以及其他富含有机质的沉积岩中硒含量也相对较高[7,8]。

硒在地壳中的丰度与硫相比低 1 000 倍以上，通常很难形成工业富集。在地壳中，硒主要和天然硫一起，呈类质同象分布于黄铁矿和硫化矿中。根据硒的工业利用情况，可将硒矿床分为独立硒矿床和伴生硒矿床。目前为止发现的伴生硒矿物有百余种，但多数由于硒含量太低而达不到工业品位，铜矿中的硒含量较高，是硒工业生产的主要来源。在自然条件下，独立硒矿物需同时具备贫硫和富硒两大条件才能形成，因此在地壳中分布极少，我国湖北恩施渔塘坝硒矿是迄今为止全球唯一探明的独立硒矿床，硒的质量分数高达 $8\ 590\times10^{-6}$[9]。

二、土壤中的硒

世界范围内土壤的硒含量范围在0.01～2 mg/kg,平均0.33 mg/kg[10]。不同国家和地区以及同一国家不同地区的土壤硒含量存在较大差别。美国、加拿大、日本、墨西哥、新西兰和澳大利亚的土壤含硒量范围分别为0.1～4.3 mg/kg、0.1～6.1 mg/kg、0.13～2.82 mg/kg、0.4～3.5 mg/kg、0.1～0.4 mg/kg和0.02～0.49 mg/kg。俄罗斯平原地带土壤硒含量低于0.01 mg/kg,我国主要土壤天然含硒量为0.05～0.8 mg/kg,平均0.25 mg/kg,低于世界土壤平均含硒量。

依据土壤含硒量是否满足人和动物对硒的需要,即从营养角度,可将土壤分为缺硒土壤(＜0.125 mg/kg)、临界缺硒土壤(0.125～0.175 mg/kg)、中等硒土壤(0.175～0.4 mg/kg)、高硒土壤(0.4～3 mg/kg)和高度硒毒性土壤(＞3 mg/kg)[10]。一般情况下,当土壤硒含量超过5 mg/kg时,通过植物吸收后极易导致动物中毒[11]。高硒或高度硒毒性土壤在美国、加拿大、墨西哥、哥伦比亚、爱尔兰、澳大利亚、委内瑞拉、以色列、印度和中国等国家的少数地区均有报道。美国部分地区高度硒毒性土壤样本的硒含量平均可达4.5 mg/kg,最高可达80 mg/kg,典型代表区为南达科他州。哥伦比亚高度硒毒性土样的平均硒含量为2～7 mg/kg,最高可达13～20 mg/kg。爱尔兰高度硒毒性土样的硒含量最高可达1 200 mg/kg。我国湖北恩施和陕西紫阳的部分地区为高硒区,土壤样本硒含量平均可达4～5 mg/kg,最高可达37 mg/kg。

世界范围内低硒或缺硒土壤的面积远远大于高硒或高度硒毒性土壤的面积,土壤缺硒是世界范围内广泛存在的问题,全世界大约有40多个国家和地区存在不同程度的缺硒。低硒或缺硒土壤的分布范围主要在北半球,包括欧洲大部,经中国、蒙古、俄罗斯、朝鲜半岛、日本、夏威夷,到太平洋彼岸的北美大陆北部和东西海岸。在南半球包括非洲南部,经澳洲大陆的西南和东南端与新西兰,到南美洲的智利、阿根廷、巴西南部和乌拉圭全境。

按照国际上公布的土壤缺硒临界值0.125 mg/kg为标准,我国有近2/3的土壤缺硒,其中有近1/3的土壤严重缺硒,硒含量低于0.02 mg/kg,典型代表区为东北克山病地区[12]。我国环境低硒带呈东北一西南走向,从东北地区的暗棕壤、黑土向西南方向走行,经黄土高原的褐土、黑垆土,到川滇地区的紫色土、红褐土、红棕壤，再到青藏高原东部和南部的高山草甸土和黑毡土，形成一条低硒带[13]。涵盖的省份(自治区)有黑龙江、吉林、辽宁、河北、河南、内蒙古、山西、山东、甘肃、四川、云南和西藏。低硒带内土壤硒含量均值仅为 0.1 mg/kg，低硒带的西北部为干旱、半干旱黑钙土、栗钙土、灰钙土和荒漠土，其硒含量逐渐递增到 0.3 mg/kg 左右，低硒带的东南部为热带、亚热带黄壤、红壤、砖红壤及相应的水稻土，其硒含量也逐渐递增到 0.3 mg/kg 左右。整个土壤硒含量分布呈中间低、两边高的马鞍形趋势面[14]。

用^{75}Se示踪技术可以把土壤中的硒区分为元素态硒(Se^0)、硒化物态硒(Se^{2-})、硒酸盐态硒(SeO_4^{2-})、亚硒酸盐态硒(SeO_3^{2-})、有机态硒和挥发态硒。根据土壤中的硒是否溶于水，可将其划分为水溶性硒和非水溶性硒两类。硒酸盐和亚硒酸盐属于水溶性硒，其他形式的硒为非水溶性硒。从世界各地土壤含硒状况可以看出，亚硒酸盐为土壤中主要的硒形式，约占40%以上，而以硒酸盐形式存在的硒不超过 10%，有机态硒主要来自聚硒植物的腐烂分解，而挥发态硒主要来自土壤微生物对硒的甲基化作用。

土壤中硒的存在形式与土壤的类型和质地、土壤中有机质含量、土壤的氧化还原电位(E_h)和 pH 以及土壤中微生物的活动直接相关，同时也受到降雨的影响。在酸性土壤($pH<6.5$)或土壤通气不良(低氧化还原电位，$E_h=500$ mV)条件下，硒主要以亚硒酸盐形式存在(图 1-1)。亚硒酸盐容易和土壤中存在的阳离子牢固结合，尤其易和铁的氧化物或氢氧化物结合形成难溶性复合物而被固定，这种作用在 pH=3～5 时最强。在碱性土壤($pH>7.5$)或土壤通气良好(高氧化还原电位，$E_h=0\sim200$ mV)条件下，即干旱和半干旱地区，土壤中的硒主要以硒酸盐形式存在(图 1-1)。硒酸盐不易被固定且易被植物吸收，但易遭受淋洗作用而迁移。在中性和微酸性土壤中，硒容易积累在富含有机质和腐殖质的土层中。在富含酸性有机质的土层中，硒酸盐还原为亚硒酸盐或硒化物而被固定。元素态硒在土壤中存在较少，适宜条件下($E_h<-200$ mV)，如在沼泽地土壤中，微生物活动可将硒酸盐或亚硒酸盐还原为元素态硒或硒化物。对于生长在高硒土壤中的聚硒植物而言，这种还原作用可降低植物对硒的吸收从而降低硒的潜在毒性。微生物活动还可将无机硒转化为有机态硒，并进一步通过甲基化作用转变为可挥发的甲基化有机硒化合物。土壤中许多细菌和真菌参与硒的甲基化过程，厌氧条件有利于硒的甲基化，但甲基化作用促进了土壤中硒的排出。

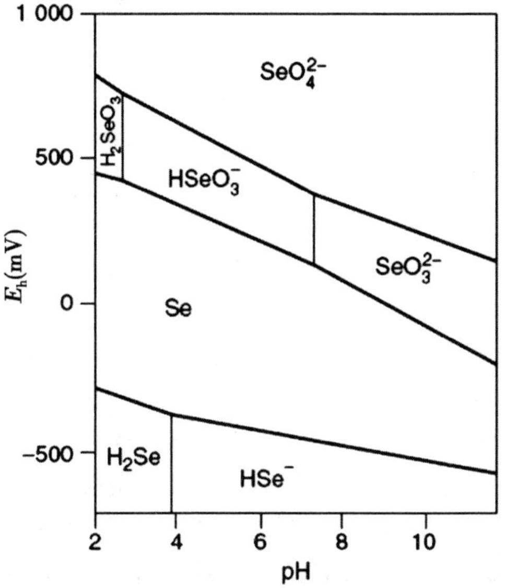

图 1-1 土壤中硒分布随 pH 和氧化还原电位的变化[15]

三、植物中的硒

(一)植物对硒的吸收

植物叶片可吸收硒,植物叶面喷施硒肥可提高植物根、茎、叶以及籽实中的硒含量,而且植物体中的硒含量与叶面施硒浓度呈显著正相关[16,17]。植物根系是植物吸收硒的主要途径,土壤硒是植物硒吸收的主要来源。

植物根系对硒酸盐的吸收为主动吸收,由于硒和硫具有相似的化学结构和性质,大部分植物在吸收过程中并不能有效区分硒酸盐和硫酸盐,因此,硒酸盐主要通过硫酸盐转运体即硫转运途径(通道)进行吸收。现已在拟南芥、大豆、大麦、玉米、还魂草和印度芥菜中克隆了几个硫转运子基因,它们的蛋白结构和真菌和动物的很相似,有 12 个跨膜伸展区。依据与硫的亲和力高低以及存在部位的不同,可将硫转运子大致分为 4 类:第一类为 SULTR1;2 和 SULTR1;1,与硫有高亲和力,主要分布在植物的根部,对硒酸盐的吸收起着重要作用;第二类为 SULTR2;1 和 SULTR2;2,与硫的亲和力较低,主要分布在维管束系统和根、叶等器官,起着迁移转运硒酸盐的作用;第三类为 SULTR3;1,主要分布在叶绿体中,起着转运硒酸盐到叶绿体的作用;第四类为 SULTR4;1 和 SULTR4;2,起着根茎之间硒酸盐转运的作用。对于非聚硒植物,植物根系对硫酸盐的吸收优先于硒酸盐。

相对于低亲和性的硫酸盐转运体，高亲和性的硫酸盐转运体对硒酸盐具有更强的选择性，因此，植物对硒酸盐的吸收可能主要受高亲和性硫酸盐转运体调控，其基因表达的提高会提高对硒酸盐的吸收。研究发现，植物对硒酸盐和硫酸盐的吸收需要三个质子的协同运输作为驱动力，然后逆电化学势穿过根系的细胞质膜进入植物体内。对拟南芥不同程度缺失 SULTR1;2 和 SULTR1;1 两种硫酸盐转运体的突变体进行硫酸盐与硒酸盐吸收的表型验证，发现两种硫载体都可进行硒酸盐的吸收转运，只是二者在硫酸盐与硒酸盐吸收转运功能上存在不对等的功能冗余[18]。而在缺失 SULTR1;2 而具有 SULTR1;1 的拟南芥突变体上表现出硒酸盐耐受性显著增强，说明植物体内不同硫酸盐转运体对硒酸盐和硫酸盐具有不同的选择性，与 SULTR1;1 相比，SULTR1;2 可能是硒酸盐吸收进入植物根细胞的主要转运体[19]。从底物角度考虑，硒酸盐和硫酸盐的吸收存在竞争性抑制作用，在生长培养基中提高硫酸盐含量能明显降低植物对硒酸盐的吸收，而对亚硒酸盐和硒代蛋氨酸(SeMet)的吸收没有影响，而当生长培养基中硫酸盐的含量降低甚至缺乏时，植物对硒的吸收明显提高[20]。对聚硒植物而言，硫转运蛋白可能更趋向于吸收硒。对沙漠王羽的研究提示，高聚硒植物在进化中可能仍然存在特异性硒酸盐转运体[21]。

很长一段时间里，人们一直认为植物对亚硒酸盐的吸收仅为被动吸收。近期研究发现，植物体还可以通过磷酸盐和硅转运体对亚硒酸盐进行主动吸收。水稻根系的磷酸盐转运子 Os PT2 在亚硒酸盐的主动吸收过程中起着重要作用，Os PT2 基因的过量表达或缺失能显著增加或减少水稻对亚硒酸盐的吸收[22]。提高培养基中磷酸盐的浓度可降低植物对亚硒酸盐的吸收，而培养基中磷酸盐缺乏可提高植物对亚硒酸盐的吸收[20]，这种磷酸盐与亚硒酸盐的竞争抑制现象间接证明了磷酸盐转运子在亚硒酸盐主动吸收过程中的重要作用。

(二)植物对硒的吸收和富集能力

不同种属植物对硒的吸收和富集能力有很大差异，根据其富硒能力的大小，可将植物分为聚硒植物和非聚硒植物，聚硒植物又可进一步分为原生聚硒植物和次生聚硒植物[23]。

目前所知有 30 多种聚硒植物。原生聚硒植物能够从土壤中吸收累积大量的硒，使其总硒含量可高达 1 000 mg/kg 以上，因此也被称为超聚硒植物。在菊科、十字花科、藜科、玉蕊科、蝶形花科、茜草科和玄参科植物中均有超聚硒植物的发现。紫云英属(*Astragalus*)部分植物中硒含量常超过 1 000 mg/kg，如生长在美国西南部的双槽紫云英(*Astragalus bisulcatus*)，硒含量可达 1 000～5 000 mg/kg。剑叶莎属(*Machaeranthera*)部分植物中硒含量可富集到 800 mg/kg,沙漠王羽(*Stanleya pinnata*)中硒含量可富集到 700 mg/kg,一些卷舌菊属(*Symphyotrichum*)植物中硒含量也可富集到几百毫克[24]。中国湖北恩施的碎米荠属植物碎米荠(*Cardamine*)是一种富硒能力极强的植物,其苗期叶片硒含量即可超过 1 000 mg/kg。玉蕊科植物巴西坚果(*Bertholletia excelsa*)是一种聚硒果树,其果实中硒含量可高达 512 mg/kg。此外,芸苔属植物芥菜的根和茎的含硒量分别为 186 mg/kg 和 133 mg/kg(干重),而叶和花的硒含量分别为 931 mg/kg 和 541 mg/kg(干重)。一些次生聚硒植物,如紫菀属(*Aster*)、滨藜属(*Atriplex*)、火焰草属(*Castilleja*)、胶菀属(*Grindelia*)、古蒂里齐亚属(*Gutierrezia*)植物,每千克含硒量一般在 50～100 mg/kg,很少超过 100 mg/kg[25]。

聚硒植物中的硒主要以非蛋白形式的甲基化的硒代蛋氨酸和硒代半胱氨酸(SeCys)形式存在,这可能是聚硒植物能够聚硒的原因之一。一般来讲,十字花科植物的聚硒能力最强,其次是豆科,谷类植物的聚硒能力最低。但同一种属作物的聚硒水平在不同地区也存在差异,这可能与不同地区土壤的理化性质以及植物对土壤中硒的可利用性不同有关。种植聚硒植物是消除或缓解土壤硒污染的有效措施之一,主要通过植物的吸收、挥发和根滤等作用净化土壤中的硒。研究发现,在 90 天的时间里,印度油菜吸收了所种植土壤中几乎全部的可摄取硒,而在 3 年的时间里,印度芥菜吸收了所种植土壤中 50%的硒,而且吸收深度可达 75 cm[26]。一些水稻和杂交白杨品种具有高的硒挥发作用,也可用于净化土壤中的硒[27]。

非聚硒植物中的硒主要以蛋白质形式存在,蛋白质中的硒以硒代蛋氨酸形式存在。相比之下,大部分植物属于非聚硒植物,组织中硒含量很少超过 50 mg/kg,绝大多数在 5 mg/kg 以下。一些动物饲粮中常用的非聚硒饲料原料,如禾本科牧草、苜蓿、稻草、玉米青贮、玉米、大麦、燕麦、小麦、豆粕、花生粕、菜籽粕、棉粕、甜菜渣等,硒含量范围约在 0.05～0.6 mg/kg 之间。

(三)影响植物吸收硒的因素

植物对土壤中硒的吸收受到多种因素的影响,包括土壤中硒的含量和存在形式、土壤的理化性质、降雨和土壤施肥等。

相比土壤中的硒含量,土壤中硒的存在形式是影响植物吸收硒的主要因素。亚硒酸盐和硒酸盐形式的硒均可被植物吸收。相比亚硒酸盐,硒酸盐的水溶性更好,不易被铁的氢氧化物固定,不易被吸附在土壤颗粒表面,因而更容易被植物吸收。研究表明,植物对土壤中硒酸盐的吸收能力大约是亚硒酸盐的10倍[28]。土壤中元素态硒、硒化物以及硒的硫化物由于低的水溶性和氧化电势，通常很难被植物吸收[29]。土壤中有机态硒，如硒代蛋氨酸和硒代半胱氨酸，可以被植物吸收。

提高土壤pH或使土壤处于氧化状态是提高土壤硒有效性的重要措施。研究表明，当土壤pH＝6时，土壤中总可吸收硒的47％可被植物吸收，而当pH提高到7时，总可吸收硒的比例提高到70％，当土壤含水量处于饱和状态(强还原状态)时，植物对土壤中硒的吸收效率显著降低[30]。当土壤中硒含量低于2 mg/kg时，土壤总硒与植物可利用硒有一定的线性相关性，而当土壤硒含量超过2 mg/kg时，二者之间的线性相关性显著降低。因此，评价土壤中植物可吸收硒的状况较土壤中硒含量更能准确反映硒的缺乏或中毒情况。研究发现，与耕作过的土壤相比，未耕作过的土壤具有高的植物硒利用率，已耕作土壤由于使用肥料，导致土壤中N、P、S的含量提高，而土壤中这些元素的提高可竞争性抑制植物对土壤中硒的吸收[31]。土壤中硫酸盐或磷酸盐的使用均可降低植物对硒的吸收，这在非聚硒植物上表现更为明显，磷酸盐对硒吸收的竞争抑制作用明显强于硫酸盐，这可能与磷酸盐带有高的负电荷数有关[32]。利用这种竞争抑制关系，生产实践中可通过给高硒土壤施肥以降低植物对硒的吸收，进而预防硒对人和动物的潜在毒性。

(四)硒在植物体内的代谢

由于硒和硫的化学相似性，植物根系从土壤中吸收的无机态硒(硒酸盐和亚硒酸盐)主要通过硫同化代谢途径进行代谢转化，这一过程可能同时发生在植物根部和地上部，一般认为主要发生在地上部的叶绿体中。

如图1-2所示，硒酸盐首先被ATP硫酸化酶(ATPS)活化生成腺苷-5′-磷酰硒酸(APSe)，APSe在APSe还原酶的作用下还原为亚硒酸盐。亚硒酸盐在还原型谷胱甘肽(GSH)作用下通过非酶促反应进一步还原为硒化物(Se^{2-})。在半胱氨酸合成酶的作用下，Se^{2-}与O-乙酰-丝氨酸(OAS)耦联生成硒代半胱氨酸。与硫化物相比，半胱氨酸合成酶对硒化物的亲和力更强[33]。硒代半胱氨酸进一步通过胱硫醚-γ-合成酶(CGS)作用生成硒代胱硫醚(SeCysth)，然后经过胱硫醚-β-裂解酶(CGL)作用分解为硒代高半胱氨酸(SeHoCys)，硒代高半胱氨酸再经甲硫氨酸合成酶作用生成硒代蛋氨酸。硒代蛋氨酸可在蛋氨酸甲基转移酶的作用下生成甲基硒代蛋氨酸(MeSeMet)，进一步甲基化转变为具有挥发性的二甲基硒化物(DMSe)。

在植物硒代谢过程中，中间产物硒代半胱氨酸也可通过硒代半胱氨酸甲基转移酶(SMT)的作用生成甲基硒代半胱氨酸(MeSeCys)，进一步甲基化转变为具有挥发性的二甲基二硒化物(DMDSe)。当硒代半胱氨酸裂解酶存在时,硒代半胱氨酸也可还原为元素态硒。

在植物硒代谢过程中,中间产物硒代蛋氨酸除了发生甲基化外,也可通过非特定的途径随意替代蛋氨酸合成到蛋白质中,非聚硒植物中的硒大约有 90%是以硒代蛋氨酸形式存在于蛋白质中的[34]。

在叶绿体细胞中,硒酸盐转化为亚硒酸盐是整个硒代谢过程的限速步骤。对印度芥菜的研究表明,与野生型相比,ATPS 过表达显著提高了硒酸盐的还原以及有机态硒的含量[35]。

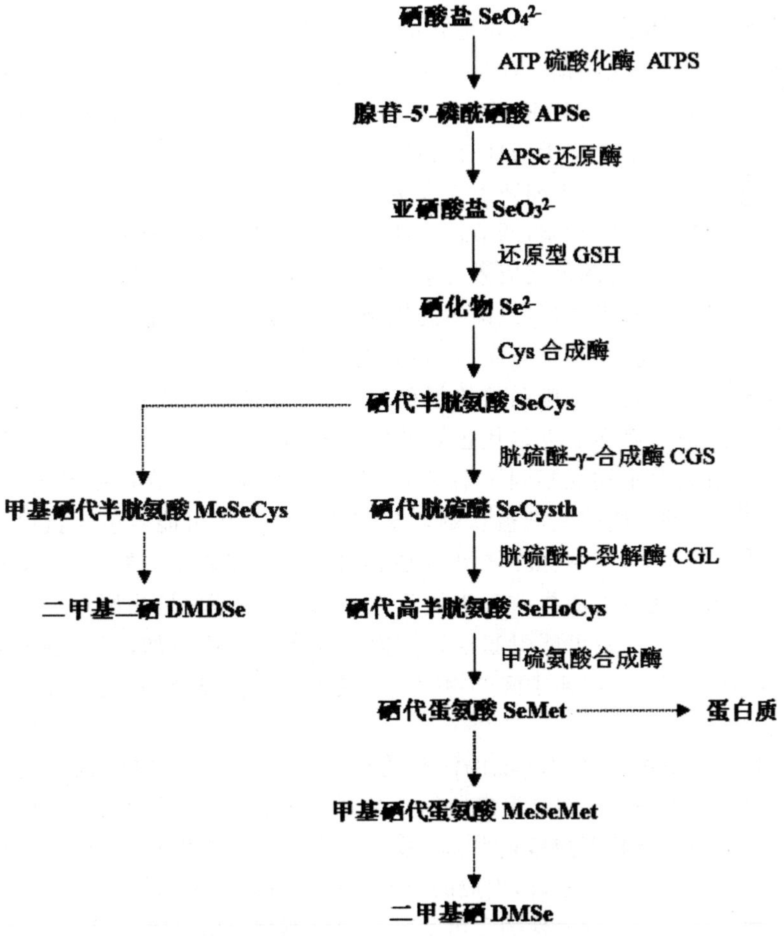

图 1-2　植物体中硒的代谢转化过程[36]

(五)植物中硒的生理作用

目前关于硒是否是植物的必需微量元素仍无定论。一些藻类植物需要硒来合成硒蛋白，但硒可能并不是高等植物的必需营养素。一些研究发现，硒缺乏会导致小麦和水稻的生长受阻以及黑麦草和莴苣对紫外线照射的敏感性提高，低浓度的硒刺激了印度芥菜、莴苣、南瓜、黑麦草和大豆的生长，提高了马铃薯的产量和淀粉含量以及加工和贮存特性，降低了高温、紫外线、低温、干旱、重金属等胁迫因子对植物的毒害作用，延缓了一些作物的衰老进程[32]。硒还会影响植物种子的萌发，土壤中高的硒添加浓度(大于 29 mg/kg)抑制了番茄、莴苣和白萝卜种子的萌发和生长[37]。低浓度的硒对植物的有益作用可能与其参与植物的抗氧化过程有关[38]。一些研究表明，硒增强了植物对活性氧引起的氧化应激的抵抗力[39-41]，在应激条件下，硒上调了抗氧化酶如超氧化物气化酶(SOD)、过氧化氢酶(CAT)、抗坏血酸盐过氧化物酶(APX)、单脱氢抗坏血酸还原酶(MDHAR)、脱氢抗坏血酸还原酶(DHAR)、谷胱甘肽还原酶(GR)、谷胱甘肽 S-转移酶(GST)和过氧化物酶(POD)的活性，同时提高了非酶抗氧化剂如抗坏血酸、还原型谷胱甘肽、类胡萝卜素和生育酚的水平[41-43]。因此，硒可能通过抑制植物体内 ROS 积累，调节植物体内促氧化和抗氧化物质的平衡，进而降低环境胁迫诱导的氧化应激对植物的不利影响，从而间接刺激植物的生长，延缓植物的衰老进程，但确切的机制还有待进一步研究证实。

大多数植物，尤其是非聚硒植物，对高硒环境比较敏感。高硒可引起植物光合效率降低、细胞膜结构损伤，最终导致植物生长受阻和减产。在非聚硒植物中，硒多数以硒代蛋氨酸的形式结合于植物蛋白质中，在黑麦草、小麦和苜蓿中均观察到了硒对蛋氨酸中硫的取代。在高硒环境下，植物蛋白质中的含硫氨基酸被含硒氨基酸的广泛取代可能会引起蛋白质结构和功能的改变。一些蛋白质的生物活性依赖于半胱氨酸侧链上—SH 的存在，当被取代后可能导致其生物活性丧失，最终导致植物硒中毒。也有一些研究认为，高硒可能引起植物体内活性氧的生成增加，高硒也可能抑制植物体内高活性代谢中间产物硒代半胱氨酸的甲基化，从而增加其毒性作用[44]。

超聚硒植物对高硒环境的耐受力较强，即使大量吸收硒也能正常生长。研究发现，SMT 在超聚硒植物中的活性很高，因而使大量吸收的硒转变为甲基硒代半胱氨酸形式[45]，甲基硒代半胱氨酸不能进入蛋白质中，因而对植物的毒性作用小，而生成的甲基硒代半胱氨酸又可进一步通过甲基化作用转变为具有挥发性的二甲基二硒化物形式而排出体外。

四、水中的硒

在大多数 pH 和氧化还原条件下，水中的硒主要以硒酸盐和亚硒酸盐两种形式存在，少数以 Se^{2-} 形式存在。硒可被水中生物吸收，也可以和水中的微粒或胶体物质以及沉积物形成复合物。

天然水体中的硒含量通常不会超过 1 μg/L。海水中硒含量范围通常为 0.1～0.35 μg/L，平均 0.2 μg/L。也有一些数据认为海洋中的平均硒含量为 0.09 μg/L[46]。通过多种途径，每年大约有 7 700～8 000 t 的硒进入海洋。在海洋表层水中，有机硒化物（主要为二甲基硒化物）大约占到总溶解硒的 80%，脱气过程可能是海水排除硒的重要机制。全球地表淡水（江、河、湖泊）中硒含量范围通常为 0.02～0.5 μg/L，平均 0.07 μg/L。如亚马孙河为 0.21 μg/L，易鲁河为 0.24 μg/L，洞庭湖为 0.13 μg/L，湘江为 0.25 μg/L，金沙江为 0.2 μg/L。然而，一些河流的含硒量会高于 0.5 μg/L，甚至超过 10 μg/L，如东湖为 1.1 μg/L，科罗拉多河为 1～4 μg/L，安尼玛斯河为 12 μg/L，密苏里河和密西西比河为 14 μg/L[1]。这些河流中高的硒含量可能主要来自工业生产，炼油可能是主要来源。

地下水中硒含量范围较大，一般为 0.06～400 μg/L，pH 的高低会直接影响地下水中硒的溶解度，进而影响其在地下水中的含量，一些地区的地下水，硒含量可高达 6 000 μg/L。

世界范围内饮用水中的硒含量一般不超过 10 μg/L，但有时可能会达到 50 μg/L。在中国的一些高硒地区，饮用水中硒含量可达 50～160 μg/L。大多数国家规定，人饮用水中最大硒含量不得超过 10 μg/L，牲畜饮用水不得超过 50 μg/L，灌溉水不得超过 20 μg/L[47]。

五、大气中的硒

大气中的硒来自海洋表层水的蒸发、火山喷发、工业排放以及植物、动物和土壤污泥中微生物作用引起的硒挥发。矿物燃料的燃烧是大气中硒的主要来源之一，全球范围内每年因矿物燃料燃烧排放到大气中的硒约为 6 000 t 以上，约占总排入大气硒量的 40%，煤炭中 53% 的硒会在燃烧过程中以挥发态或飞尘形式散逸到大气中[47]。

大气中硒含量差异极大，南极上空大气中硒含量较低，约为 0.06 ng/m^3，世界范围内农村大气平均含硒量为 0.2 ng/m^3，但一些硒污染地区可达到 4 ng/m^3，城市与农村相比，大气硒含量较高，约为 0.1～10 ng/m^3，但在一些铜冶炼厂附近硒含量往往超过 10 ng/m^3。

第四节　硒的必要性和对动物的毒性

微量元素研究最重要的一个方面就是了解其对动物和人体健康的益处和害处，并知道如何去控制。硒是人和动物体的必需微量元素，但过量时会产生毒性作用。当家畜日粮干物质(DM)中硒含量低于 0.1 mg/kg 时，表现为硒缺乏效应，而硒含量超过 5 mg/kg DM 时，表现为中毒效应。对于人而言，当每天的硒摄入量低于 40 μg 时表现为硒缺乏效应，大于 400 μg 时表现为硒中毒效应。由于硒在人和动物体内从缺乏到中毒的反应剂量范围很窄，人们常把硒形象地称为“必需的毒素”“双刃剑”，就像以“月亮”来命名硒一样，既具有“黑暗”的一面，同时也具有“光明”的一面。相对于硒中毒，人和动物硒缺乏更易发生，影响范围几乎遍布全球，主要原因是由于全球范围内的土壤普遍表现为缺硒。

一、硒对动物的毒性

关于硒对动物的毒性作用，最早可追溯到发现硒元素之前 500 年。大约在 1295 年，马可波罗在中国西北地区记录到马采食某种植物后出现蹄子脱落的现象[48]。1560 年，传教士 Pedro Simon 在哥伦比亚也记录到类似的人的中毒现象。1860 年，美国军医 Madison 在南达科他州和内布拉斯加州附近的密苏里河河畔的兰德尔堡记录到军马毛发和蹄子脱落的现象，在这些地区及邻近的北部平原的居民也经常观察到家畜出现类似的症状，由于当地居民认为这种病与家畜饮用当地的一种盐碱水有关，因此形象地称之为“碱毒病”，但“碱毒病”的发病原因一直以来是一个谜[49]。在之后的 75 年里，美国农业部组织专家致力于寻找该病的发病原因，直到 1937 年(硒元素被发现之后 100 年)，Moxon 首次证实了“碱毒病”的发生是由于家畜采食大量含硒量高的植物(紫云英属的野豌豆)引起的[50]。在接下来的 15 年里，美国农业部和地质勘查局集中调研了土壤中硒的含量和分布情况、植物对硒的吸收和富集过程以及硒对动物的毒性作用，积累了大量的事实证据。

现已明白，硒中毒包括急性硒中毒和慢性硒中毒两类。历史上，在爱尔兰、以色列、澳大利亚、俄罗斯、委内瑞拉、美国、印度和南非等国家的高土壤硒含量地区均发生过人的硒中毒事件，我国湖北恩施和陕西紫阳两大富硒区，因不当利用高硒石煤和食用高硒作物，也曾爆发过人的硒中毒事件。硒中毒事件发生地区土壤中硒的含量往往超过 5 mg/kg，一些植物中硒的含量往往超过 3 mg/kg 干物质[51]。

（一）急性硒中毒

自由放牧条件下引起的家畜急性硒中毒常发生于短时间内采食大量富硒植物，但发生几率一般较小，原因在于这些植物通常适口性很差（有异味），家畜通常会有意识地避开它们。只有在气候恶劣、过度放牧或牧草匮乏的情况下，家畜才有可能被迫采食这些植物而导致中毒。事实上，家畜急性硒中毒常因日粮中过量添加硒、注射过量硒制剂或日粮配制不当或计算错误引起。

家畜急性硒中毒的症状包括运动失调、厌食、腹泻、腹痛、虚脱、呼吸困难、体温升高、心率加快、瞳孔散大、鼻孔有泡沫、黏膜发绀、呼出气体有明显的大蒜气味，常见于反刍动物和马。血液学改变包括纤维蛋白原水平和凝血酶原活性降低，还原型谷胱甘肽水平降低，碱性磷酸酶、转氨酶和琥珀酸脱氢酶活性提高，血硒水平提高（高于 100 μg/dL）。病理变化表现为肝坏死、肝细胞水肿变性、肾组织充血、肾小管上皮细胞变性或坏死、肺充血水肿及渗液、肺泡内出血、多病灶心肌纤维变性和坏死并伴有弥散性心肌炎，有时还会出现中枢神经系统病变。急性硒中毒动物常因肺部水肿和充血导致呼吸衰竭而死亡，从出现明显症状到死亡的时间间隔可以是几小时，也可以是几天，主要和硒的摄入率有关，摄入率越高，死亡速度越快。

当给牛按每千克体重饲喂 10～20 mg 硒或注射超过 0.5 mg 的硒会导致急性硒中毒。给猪按每千克体重饲喂超过 20 mg 硒或注射超过 1.65 mg 硒会引起急性硒中毒。与饲喂硒适宜日粮的猪相比，饲喂硒缺乏日粮的猪在高硒日粮环境中更易发生硒中毒。给实验动物（兔、小鼠、猫）按每千克体重注射超过 3 mg 硒会引起急性硒中毒，呼出的气体有强烈的大蒜气味。以每千克活体重计，亚硒酸钠对动物的最小致死剂量分别为：马：1.5 mg/kg，牛：4.5～5.0 mg/kg，猪：6～8 mg/kg，大鼠：3.0～5.7 mg/kg，兔：0.9～1.5 mg/kg，狗：2 mg/kg。

（二）慢性硒中毒

虽然在放牧条件下家畜急性硒中毒的发生并不多见，但慢性硒中毒却屡有发生。慢性硒中毒主要影响动物的脾脏、肝脏、心脏、肾脏和胰腺。依据所采食植物的含硒量不同，家畜慢性硒中毒可分为碱毒病和晕倒病（也称蹒跚病）两种。

碱毒病主要发生在马、牛和猪身上，放牧条件下常因持续数周甚至数月采食含硒量高于 5 mg/kg 但低于 40 mg/kg 的植物引起，这些植物中的硒常以非水溶性的蛋白形式存在[23]。实验条件下饲喂硒酸钠或亚硒酸钠等水溶性无机硒盐也可诱导碱毒病发生。碱毒病的症状包括反应迟钝、精神萎靡、食欲减退、消瘦、被毛粗糙、脱毛(尤其是鬃毛和尾毛)、蹄和关节损伤、跛行及蹄裂。与牛和猪不同，碱毒病在羊身上并不表现明显症状，既没有羊毛的脱落，也没有蹄的损伤[23]。碱毒病的病理表现主要为心脏和肝脏损害。碱毒病对家畜最显著的影响是导致生殖功能降低[52]，在碱毒病典型症状出现之前即表现为明显的生殖功能降低。血液和毛发中硒水平可作为碱毒病的临床诊断指标，当血硒水平超过 2 mg/L 或毛发硒水平超过 2 mg/kg 时，表现为明显的碱毒病临床症状。

晕倒病常由少量但长时间食用含硒量超过 100 mg/kg 的硒指示植物引起，这些植物中的硒常以水溶性的非蛋白形式存在[23]。晕倒病主要发生在牛和羊身上，在猪、马和家禽身上通常不发生。对于牛而言，晕倒病的发生通常经历三个阶段：在第一阶段，家畜的体温和呼吸正常，但视力会减退，表现为转圈行走、无视前方障碍物，还表现出明显的厌食症状；在第二阶段，除第一阶段的症状更加明显外，前肢越来越无力，无法支撑整个身体；在第三阶段，舌肌部分或全部麻痹，吞咽功能丧失，动物几乎完全失明，呼吸加快和呼吸困难，腹痛明显，体温降低，磨牙，流涎，消瘦，多因呼吸衰竭而死亡。对于羊而言，晕倒病的发生并无明显的阶段区分，通常在无明显症状的情况下几小时内突然死亡。晕倒病的病理变化表现为肝坏死、肝硬化、肾炎、肾充血和胃肠道积食。家禽慢性硒中毒常表现为产蛋率下降、孵化率降低及胚胎畸形。

(三)硒化合物的毒性

硒化合物的形态不同，其毒性也不同。元素硒和金属硒化物因几乎不能被动物利用基本上是无毒的，硒酸盐和亚硒酸盐的毒性相对较大，而硒化氢是已知毒性最强的硒化合物，经生物代谢产生的有机硒化合物如硒代半胱氨酸、硒代蛋氨酸和甲基硒化合物的毒性相对较低，但剂量过高也会引起中毒。在已知的硒化合物中，二甲基硒的毒性最低，亚硒酸盐的毒性略大于硒酸盐，大约是硒代蛋氨酸的 4 倍，二甲基硒的 500～700 倍。

(四)硒中毒的预防和控制

预防和控制硒中毒最直接有效的途径是避免过量硒的摄入。放牧条件下，应避免家畜采食大量高硒植物，也可考虑通过治理土壤，降低植物对硒的吸收，使土壤和植物中的硒含量维持在适当水平。集约化饲养条件下，应特别注意避免日粮配制不当或计算错误，可考虑将高硒植物和其他低硒饲料原料按一定比例搭配使用，也可考虑在动物日粮中使用低毒的硒添加剂，如纳米硒、硒酵母、富硒苜蓿等。

当动物发生硒中毒时，一般没有特效的解毒药物。高蛋白日粮能减轻动物的硒中毒症状，其中的蛋氨酸起主要作用，可能与蛋氨酸和硒在动物肠道的吸收存在竞争抑制作用有关。日粮中添加亚麻籽粉可有效缓解硒对动物的毒性，可能与亚麻籽中含有生氰糖苷、龙胆二糖丙酮氰醇、龙胆二糖甲乙酮氰醇等物质有关。有机砷化合物如对氨基苯胂酸、洛克沙胂也可降低硒对动物的毒性，而且效果显著，其原理是有机砷制剂可与吸收的硒螯合并将其转移到胆道系统，但有机砷本身也是一种有毒物质，应在专业指导下使用。此外，口服维生素 C 和维生素 E 可加速硒的排泄，进而降低硒的毒性。

二、硒对动物的必要性

早在 1916 年，人们就发现硒存在于正常的人体组织和细胞中[53]，暗示硒可能是人和动物的必需微量元素，然而由于没有直接的证据，科学家研究的侧重点仍然是硒的毒性作用。20 世纪 40 年代，由于战争和食物缺乏，许多德国人患上了肝病，于是，德国政府邀请一位叫施瓦茨（Schwarz）的科学家研究营养与肝病的关系。最初，施瓦茨发现蛋白质缺乏是导致肝坏死的主要原因，进一步研究发现蛋白质中的含硫氨基酸以及维生素 E 对肝脏有很好的保护作用。施瓦茨还发现，用酵母饲料饲喂大鼠一个月就能引起肝坏死，进一步发现用造纸厂生产出来的酵母（torula yeast）饲喂的大鼠 4 周后都出现了肝坏死，而用啤酒厂生产出来的酵母（brewer's yeast）饲喂的大鼠 4 周后都未出现肝坏死。这一研究说明在啤酒酵母中存在一种可以预防肝坏死的物质，这种物质的生物活性比含硫氨基酸和维生素 E 都强，因此，他把这种未知物质称为“第三因子”。施瓦茨进一步从啤酒酵母的络氨蛋白乙醇萃取液中分离纯化了该种物质，当滴一滴碱时会发出很强的大蒜气味，很像是吃了高硒饲料的牛呼出的气味。经过多次检测和验证，1957 年 5 月 17 日，确定这种能强烈保护肝脏的物质是一种硒化合物[54]。这是人类第一次发现硒是营养性肝坏死的重要保护因子，也是人类第一次证实硒具有营养作用，从此拉开了研究硒与健康的序幕。

早在 20 世纪 30 年代，中国黑龙江省克山县就曾发生过以多病灶的心肌坏死为主要症状的疾病，主要发生在农村地区，易感人群为育龄妇女和 2～6 岁儿童，死亡率较高，因在克山县发病率高且症状典型，故被命名为

“克山病”。克山病是由于严重硒缺乏(硒摄入量低于 12 μg/d)引起[55,56]。给克山病病人口服亚硒酸钠可有效防止该病的进一步发展,这是人类第一次证实了硒对人的营养重要性。新西兰和芬兰的土壤低硒地区也曾报道过有该病的发生[57]。研究发现,硒缺乏还可导致人类患大骨节病,一种类似于类风湿性关节炎但比类风湿性关节炎更严重的疾病,又叫矮人病、算盘珠病。该病还与碘缺乏有关,中国西藏和从东北到西南的缺硒地区、西伯利亚以及朝鲜均有该病的报道。之后对动物的研究还发现,硒缺乏可导致猪患营养性肝坏死病[58]、犊牛和羔羊患白肌病[59]以及家禽患渗出性素质病[60](表 1-2)。动物硒缺乏还可导致黄疸、水肿、小动脉管壁透明样变化以及精子形态异常。

表 1-2　不同动物硒缺乏的临床症状[61]

动物	临床症状
犊牛和羔羊	白肌病(骨骼肌和心肌营养不良、变性坏死)、羔羊僵硬病
成年反刍动物	生长发育不良、生殖功能障碍(早期胚胎死亡)、免疫力低下、胎衣不下
马	白肌病
猪	营养性肝坏死、白肌病、桑葚心病
家禽	渗出性素质病(毛细血管通透性增加引起血浆成分漏出)、胰腺萎缩

尽管人们认识到人和动物的许多疾病与硒缺乏有关,但此时关于硒的生物学功能还知之甚少。1973 年,Rotruck 等[61]首次证实硒是谷胱甘肽过氧化物酶(GPX)的必需结构组分,以硒代半胱氨酸的形式存在于该酶的活性中心。GPX 是生物体内广泛存在的一种重要的过氧化物分解酶,在机体抗氧化防御系统中起着关键作用,硒缺乏会导致该酶的基因表达和活性显著降低,进而导致细胞甚至组织因氧化应激而损伤。就在 1973 年,世界卫生组织(WHO)宣布,硒是人和动物生命活动所必需的微量元素,关于硒的研究也涉足分子生物学领域。对小鼠的研究发现,吸收的硒仅有 1/3 存在于 GPX 中[62],暗示机体中可能存在其他的硒蛋白或硒酶。大量基于硒的代谢标记研究表明,大约吸收硒的 80%以硒代半胱氨酸的形式存在于蛋白质中[63]。迄今为止,在生物体内已发现 30 种硒蛋白,其中 25 种存在于人和动物体内,包括 GPX 家族、硫氧还蛋白还原酶(TrxR)家族、脱碘酶(DIO)家族以及硒蛋白 P、K、S 和 W 等[61,64]。随着人们对各种硒蛋白生物

学功能的认识，硒对人和动物健康的重要性已成为不争的事实，具有调节抗氧化功能、免疫功能、甲状腺功能、营养物质代谢和生殖功能以及抗病甚至预防癌症发生的重要作用。

第五节　人和动物对硒的需要量

1989 年，首次确定了人对硒的需要量：成年男性 70 μg/d，成年女性 55 μg/d。1996 年调整为：成年男性 34 μg/d，成年女性 26 μg/d。2000 年调整为：成年男性和女性均为 55 μg/d。

基于硒缺乏对家畜生长、生产和健康的影响以及世界范围内绝大多数土壤处于低硒或缺硒状态，在动物饲养实践中使用硒添加剂已成必然。动物日粮硒添加不仅是为了预防动物因硒缺乏而导致的健康问题，而且还是通过动物源性食品为人类提供硒的一项有效措施。1979 年，美国食品和药品管理局（FDA）正式通过法案，允许在家畜日粮中添加微量元素硒，添加形式为无机硒盐（亚硒酸盐或硒酸盐），推荐的添加剂量以硒计为 0.1 mg/kg DM。1987 年，最大允许硒添加剂量上调为 0.3 mg/kg DM，并且允许在反刍动物日粮中添加，添加形式仍然为无机硒盐。2003 年，允许在反刍动物日粮中使用有机硒源（如硒酵母），最大允许添加量以硒计为 0.3 mg/kg DM[65]。动物的种类以及相同动物在不同生长阶段推荐的硒添加量有所区别，详见表 1-3。

表 1-3　动物日粮中推荐的硒添加量

动物	推荐添加量	参考文献
猪	0.15～0.3 mg/kg DM	[66]
肉牛	0.1 mg/kg DM	[65]
奶牛	0.3 mg/kg DM	[65]
犊牛	0.1 mg/kg DM	[65]
绵羊	0.1～0.2 mg/kg DM	[67]
山羊	0.1 mg/kg DM	[68]
马	役用：0.3 mg/kg DM；非役用：0.1 mg/kg DM	[69]
猴子	2 mg/d；0.1～0.15 mg/100 kg BW	[70]
鸡（0～8 周龄）	0.3 mg/kg DM	[71]
火鸡（0～8 周龄）	0.2～0.3 mg/kg DM	[72]

第六节　硒在动物体内的吸收、代谢和分布

一、动物对硒的吸收

动物对硒的吸收一方面因动物种类不同而有所不同，另一方面和摄入硒的化学形式关系极大，动物对硒的吸收程度还取决于硒的摄入量，日粮中钙、砷、钴以及硫的含量也会影响硒的小肠吸收。总的来说，可溶性无机硒盐（硒酸盐、亚硒酸盐）和硒代氨基酸（SeMet、SeCys）最易吸收，而金属硒化物以及硒代二乙酸、硒代丙酸和硒代嘌呤等有机形式硒吸收缓慢，元素硒则几乎完全不被吸收。大多数植物来源的硒较易吸收，而动物来源的硒则相对较难吸收。

（一）单胃动物对硒的吸收

单胃动物的胃不吸收硒，小肠是硒吸收的主要场所。猪对硒的最大吸收部位为盲肠和结肠，小鼠对硒的最大吸收部位在十二指肠。早期的研究认为，硒酸钠在小肠中主要通过被动转运（经载体的易化扩散）的方式吸收[73,74]。后来的研究发现，硒酸钠在小肠中的吸收是耗能的，是一种主动转运过程[75]。亚硒酸钠在小肠中主要通过单纯扩散的方式吸收，吸收效率低于硒酸钠。对小鼠的研究表明，硒酸钠在小肠中的吸收率约为80%，而亚硒酸钠在小肠中的吸收率仅为35%[75]。硒代蛋氨酸在小肠中的吸收机制和蛋氨酸相同，主要通过依赖Na^+的氨基酸跨膜转运系统主动吸收，吸收率较高，约为80%[75]。由于硒代蛋氨酸和蛋氨酸在小肠中利用相同的转运系统，因而存在竞争关系，增加蛋氨酸的小肠流量会降低对硒代蛋氨酸的吸收[75]。增加日粮中硫酸盐的含量会降低硒酸钠在小肠中的吸收，但增加相应的硫化合物不影响亚硒酸钠和硒代半胱氨酸在小肠中的吸收。

（二）反刍动物对硒的吸收

反刍动物的瘤胃和皱胃不吸收硒，小肠和盲肠是硒吸收的主要场所。与单胃动物相比，反刍动物小肠对相同化学形式硒的吸收率以及吸收机制可能没有本质上的区别。尽管瘤胃壁不吸收硒，但通过日粮摄入的硒在进入小肠之前会被瘤胃微生物进行代谢，瘤胃微生物对摄入硒的代谢

会改变进入瘤胃后消化道的硒的化学形式，从而影响硒在小肠中的吸收。口服^{75}Se（亚硒酸钠）的研究表明，猪的小肠吸收率为85%，而绵羊的小肠吸收率仅为34%[76]，暗示瘤胃微生物对硒的代谢会降低硒在小肠中的吸收。

摄入硒的形式不同，其在瘤胃中的代谢也不尽相同。无机硒盐（硒酸钠、亚硒酸钠）是反刍动物日粮中添加的主要硒源。总的来讲，由于瘤胃的强还原环境，硒酸钠首先会被还原为亚硒酸钠，尽管没有具体的数据，但可以肯定的是，绝大多数硒酸钠在瘤胃中会被还原为亚硒酸钠，仅有少数硒酸钠可逃过瘤胃微生物的作用直接进入小肠[77]。因此，对于反刍动物，硒酸钠和亚硒酸钠的生物利用效率基本相近[78]。日粮中的亚硒酸钠以及瘤胃中由硒酸钠还原生成的亚硒酸钠在瘤胃中会进一步被瘤胃微生物代谢：一部分亚硒酸钠被瘤胃微生物还原为不溶形式的元素硒和硒化物，一部分被瘤胃微生物利用以硒代氨基酸的形式合成到菌体蛋白中，还有一部分逃过瘤胃微生物的作用，以亚硒酸钠的形式离开瘤胃进入小肠[79]。Serra等[80]通过离心分离绵羊的瘤胃液，认为当饲喂亚硒酸钠时，约30%～40%的硒转变为不溶形式，10%～15%的硒合成到菌体蛋白中，40%～60%的硒以亚硒酸钠的形式离开瘤胃。反刍动物粪便中的硒主要以元素硒和硒化物形式存在，溶解度低，酸中溶解率为60%，水中溶解率仅为10%，用反刍动物粪便种植牧草75天后，只有0.3%的硒被植物吸收[81]。很早以前，科学家就曾推测瘤胃微生物可以利用无机硒合成含硒氨基酸，之后很多研究证实了这一推测。Whanger等[79]用亚硒酸钠和硒酸钠饲喂绵羊，发现瘤胃微生物蛋白中存在硒，且主要以硒代半胱氨酸形式存在。Paulson等[77]用亚硒酸钠和硒酸钠对瘤胃微生物进行体外培养，在瘤胃微生物蛋白中未检测出硒代蛋氨酸形式的硒，瘤胃上清液中也未检测出存在该种形式的硒。近年来的研究发现，当以硒酸钠和亚硒酸钠作为硒源时，二者合成到菌体蛋白中的硒无明显差别，约为3.3%～5.5%，其中既有硒代半胱氨酸形式（42%），也有硒代蛋氨酸形式（20%）[82]。瘤胃中硒酸钠还原为亚硒酸钠以及亚硒酸钠还原为元素硒和硒化物均降低了硒在小肠的吸收利用。研究表明，经瘤胃微生物代谢后，奶牛、山羊和绵羊对日粮亚硒酸钠和硒酸纳的小肠吸收率约为50%[83]。日粮精粗比会影响反刍动物对无机硒的吸收利用，在高粗料条件下，奶牛对硒的吸收率较低，而在高精料条件下，奶牛对硒的吸收率提高。Hudman和Glenn用亚硒酸钠分别对反刍兽新月形单孢菌（*Selenomonas ruminantium*）、溶纤维丁酸弧菌（*Butyrivibrio fibrisolven*）和栖瘤胃普雷沃氏菌（*Prevotella ruminicola*）进行体外培养后发现，反刍兽新月形单孢菌和溶纤维丁酸弧菌能够利用无机硒合成含硒氨基酸进而合成到菌体蛋白中，反刍兽新月形单孢菌对硒的利用速度很快，在利用亚硒酸钠合成含硒氨基酸的同时，仅有少部分的硒被还原为元素硒，而栖瘤胃普雷沃氏菌会把大部分的硒还原为元素硒[84]。在高粗料条件下，溶纤维丁酸弧菌和栖瘤胃普雷沃氏菌是瘤胃中的优势菌，而在高精料条件下，瘤胃中占主导地位的菌群为嗜淀粉拟杆菌（*Bacterorides amylophilus*）、乳酸杆菌（*Lactobacillus acidophilus*）、反刍兽新月形单孢菌和链球菌。

目前，在反刍动物日粮中添加的有机硒源主要为硒酵母。硒酵母产品中90%以上的硒是以硒代蛋氨酸形式存在的，少量以硒代半胱氨酸形式存在，无机硒的含量不超过2%。因此，酵母硒在瘤胃中的代谢可看作是硒代蛋氨酸在瘤胃中的代谢。用硒代蛋氨酸对瘤胃液进行体外培养发现，60%的硒存在于菌体蛋白中[77]。用硒代蛋氨酸饲喂绵羊的研究发现，菌体蛋白中既有硒代半胱氨酸形式的硒，又有硒代蛋氨酸形式的硒[79]。上述研究暗示，在瘤胃中，硒代蛋氨酸既可以通过非特定方式随机取代蛋氨酸合成到菌体蛋白中，也可以通过蛋白质生物合成途径进入到菌体蛋白中。分别用亚硒酸钠、硒酸钠和硒代蛋氨酸对瘤胃液进行体外培养发现，以硒代蛋氨酸作为硒源时，有23%的硒合成到菌体蛋白中，菌体蛋白中79%的硒以硒代蛋氨酸形式存在，是以亚硒酸钠和硒酸钠作为硒源时菌体蛋白中硒含量的3.8～4倍[82]，说明饲喂酵母硒可提高瘤胃细菌硒含量[85]。有限的研究资料表明，硒代蛋氨酸经瘤胃代谢后在山羊小肠的吸收率约为65%[86]，硒酵母经瘤胃代谢后在绵羊小肠的吸收率约为44%[87]。

以酵母硒中硒代蛋氨酸的含量为90%计，且60%的硒代蛋氨酸逃过瘤胃微生物的代谢进入小肠，则酵母硒中55%的硒以硒代蛋氨酸形式在小肠中吸收，按其80%的小肠吸收率计算，这部分硒的小肠吸收率为45%，剩余45%的硒以非硒代蛋氨酸形式在小肠吸收，按其50%的小肠吸收率计算，这部分硒的吸收率为22.5%。由此计算，酵母硒经瘤胃代谢后在反刍动物小肠的吸收率约为66%，其吸收率大约比亚硒酸钠高30%[83]。绝大多数研究表明，与亚硒酸钠相比，酵母硒有更高的生物利用率，更易提高反刍动物的血硒、乳硒和组织硒含量[88-92]。与亚硒酸钠相比，日粮添加酵母硒后全血硒含量平均可提高20%，全血GPX活性平均可提高16%，乳硒含量平均可提高90%，硒由胎盘向胎儿的转运效率平均可提高42%[93]。

二、硒在动物体内的代谢

日粮硒经小肠吸收后被转运到肝脏，肝脏是硒代谢的关键器官，可以合成绝大多数硒蛋白。其中硒蛋白P（Selenoprotein P，SelP，SEPP1，

SEPP)是肝脏合成的重要硒蛋白,负责将硒通过血液循环转运到机体的其他组织[94]。对小鼠的研究表明,肝脏 SelP 的基因表达水平最高,肾脏、骨骼肌、心脏、睾丸和脑组织 SelP 的基因表达水平分别相当于肝脏的 38%、10%、6%、6%和 2%[95]。人血浆中 SelP 的含量平均约为 5.6 mg/L。无论何种形式的硒,想要合成硒蛋白,吸收后在细胞内均需代谢生成中间产物 H_2Se(图 1-3)[96],关于硒蛋白的生物合成过程将在第二章中详细阐述。硒酸盐在被作为硒源利用前需还原为亚硒酸盐,还原途径类似于硝酸盐或硫酸盐的还原,亚硒酸盐在还原型谷胱甘肽或硫氧还蛋白还原酶(TrxR)作用下进一步还原为硒化物。硒代蛋氨酸需通过转硫途径生成硒代半胱氨酸,硒代半胱氨酸在硒代半胱氨酸裂解酶作用下生成硒化物和丙氨酸。体内硒蛋白分解产生的硒代半胱氨酸也可在硒代半胱氨酸裂解酶作用下生成硒化物。因此,硒代半胱氨酸裂解酶在细胞内硒的循环利用中起着重要的作用。尤其是当摄入硒不足时,这种酶对于维持硒蛋白的合成至关重要。此外,该酶的存在还可以最大限度地降低游离硒代半胱氨酸对细胞的毒副作用。细胞并不能有效区分硒代蛋氨酸和蛋氨酸,因此,硒代蛋氨酸除了代谢生成硒化物合成硒蛋白外,还可通过非特定方式随机取代蛋氨酸合成到蛋白质中。为了与硒蛋白加以区分,通常把这类蛋白称为含硒蛋白,机体所有含有蛋氨酸的蛋白质中均可见硒代蛋氨酸的存在,含硒蛋白的合成有利于机体对硒的沉积或存留。

三、硒的排泄

动物体内的硒可通过粪便、尿液和呼吸排出体外,也可通过毛发和皮肤细胞的脱落排出体外。排出的比例取决于摄入硒的化学形式、摄入水平和摄入方式,同时也受物种差异和日粮中其他因素的影响。对于单胃动物,无论硒摄入方式是口服还是注射,尿液是硒排泄的主要途径。对于反刍动物,当硒摄入方式是口服时,主要通过粪便排出,而当硒摄入方式为注射(皮下或静脉)时,通过尿液排出的硒要高于粪便。

正常生理情况下,日粮中未被消化吸收的硒以及少量通过胆汁、胰腺和肠排泄的内源硒主要通过粪便排出,而消化吸收的硒经代谢生成硒化物,过量的硒化物主要通过尿液排泄(图 1-3)。消化吸收的硒代蛋氨酸和硒代半胱氨酸除了代谢生成硒化物经尿液排泄外,也可在胱硫醚-γ-裂解酶的作用下生成甲基硒,并进一步代谢生成离子形式的三甲基硒阳离子(TMSe)经尿液排出。

尿液中至少有 5 种不同的硒代谢产物,对小鼠的研究发现,TMSe 是尿液中硒的主要存在形式，当硒摄入过量时，尿液中 TMSe 的浓度明显提高[97]，在正常生理剂量到低毒性剂量范围内，尿液中还存在一种叫作 1β-甲硒基-N-乙酰基-D-半乳糖胺的硒糖类物质，这种物质在肝脏中合成，通过尿液排出[98]。

当硒摄入过量时，除了粪尿硒排泄增加外，呼吸也成为硒排泄的重要途径，挥发性的二甲基硒(DMSe)是硒呼吸排泄的主要形式(图 1-3)[99]。

从物质代谢的角度讲，毛发和皮肤细胞中的硒是硒沉积或存留的结果，并不属于硒排泄的范畴，但毛发和皮肤细胞的脱落可导致体内硒的丢失。

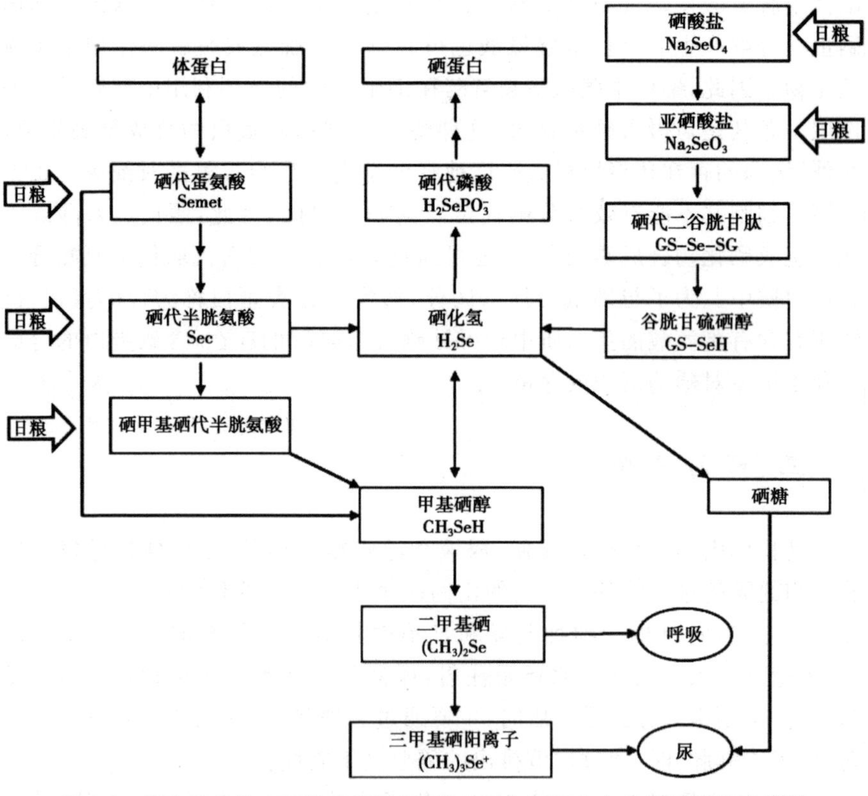

图 1-3　硒在动物体内的代谢通路(硒蛋白及含硒蛋白合成及硒的排泄)[100]

四、硒的组织分布

在日粮硒摄入充足情况下，肾脏硒水平最高，其次为肝脏、脾脏和胰脏。肠和肺组织硒水平相对较高，心肌硒水平明显高于骨骼肌。毛发硒水平相对较高，而神经组织中硒含量相对较低，但每单位组织的硒含量在大脑和内分泌器官中最高。日粮硒摄入水平和形式均会影响组织硒水平，一定范围内，组织硒水平会随着日粮硒摄入水平的提高而增加。相同剂量的有机硒要比无机硒更易提高组织硒水平，这主要与有机硒有较高的生物利用效率有关。

第七节　动物硒营养状况评价

动物的营养状况与动物的健康水平以及动物的生产和生殖密切相关，营养素缺乏或过量均会对动物的健康和生产产生不利影响，适宜的营养状况是维持动物健康并在健康状况下发挥最大生产性能的重要保证，同时也是确定动物对各种营养素需要量的重要参考依据。硒是动物的必需微量营养素，在动物体内通过合成硒蛋白行使其广泛的生物学功能，同其他必需营养素一样，适宜的硒营养状况同样是维持动物健康、使动物发挥最优生产的重要保证。硒缺乏引起的疾病通常在严重硒缺乏时才会表现出来，因此，寻找一种在生产实践中可操作性强且能准确诊断亚临床代谢紊乱或硒供应不足的方法，以能够敏感、准确反应动物硒营养状况的生物学指标作为指示物（marker），通过大量的临床试验确定其在生物样本中的正常参考范围，进而指导在动物日粮中的硒适宜添加量，这对于保证动物健康以及提高或达到预期的生产性能具有重要的意义。

目前，硒营养状况的评价方法主要包括两类。一类是直接评价法，主要是通过测定生物体体液和组织中的硒含量来反应机体的硒营养状况；另一类是间接评价法，主要是通过测定生物体体液或组织中能够敏感反应硒营养状况的功能性生物学指标进行评价。

一、直接评价法

动物硒营养状况的直接评价指标主要包括血液和组织中的总硒含量。血硒包括全血或红细胞硒以及血清或血浆硒。组织硒主要包括肝脏、肾脏和骨骼肌硒。对于人而言，直接评价指标也可以是头发和指甲中的硒含量。对于产奶母畜而言，也可通过测定奶中的硒含量来评价硒营养状况。

（一）血浆或血清硒

动物血清或血浆硒含量与日粮硒摄入量以及肌肉硒注射剂量存在高度的正相关关系，对当前的硒摄入水平反应迅速且准确，通常反映的是动物的短期硒营养状况[101]，血清或血浆硒被誉为是评价动物硒营养状况的黄金标准。早期研究认为，成年牛血清或血浆中硒的正常水平为40～120 μg/L[102,103]。后来的研究认为，成年牛血清或血浆中硒的正常水平为70～100 μg/L[104,105]。根据大量的临床试验观察，Scholz 和 Fleischer 对奶牛血清硒水平制定了6级评价参考标准，极度硒缺乏：血清硒水平低于10 μg/L；重度硒缺乏：血清硒水平10～30 μg/L；轻度硒缺乏：血清硒水平30～50 μg/L；临界硒缺乏：血清硒水平50～70 μg/L；硒适宜：血清硒水平70～90 μg/L；硒充足：血清硒水平100 μg/L 及以上[106,107]。依据该参考范围，对1 365头成年荷斯坦奶牛为期一年的硒营养状况评价结果表明，血清硒含量低于70 μg/L 的奶牛占33.7％，其日粮硒摄入水平低于0.3 mg/kg DM，其临床表现为免疫功能降低，易患乳腺炎，而其余奶牛的血清硒水平均在70 μg/L 以上，平均75 μg/L，处于硒适宜水平范围内[105]。

相同动物在不同年龄段的血清硒水平存在较大差异，而不同动物的血清硒水平存在着相当大的差异，成年后这种差异表现最为明显。例如，成年绵羊和山羊的血清硒水平为120～160 μg/L，而成年马的血清硒水平为130～160 μg/L，成年猪的血清硒水平为180～220 μg/L（表1-4）。因此，当用血清或血浆硒评价动物的硒营养状况时，一定要考虑动物及其年龄。此外，当用血清或血浆硒评价动物的硒营养状况时，还需要考虑日粮硒的存在形式[108,109]。当硒摄入形式为无机硒时，血浆或血清中的硒主要以 SelP 和血浆 GPX（PGPX，GPX3）形式存在，随着硒摄入量的提高，血清或血浆 SelP 和 GPX3 水平逐渐提高，但当硒摄入过量时，由于受到稳态调节作用的影响，血清或血浆 SelP 和 GPX3 水平达到平台期而不再明显提高，过量摄入的硒主要通过增加硒的排泄而排出体外。当硒摄入形式为酵母硒或硒代蛋氨酸时，血浆或血清中的硒除了以 SelP 和 GPX3 形式存在外，还存在血浆蛋白硒，因为硒代蛋氨酸可以以非特定方式随机取代蛋氨酸合成到蛋白质中。因此，当硒代蛋氨酸形式的硒摄入超过正常水平时，血浆或血清硒仍可提高，与摄入超营养的无机硒相比，受稳态调节作用的影响较小，在一定程度上对机体硒的存留、胎儿对硒的利用以及牛奶中硒含量的提高有积极作用，但蛋白质中蛋氨酸被过多取代会影响蛋白质正常功能的发挥。

表 1-4　动物血清硒水平参考值[104]

年龄(天)	动物血清硒含量参考范围(μg/L)				
	牛	绵羊	山羊	猪	马
<1	50～70	50～80	50～80	70～90	70～90
1～9	50～70	60～90	60～90	70～120	70～90
10～29	55～75	70～100	70～100	70～120	80～100
30～70	60～80	80～110	80～110	100～160	90～110
71～180	60～80	80～110	80～110	140～190	90～110
181～300	60～80	80～110	80～110	180～220	90～110
301～700	65～90	90～120	90～120	180～220	100～130
>700	70～100	120～160	120～160	180～220	130～160

(二)全血硒

全血硒包括血清硒和红细胞硒两部分，全血硒含量与日粮硒摄入水平也存在明显正相关关系。然而，与血清或血浆硒相比，全血硒对日粮硒摄入水平的变化反应缓慢。原因在于红细胞中的硒主要以胞质 GPX(cGPX，GPX1)形式存在，红细胞中的 GPX1 是在红细胞生成过程中合成的。因此，红细胞硒对日粮硒摄入的完全反应需要一定的时间，这个时间等同于红细胞的平均寿命。对于牛而言，红细胞的平均寿命较长，约为 90～120 天，因此，红细胞或全血硒反映的是早期而不是当下的硒摄入水平，故红细胞或全血硒通常主要反映的是长期硒营养状况[101]。

从数据准确性的角度讲，全血硒要优于血清或血浆硒，因为全血在运输和贮存过程中红细胞的部分溶血现象会造成血清或血浆硒含量的假性提高。然而，由于全血硒对日粮硒摄入量的提高和降低的反应都比较慢，获得直接的临床数据需要更长的时间，在生产实践中，人们更趋向于用血浆或血清硒评价动物的硒营养状况。尽管没有用全血硒评价动物硒营养状况的直接数据，但在许多相关的研究中测定了全血硒，在利用这些数据进行动物硒营养状况评价时，可在已知血清或血浆硒水平正常范围的基础上，通过二者的比值估计全血硒含量的正常范围，以此作为用全血硒评价动物硒营养状况的参考标准。猪、马、奶牛以及羊全血硒和血清硒的比值分别大约为 1、1.4～1.5、2.5 和 4，以此计算，上述动物全血硒的参考范围分别为 180～220 μg/L(猪)、195～240 μg/L(马)、175～250 μg/L(牛)、480～640 μg/L

(羊)。对奶牛的研究表明,当全血硒水平低于 60 μg/L 时表现为重度硒缺乏,60~200 μg/L 时处于临界硒缺乏状态[110-112]。需要注意的是,全血硒和血清硒的比值会因动物的年龄不同而有所差异,而且全血硒和血清硒的比值会随着硒摄入量的变化而变化,当硒摄入水平较低时,比值会变大,而当硒摄入水平过高时,比值会减小。对美洲驼的研究表明,在血红蛋白浓度不变的情况下,当日粮硒摄入量缓慢提高到 20 mg/kg 时,全血硒和血清硒的比值从通常的 1.4~1.5 降低到 0.8[104]。此外,由于血清硒水平的参考值范围较宽,会很大程度上影响利用比值估算的准确性。一些研究表明,遗传多样性也是影响比值的一个因素[113]。当用全血硒评价动物的硒营养状况时,同样需要考虑硒的摄入形式,因为与无机硒相比,硒代蛋氨酸形式的硒同样可以以非特定方式随机替代蛋氨酸进入到红细胞中的血红蛋白中。大量研究表明,与亚硒酸钠相比,相同摄入水平的硒酵母可显著提高奶牛全血的硒含量[92,114,115],平均提高 20%左右[83]。

(三)组织硒

动物组织中硒含量由高到低的顺序为肝脏、肾脏、心脏、骨骼肌和脂肪组织[116]。对猪的研究表明,组织中硒含量与血浆硒水平存在显著的正相关关系,相关系数分别为 0.81(骨骼肌)、0.91(肝脏)和 0.71(肾脏)[117]。对牛的研究表明,肝脏和骨骼肌硒含量与全血硒水平呈显著的正相关,相关系数分别为 0.78 和 0.83[118]。肝脏是机体硒代谢的重要器官,因此在组织中,通常主要以肝脏硒水平来评价动物的硒营养状况。Kincaid 研究认为,动物肝脏中硒的正常参考范围为 1.2~2.0 μg/g 干重,肝脏中硒含量范围不因动物种类和年龄的不同而不同[111]。当肝硒含量(干重)高于 2.5 μg/g 时,常提示动物处于硒中毒状态,同时会表现出明显的临床症状,当肝硒含量(干重)低于 1.2 μg/g 但高于 0.6 μg/g 时,常提示动物处于轻度硒缺乏状态,而当肝硒含量(干重)低于 0.6 μg/g 时,则动物处于重度硒缺乏状态,常会表现出明显的临床硒缺乏症状。当以湿重基础表示肝脏硒含量时,可以以干重基础肝脏硒含量的 1/3 估算。当用组织硒评价动物的硒营养状况时,同样需要考虑硒的摄入形式,同样也是因为硒代蛋氨酸形式的硒可以以非特定方式随机替代蛋氨酸进入到组织蛋白中。

(四)乳硒

对于产奶母畜而言,乳硒含量是评价硒营养状况的潜在指标。与血液和组织硒相比,乳硒评价指标具有取样简单、对动物伤害小、评价范围广等优点。对奶牛的研究表明,乳硒含量与血清和全血硒含量以及全血 GPX

活性均存在明显的相关性，相关系数分别为0.92、0.57和0.82[103,119,120]，相比较而言，乳硒含量与血清硒水平的相关性最好，呈S形曲线相关。如图1-4所示，当乳硒含量达到20 μg/L时，此时的血清硒含量达到70 μg/L，提示动物处于硒适宜状态，而当乳硒含量低于10 μg/L时，此时血清硒含量低于30 μg/L，提示动物处于重度硒缺乏状态。与血清和血浆硒一样，乳硒也可对日粮硒的摄入作出快速反应，因而反映的也是短期硒营养状况。

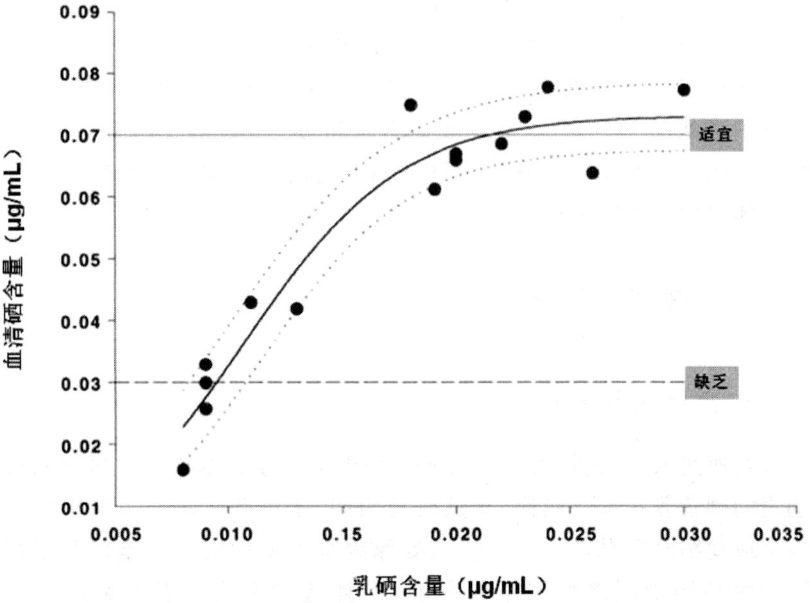

图1-4 血清硒含量与桶装乳硒含量的相关关系[119]

当用乳硒作为硒营养状况的评价指标时，一定要考虑胎次、泌乳阶段、乳产量以及乳蛋白和乳脂水平。早期泌乳阶段的乳硒含量要高于泌乳后期，原因可能主要与乳的稀释有关，因为泌乳阶段对血清硒含量无显著影响。乳脂和乳蛋白水平较高时往往乳硒含量也较高。同样，当用乳硒评价动物的硒营养状况时，需要考虑硒的摄入形式，因为硒代蛋氨酸形式的硒可以以非特定方式随机替代蛋氨酸进入到乳蛋白中。大量研究表明，与亚硒酸钠相比，相同摄入水平的硒酵母可显著提高牛奶中的硒含量[89,121,122]，乳硒含量平均提高90%左右[83]。

（五）毛发硒

毛发中硒的积累通常需要几个月的时间，通常反映的是长期硒营养状况。用毛发硒含量评价硒营养状况具有样本易于获得、保存简单，采样技术要求不高，对动物无危害等优点，适合于偏远地区和某一地区畜群硒营养状况的整体评价。研究表明，当奶牛毛发硒含量在0.06～0.23 mg/kg DM时，犊牛出生的死亡率较高，患肌肉营养不良的几率提高，而当毛发硒含量高于0.25 mg/kg DM时，犊牛出生后健康状况良好[123]。Chrstodoulopoulos等[124]对希腊40个牧场400头奶牛的全血和毛发硒含量的分析结果表明，在饲喂制度和日粮硒含量相同的条件下，毛发硒含量和全血硒含量存在显著的正相关，其中黑色毛发的相关系数为0.61，白色毛发的相关系数为0.77，暗示毛发硒可作为评价奶牛硒营养状况的指标。当日粮硒摄入量超过0.3 mg/kg DM时，即使全血硒含量达到饱和状态(大约300 μg/L)，但毛发硒含量仍可提高，暗示毛发硒含量不仅可以指示硒营养状况，还可以指示硒中毒状况[103,107,123]。当毛发硒含量超过5 mg/kg DM时，往往指示动物处于硒中毒状态。毛发硒对过量硒摄入的反应很可能是因为毛发参与了机体对硒的脱毒过程[125]。

二、间接评价指标

(一)土壤或牧草硒

针对放牧家畜，最初人们试图通过分析土壤或牧草中硒含量来间接评价动物的硒营养状况。然而，大量研究表明，土壤硒含量并不能很好地反映牧草或动物的硒状况[126-128]。主要原因是土壤中的硒含量以及植物对土壤硒的利用受到多种因素的影响，而且这些影响因素很难进行人为控制。

(二)血液GPX

随着人们对硒的生理功能的逐渐了解，无论是放牧家畜还是集约化饲养家畜，硒的补给已成必然。在日粮硒添加的基础上，Smith等[112]建议用血液中GPX的活性评价动物的硒营养状况，其出发点在于GPX是硒的生理功能形式，从功能生物学的角度讲，GPX可作为机体硒营养状况的重要功能指示物，要比用血液硒含量评价更能准确反映动物的硒营养状况。大量研究表明，硒的添加水平与血液中GPX活性存在明显的正相关[129]，从而从理论上解释了用血液GPX活性评价动物硒营养状况的可行性。

机体内总硒的大约12%存在于GPX中[130]，血液中的GPX主要存在于红细胞中，占到全血GPX的几乎98%[107]，由于红细胞的寿命较长(90～120天)，红细胞或全血GPX活性通常反映长期硒营养状况，而血浆和血小板GPX活性通常反映短期硒营养状况。研究表明，随着硒摄入量的提高，血小板 GPX 活性在 2 周内明显提高，原因在于血小板的寿命较短，平均 8～14 天[131]。就目前而言，主要采用全血 GPX 活性来评价动物的硒营养状况。研究表明，牛、羊、猪、马以及人的全血 GPX 活性与血硒含量呈显著的正相关，相关系数最高接近 1.0，最低也可达到 0.81[106]。根据血硒的参考范围，一些研究给出了部分动物全血中适宜 GPX 活性的下限：绵羊为 600 μkat/L[132]，牛为 600～700 μkat/L[106]，马为 200～300 μkat/L[133]，山羊为 700 μkat/L[134]。也有一些研究给出了正常参考范围：猪为 100～200 μmol/g 血红蛋白（红细胞），牛为 19～36 μmol/g 血红蛋白（红细胞）、472.2～665.4 μkat/L（全血），绵羊为 60～180 μmol/g 血红蛋白（红细胞），马为 30～150 μmol/g 血红蛋白（红细胞）[135]。

与硒含量相比，用全血 GPX 活性评价动物硒营养状况具有测定简单方法简单、成本低等优点。但是，当日粮硒摄入水平或全血硒含量过高时，不适宜用全血 GPX 活性评价硒营养状况，因为当机体硒含量超过一定水平时，GPX 活性达到饱和而不再增加。此外，GPX 活性的度量单位很多，包括 μkat/L、mmol/L、IU/L、U/mL、μmol/g 血红蛋白、U/g 血红蛋白、IU/mg 血浆蛋白等，由于没有统一的标准，各个实验室使用的度量单位不尽相同，因而无法对实验室间的数据进行比较。用全血 GPX 活性评价动物的硒营养状况时，也应该考虑日粮硒的摄入形式。尽管相同水平的无机硒和有机硒对 GPX 的合成量无明显影响，但会影响其合成速度。对山羊的研究发现，亚硒酸钠与乳蛋白硒对机体的供硒能力无明显区别，但亚硒酸钠可明显增加 GPX 活性的提高速度[136]。

（三）血浆 SelP

SelP 是一种分泌型硒蛋白，与其他硒蛋白的一个显著的不同点是 SelP 中的含硒量很高。SelP 在肝脏合成，分泌到血浆，通过血液循环将硒转运到机体的其他组织以供组织利用。关于血浆 SelP 作为评价硒营养状况的研究主要集中在人上。研究表明，人的血浆中 60%～80%的硒以 SelP 形式存在，血浆 SelP 与其他反映硒营养状况的指标如血清硒、血浆 GPX 和红细胞 GPX 呈显著的正相关[137-139]。对肝硬化严重程度与血浆 SelP 水平关系的研究发现，肝硬化疾病越严重，血浆 SelP 水平越低，严重肝硬化病人血浆 SelP 水平只有健康人的 50%，而血浆 GPX 活性不降反升，血浆 SelP 水平的降低可能主要与肝损伤有关，血浆 GPX 活性的升高可能是一种补偿性反应[140]。对克山病与血浆 SelP 水平的关系的研究发现，克山病病人的血浆 SelP 水平显著低于正常人，补充一定量的硒后发现，当血浆 GPX 活性达到正常时，血浆 SelP 水平仅为正常水平的 75%，说明 SelP 达到正常水平所需的硒要比 GPX 达到正常水平所需的硒要多，暗示 SelP 可作为评价硒营养状况的指标，而且要比 GPX 更灵敏[141]。然而，另一些研究发现，当硒摄入量较正常水平降低 50%时，血清硒降低 11%，红细胞 GPX 活性也显著降低，但血浆 SelP 水平无明显改变，而当硒摄入量回归正常后，血清硒含量和红细胞 GPX 活性显著提高，但血浆 SelP 水平无明显改变[139,142]。还有一些研究发现，达到最大血浆 SelP 水平所需的硒摄入量要比达到最大 GPX 活性所需的硒摄入量低[142,143]。上述研究结果存在明显的不一致甚至是相反的，因此，血浆 SelP 能否作为评价硒营养状况的指示物还有待进一步研究。

（四）其他评价指标

乳体细胞数通常用来估计乳腺的健康状况，当乳腺受到细菌感染或处于炎症状态时，白细胞迁移到乳中导致细胞数量增加。健康乳腺分泌的乳中体细胞数一般不超过 1×10^5 个/mL。当血液中 GPX 活性低于 3.3 mkat/g 血红蛋白时，乳腺易受到细菌侵染。当乳体细胞数量较低时，通常血清硒含量和血液 GPX 活性较高，乳体细胞数与动物硒营养状况呈明显的负相关[144-146]。因此，乳体细胞数也可作为奶牛硒营养状况的评价指标，但应结合其他评价指标特别是血液 GPX 活性进行评价。

参考文献

[1] 黄开勋，徐辉碧. 硒的化学、生物化学及其在生命科学中的应用[M]. 武汉：华中科技大学出版社，2009.

[2] Zayed A, Terry N. Selenium volatilization in roots and shoots: Effects of shoot removal and sulfate level[J]. Journal of Plant Physiology, 1994, 143(1): 8-14.

[3] Winkel L H E, Johnson C A, Lenz M, et al. Environmental Selenium Research: From Microscopic Processes to Global Understanding[J]. Environmental Science & Technology, 2012, 46(2): 571-579.

[4] Mosher B W, Duce R A. A global atmospheric selenium budget[J]. Journal of Geophysical Research Atmospheres, 1987, 92(D11): 13289-13298.

[5] Nakamaru Y, Tagami K, Uchida S. Distribution coefficient of selenium in Japanese agricultural soils[J]. Chemosphere, 2005, 58(10): 1347-1354.

[6] Taylor S R, Mclennan S M. The Continental Crust: Its Composition and Evolution, An Examination of the Geochemical Record Preserved in Sedimentary Rocks [M]. Blackwell Scientific Pub., 1985.

[7] Mehdi Y, Hornick J L, Istasse L, et al. Selenium in the environment, metabolism and involvement in body functions [J]. Molecules, 2013, 18(3): 3292-3311.

[8] Natasha, Shahid M, Niazi N K, et al. A critical review of selenium biogeochemical behavior in soil-plant system with an inference to human health[J]. Environmental Pollution, 2017, 234: 915-934.

[9] 田欢. 典型富硒区岩石-土壤-植物中硒的赋存状态及环境行为研究[D]. 武汉：中国地质大学，2017.

[10] Kabata P A, Pendias H. Trace elements in soils and plants [M]. Washington D C: CRC Press, 1992.

[11] Tan J A, Zhu W, Wang W, et al. Selenium in soil and endemic diseases in China[J]. Science of the Total Environment, 2002, 284(1): 227-235.

[12] Rayman M P. Selenium and human health : The Lancet[J]. Lancet, 2000, 356: 942-943.

[13] Wang Z, Gao Y. Biogeochemical cycling of selenium in Chinese environments[J]. Applied Geochemistry, 2001, 16(11): 1345-1351.

[14] 王子健. 中国低硒带生态环境中硒的环境行为研究进展[J]. 环境化学, 1993, 12(3): 237-243.

[15] Jezek P, Skarpa P, Losak T, et al. Selenium-An Important Antioxidant in Crops Biofortification [M]. UK: InTech, 2012.

[16] 高新楼，秦中庆，苏利，等. 喷施富硒液对富硒小麦籽粒硒含量及产量的影响[J]. 安徽农业科学，2007，35(18)：5498-5499.

[17] 程兆东，康怀启，王永刚，等. 叶面喷施硒肥对梨果实和叶片的影响[J]. 中国果菜，2017，37(4)：30-33.

[18] Barberon M, Berthomieu P, Clairotte M, et al. Unequal functional redundancy between the two Arabidopsis thaliana high-affinity sulphate transporters SULTR1; 1 and SULTR1; 2 [J]. New Phytologist, 2008, 180(3): 608-619.

[19] E E K, N C, H R, et al. Characterization of a selenate-resistant Arabidopsis mutant. Root growth as a potential target for selenate toxicity [J]. Plant Physiology, 2007, 143(3): 1231-1241.

[20] Li H F, Mcgrath S P, Zhao F J. Selenium uptake, translocation and speciation in wheat supplied with selenate or selenite[J]. New Phytologist, 2010, 178(1): 92-102.

[21] Schiavon M, Pilon M, Malagoli M, et al. Exploring the importance of sulfate transporters and ATP sulphurylases for selenium hyperaccumulation a comparison of Stanleya pinnata and Brassica juncea (Brassicaceae)[J]. Frontiers of Plant Science, 2015, 6(6): 2.

[22] Zhang L, Hu B, Li W, et al. OsPT2, a phosphate transporter, is involved in the active uptake of selenite in rice[J]. New Phytologist, 2014, 201(4): 1183-1191.

[23] Rosenfeld I, Beath O A. Selenium. Geobotany, biochemistry, toxicity, and nutrition[J]. Bulletin of the Torrey Botanical Club, 1964, 92 (5): 414.

[24] Surade J M, Reynolds R J, Richterova K, et al. Selenium hyperaccumulators harbor a diverse endophytic bacterial community characterized by high selenium resistance and plant growth promoting properties [J]. Frontiers of Plant Science, 2015, 6: 113.

[25] Mayland H F, James L F, Panter K E, et al. Selenium in Seleniferous Environments[J]. Selenium in Agriculture & the Environment, 1989, 23(seleniuminagric): 15-50.

[26] Baä±Uelos G S, Ajwa H A, Wu L, et al. Selenium Accumulation by Brassica Napus Grown in Se-Laden Soil From Different Depths of Kesterson Reservoir[J]. Journal of Soil Contamination, 1998, 7(4): 481-496.

[27] Zayed A, Gowthaman S, Terry N. Phytoaccumulation of Trace Elements by Wetland Plants: I. Duckweed[J]. Journal of Environmental Quality, 1998, 27(3): 715-721.

[28] Wang S, Xiongping W U, Liang D, et al. Transformation and bioavailability for Pak choi(Brassica chinensis) of different forms of selenium added to calcareous soil[J]. Acta Scientiae Circumstantiae, 2010, 30 (12): 2499-2505.

[29] Fernandezmartinez A, Charlet L. Selenium environmental cycling and bioavailability: a structural chemist point of view[J]. Reviews in Environmental Science and Biovtechnology, 2009, 8(1): 81-110.

[30] Eichgreatorex S, Sogn T A, Ogaard A F, et al. Plant availability of inorganic and organic selenium fertiliser as influenced by soil organic

matter content and pH[J]. Nutrient Cycling in Agroecosystems, 2007, 79(3): 221-231.

[31] Lessa J H L, Araujo A M, Silva G N T, et al. Adsorption-desorption reactions of selenium (VI) in tropical cultivated and uncultivated soils under Cerrado biome[J]. Chemosphere, 2016, 164: 271-277.

[32] Saha U. Selenium in the Soil-Plant Environment: A Review[J]. International Journal of Applied Agricultural Sciences, 2017, 3(1): 1.

[33] Wallenberg M, Olm E, Hebert C, et al. Selenium compounds are substrates for glutaredoxins: a novel pathway for selenium metabolism and a potential mechanism for selenium-mediated cytotoxicity[J]. Biochemical Journal, 2010, 429(1): 85-93.

[34] Cubadda F, Aureli F, Ciardullo S, et al. Changes in selenium speciation associated with increasing tissue concentrations of selenium in wheat grain[J]. Journal of Agricultural and Food Chemistry, 2010, 58(4): 2295-2301.

[35] Pilonsmits E A, Hwang S, Mel L C, et al. Overexpression of ATP sulfurylase in indian mustard leads to increased selenate uptake, reduction, and tolerance[J]. Plant Physiology, 1999, 119(1): 123-132.

[36] Sors T G, Ellis D R, Salt D E. Selenium uptake, translocation, assimilation and metabolic fate in plants[J]. Photosynthesis Research, 2005, 86(3): 373-89.

[37] Carvalho K M, Gallardo-Williams M T, Benson R F, et al. Effects of selenium supplementation on four agricultural crops[J]. Journal of Agricultural and Food Chemistry, 2003, 51(3): 704-709.

[38] 姜英，曾昭海，杨麒生，等. 植物硒吸收转化机制及生理作用研究进展[J]. 应用生态学报，2016，27(12)：4067-4076.

[39] Xue T, Hartikainen H, Piironen V. Antioxidative and growth-promoting effect of selenium on senescing lettuce[J]. Plant & Soil, 2001, 237(1): 55-61.

[40] Djanaguiraman M, Devi D D, Shanker A K, et al. Selenium-an antioxidative protectant in soybean during senescence[J]. Plant & Soil, 2005, 272(1/2): 77-86.

[41] Hasanuzzaman M, Fujita M. Selenium pretreatment upregulates the antioxidant defense and methylglyoxal detoxification system and confers enhanced tolerance to drought stress in rapeseed seedlings[J]. Biolog-

ical Trace Element Research, 2011, 143(3): 1758-1776.

[42] Hasanuzzaman M, Hossain M A, Fujita M. Selenium in higher plants: physiological role, antioxidant metabolism and abiotic stress tolerance[J]. Journal of Plant Sciences, 2010, 5(4): 354-375.

[43] Hasanuzzaman M, Hossain M, Fujita M. Selenium-induced up-regulation of the antioxidant defense and methylglyoxal detoxification system reduces salinity-induced damage in rapeseed seedlings[J]. Biological Trace Element Research, 2011, 143(3): 1704-1721.

[44] Spallholz J E, Hoffman D J. Selenium toxicity: cause and effects in aquatic birds[J]. Aquatic Toxicology, 2002, 57(1): 27-37.

[45] Freeman J L, Zhang L H, Marcus M A, et al. Spatial imaging, speciation, and quantification of selenium in the hyperaccumulator plants Astragalus bisulcatus and Stanleya pinnata[J]. Plant Physiology, 2006, 142(1): 124-134.

[46] Schutz D F, Turekian K K. The investigation of the geographical and vertical distribution of several trace elements in sea water using neutron activation analysis[J]. Geochimica et Cosmochimica Acta, 1965, 29(4): 259-313.

[47] Kabata-Pendias H A, Mukherjee A B. Trace Elements from Soil to Human[J]. Springer Berlin, 2007: 1-550.

[48] Maplesden D C. Selenium. Geobotany, Biochemistry, Toxicity and Nutrition[J]. Canadian Veterinary Journal, 1965, 6(2): 37.

[49] James L F. Selenium Poisoning in Livestock[J]. Rangelands, 1984, 6(2): 64-67.

[50] Moxon A L. Alkali disease or selenium poisoning[J]. SD Agric Exp Stn Bul, 1937, 311(1): 1-84.

[51] Rogers P a M, Arora S P, Fleming G A, et al. Selenium toxicity in farm animals: treatment and prevention[J]. Irish Veterinary Journal, 1990, 43(6): 151-153.

[52] Olson O E. selenium in plants as a cause of livestock poisoning//Keeler R F. Effects of Poisonous Plants on Livestock[M]. New York: Academic Press,1978: 121-133.

[53] Gaßmann T. Der Nachweis des Selens im Knochen-und Zahngewebe [J]. Hoppe-Seyler's Zeitschrift für physiologische Chemie, 1916, 97(6): 307-310.

[54] Schwarz K, Foltz C M. Selenium as an Integral Part of Factor 3 against Dietary Necrotic Liver Degeneration[J]. Nutrition, 1978, 36(11): 255.

[55] Listed N. Epidemiologic studies on the etiologic relationship of selenium and Keshan disease[J]. Chinese Med J, 1979, 92(7): 471-476.

[56] Li Q, Liu M, Hou J, et al. The prevalence of Keshan disease in China[J]. International Journal of Cardiology, 2013, 168(2): 1121-1126.

[57] Van A R, Thomson C D, Mckenzie J M, et al. Selenium deficiency in total parenteral nutrition[J]. American Journal of Clinical Nutrition, 1979, 32(10): 2076-2085.

[58] 祁周约，张碧霞，许彩萍，等. 猪营养性肝坏死病的调查研究[J]. 陕西农业科学，1980(6)：30-34.

[59] 蒋守群. 有机硒在动物营养上的研究与应用[J]. 饲料工业，2005，26(20)：43-45.

[60] 顾履珍，荫士安，周瑞华，等. 膳食蛋白质水平对硒利用的影响——Ⅲ、蛋白质水平对低硒饲料小鸡渗出性素质发病率的影响[J]. 营养学报，1987，(1)：8-13.

[61] Saha U. Selenium in Animal Nutrition: Deficiencies in Soils and Forages, Requirements, Supplementation and Toxicity[J]. 2016, 2(6): 112.

[62] Behne D, Wolters W. Distribution of selenium and glutathione peroxidase in the rat[J]. Journal of Nutrition, 1983, 113(2): 456-461.

[63] Hawkes W C, Wilhelmsen E C, Tappel A L. Abundance and tissue distribution of selenocysteine-containing proteins in the rat[J]. Journal of Inorganic Biochemistry, 1985, 23(2): 77-92.

[64] Mehdi Y, Dufrasne I. Selenium in Cattle: A Review[J]. Molecules, 2016, 21(4): 545.

[65] Nrc. Nutrient requirements of dairy cattle [M]. 7th ed. Washington DC, USA: The National Academies Press, 2001.

[66] Surai P F, Fisinin V I. Selenium in Pig Nutrition and Reproduction: Boars and Semen Quality—A Review [J]. Asian-australasian Journal of Animal Sciences, 2015, 28(5): 730-746.

[67] Nrc. Nutrient requirements of sheep [M]. 6th ed. Washington DC, USA: National Academy Press, 1985.

[68] Papazafeiriou A Z, Lakis C, Stefanou S, et al. Trace elements content of plant material growing on alkaline organic soils and its suitability for small ruminant extensive farming[J]. Bulgarian Journal of Agricul-

tural Science, 2016, 22(5): 733-739.

[69] Pagan J D, Karnezos P, Kennedy M a P, et al. Effect of selenium source on selenium digestibility and retention in exercised Thoroughbreds. in proceedings of the Equine Nutrition and Physiology Society [C]. Raleigh, NC, USA, 1999.

[70] Nrc. Minerals in nutrient requirements of horses [M]. Washington DC, USA: National Academies Press, 2007.

[71] Nrc. Nutrient requirements of poultry [M]. 6th ed. Washington DC, USA: National Academy Press, 1984.

[72] Nrc. Nutrient requirements of poultry [M]. 9th ed. Washington DC, USA: National Academy Press, 1994.

[73] Wolffram S, Ardüser F, Scharrer E. In vivo intestinal absorption of selenate and selenite by rats[J]. Journal of Nutrition, 1985, 115(4): 454-459.

[74] Turner J C, Osborn P J, Mcveagh S M. Studies on selenate and selenite absorption by sheep ileum using an everted SAC method and an isolated, vascularly perfused system[J]. Comparative Biochemistry and Physiology A: Comparative Physiology, 1990, 95(2): 297-301.

[75] Vendeland S C, Butler J A, Whanger P D. Intestinal absorption of selenite, selenate, and selenomethionine in the rat[J]. Journal of Nutritional Biochemistry, 1992, 3(7): 359-365.

[76] Cousins F B, Cairney I M. Some aspects of selenium metabolism in sheep[J]. Australian Journal of Agricultural Research, 1961, 12(5): 927-943.

[77] Paulson G D, Baumann C A, Pope A L. Metabolism of 75Se-selenite, 75Se-selenate, 75Se-selenomethionine and 35S-sulfate by rumen micro-organisms in vitro[J]. Journal of Animal Science, 1968, 27(2): 497.

[78] Podoll K L, Bernard J B, Ullrey D E, et al. Dietary selenate versus selenite for cattle, sheep, and horses[J]. Journal of Animal Science, 1992, 70(6): 1965-1970.

[79] Whanger P D, Weswig P H, Muth O H. Metabolism of Se-75-selenite and Se-75-selenomethionine by rumen microorganisms[J]. Federation of America Societies Experimental Biology, 1968, 27(2): 418.

[80] Serra A B, Nakamura K, Matsui T, et al. Inorganic selenium for sheep. I. Selenium balance and selenium levels in the different ruminal

fluid fractions[J]. Asian Australasian Journal of Animal Sciences, 1994, 7 (1): 83-89.

[81] Peterson P J, Spedding D J. The excretion by sheep of 75Selenium incorporated into red clover (Trifolium pratense L.): The chemical nature of the excreted selenium and its uptake by three plant species[J]. New Zealand Journal of Agricultural Research, 1963, 6(1-2): 13-23.

[82] Galbraith M L, Vorachek W R, Estill C T, et al. Rumen Microorganisms Decrease Bioavailability of Inorganic Selenium Supplements[J]. Biological Trace Element Research, 2016, 171(2): 338-343.

[83] Weiss W P, Eastridge M L. Selenium sources for dairy cattle. in proceedings of the Tri-State Dairy Nutrition Conference [C]. Columbus: The Ohio State University, 2005.

[84] Hudman J F, Glenn A R. Selenium uptake by Butyrivibrio fibrisolvens and Bacteroides ruminicola [J]. FEMS Microbiology Letters, 1985, 27(2): 215-220.

[85] Ryssen J B J V, Deagen J T, Beilstein M A, et al. Comparative metabolism of organic and inorganic selenium by sheep[J]. Journal of Agricultural and Food Chemistry, 1989, 37(5): 1358-1363.

[86] Aspila P. METABOLISM OF SELENITE, SELENOMETHIONINE AND FEED-INCORPORATED SELENIUM IN LACTATING GOATS AND DAIRY-COWS[J]. Journal of Agriculturalence in Finland, 1991, 63(1): 9-73.

[87] Koenig K M, Rode L M, Cohen R D, et al. Effects of diet and chemical form of selenium on selenium metabolism in sheep[J]. Journal of Animal Science, 1997, 75(3): 817-827.

[88] Wolffram S. Absorption and metabolism of selenium: differences between inorganic and organic sources. in proceedings of the Alltech's 15 the Annual Symposium [C]. Nottingham: Nottingham University Press, 1999.

[89] Juniper D T, Phipps R H, Jones A K, et al. Selenium Supplementation of Lactating Dairy Cows: Effect on Selenium Concentration in Blood, Milk, Urine, and Feces[J]. Journal of Dairy Science, 2006, 89 (9): 3544-3551.

[90] Heard J W, Stockdale C R, Walker G P, et al. Increasing selenium concentration in milk: effects of amount of selenium from yeast and cereal

grain supplements[J]. Journal of Dairy Science, 2007, 90(9): 4117-4127.

[91] Behne D, Kyriakopoulos A, Scheid S, et al. Effects of chemical form and dosage on the incorporation of selenium into tissue proteins in rats[J]. Journal of Nutrition, 1991, 121(6): 806.

[92] Knowles S O, Grace N D, Wurms K, et al. Significance of amount and form of dietary selenium on blood, milk, and casein selenium concentrations in grazing cows[J]. Journal of Dairy Science, 1999, 82(2): 429.

[93] Weiss W P, Hogan J S. Effect of Selenium Source on Selenium Status, Neutrophil Function, and Response to Intramammary Endotoxin Challenge of Dairy Cows * [J]. Journal of Dairy Science, 2005, 88(12): 4366-4374.

[94] Hill K E, Zhou J, Mcmahan W J, et al. Deletion of Selenoprotein P Alters Distribution of Selenium in the Mouse[J]. Journal of Biological Chemistry, 2003, 278(16): 13640-13646.

[95] Hill K E, Wu S, Motley A K, et al. Production of Selenoprotein P (Seppl) by Hepatocytes is Central to Selenium Homeostasis[J]. Journal of Biological Chemistry, 2012, 287(48): 40414-40424.

[96] Weekley C M, Harris H H. Which form is that? The importance of selenium speciation and metabolism in the prevention and treatment of disease[J]. Chemical Society Reviews, 2013, 42(23): 8870-8894.

[97] Suzuki K T, Kurasaki K, Okazaki N, et al. Selenosugar and trimethylselenonium among urinary Se metabolites: dose-and age-related changes[J]. Toxicology and Applied Pharmacology, 2005, 206(1): 1-8.

[98] Kobayashi Y, Ogra Y, Ishiwata K, et al. Selenosugars are key and urinary metabolites for selenium excretion within the required to low-toxic range[J]. Proceedings of the National Academy of Sciences of the United States of America, 2002, 99(25): 15932-15936.

[99] Mcconnell K P, Portman O W. Excretion of dimethyl selenide by the rat[J]. Journal of Biological Chemistry, 1952, 195(1): 277-282.

[100] Wrobel J K, Power R, Toborek M. Biological activity of selenium: Revisited[J]. IUBMB Life, 2015, 68(2): 97-105.

[101] Saun R J V. Rational approach to selenium supplementation essential[J]. Feedstuffs, 1990, 62(3): 15-17.

[102] De Toledo L R, Perry T W. Distribution of supplemental selenium in the serum, hair, colostrum, and fetus of parturient dairy cows

[J]. Journal of Dairy Science, 1985, 68(12): 3249-3254.

[103] Maus R W, Martz F A, Belyea R L, et al. Relationship of dietary selenium to selenium in plasma and milk from dairy cows[J]. Journal of Dairy Science, 1980, 63(4): 532-537.

[104] Stowe H D, Herdt T H. Clinical assessment of selenium status of livestock[J]. Journal of Animal Science, 1992, 70(12): 3928-3933.

[105] Gerloff B J. Effect of selenium supplementation on dairy cattle [J]. Journal of Animal Science, 1992, 70(12): 3934-3940.

[106] Pavlata L, Pechová A, Illek J. Direct and indirect assessment of selenium status in cattle-a comparison [J]. Acta Veterinaria Brno, 2000, 69(4): 281-287.

[107] Scholz R, Hutchinson L. Distribution of glutathione peroxidase activity and selenium in the blood of dairy cows[J]. American Journal of Veterinary Research, 1979, 40(2): 245-249.

[108] Fortier M E, Audet I, Giguère A, et al. Effect of dietary organic and inorganic selenium on antioxidant status, embryo development, and reproductive performance in hyperovulatory first-parity gilts[J]. Journal of Animal Science, 2012, 90(1): 231-240.

[109] Kim Y Y, Mahan D C. Prolonged feeding of high dietary levels of organic and inorganic selenium to gilts from 25 kg body weight through one parity[J]. Journal of Animal Science, 2001, 79(4): 956-966.

[110] Ryssen J V, O'dell E D. Selenium (Se) supplementation on the Se status of dairy cows in the Midlands of KwaZulu-Natal[J]. South African Journal of Animal Science, 2006, 36(5): 18-21.

[111] Kincaid R L. Assessment of trace mineral status of ruminants: A review[J]. Journal of Animal Scienc, 1999, 77: 1-10.

[112] Smith K. Selenium in dairy cattle : Its role in disease resistance [J]. Veterinarna Meditsina, 1988, 83(1): 72-78.

[113] Maas J, Galey F D, Peauroi J R, et al. The correlation between serum selenium and blood selenium in cattle[J]. Journal of Veterinary Diagnostic Investigation Official Publication of the American Association of Veterinary Laboratory Diagnosticians Inc, 1992, 4(1): 48-52.

[114] Ortman K, Pehrson B. Effect of selenate as a feed supplement to dairy cows in comparison to selenite and selenium yeast[J]. Journal of Animal Science, 1999, 77(12): 3365-3370.

[115] Gunter S A, Beck P A, Phillips J M. Effects of supplementary selenium source on the performance and blood measurements in beef cows and their calves[J]. Journal of Animal Science, 2003, 81(4): 856-864.

[116] Ullrey D E. Biochemical and physiological indicators of selenium status in animals [J]. Journal of Animal Science, 1987, 65 (6): 1712-1726.

[117] Meyer W R, Mahan D C, Moxon A L. Value of dietary selenium and vitamin E for weanling swine as measured by performance and tissue selenium and glutathione peroxidase activities[J]. Journal of Animalence, 1981, 52(2): 302-311.

[118] Pavlata L, Pechova A, Becvar O. Selenium status in cattle at slaughter: analyses of blood, skeletal muscle, and liver[J]. Acta Veterinaria Brno, 2001, 70(3): 277-284.

[119] Wichtel J J, Keefe G P, Leeuwen J a V, et al. The selenium status of dairy herds in Prince Edward Island[J]. Canadian Veterinary Journal-revue Veterinaire Canadienne, 2004, 45(2): 124-132.

[120] Lean I J, Troutt H F, Boermans H, et al. An investigation of bulk tank milk selenium levels in the San Joaquin Valley of California[J]. Cornell Veterinarian, 1990, 80(1): 41-51.

[121] Phipps R H, Grandison A S, Jones A K, et al. Selenium supplementation of lactating dairy cows: effects on milk production and total selenium content and speciation in blood, milk and cheese[J]. Animal, 2008, 2(11): 1610-1618.

[122] Muñiz-Naveiro , Domínguez-González R, Bermejo-Barrera A, et al. Selenium Content and Distribution in Cow's Milk Supplemented with Two Dietary Selenium Sources[J]. Journal of Agricultural and Food Chemistry, 2005, 53(25): 9817-9822.

[123] Underwood E J. Trace elements in human and animal nutrition [M]. 4th ed. New York: Academic, 1977.

[124] Christodoulopoulos G, Roubies N, Karatzias H, et al. Selenium concentration in blood and hair of holstein dairy cows[J]. Biological Trace Element Research, 2003, 91(2): 145-150.

[125] Wahlstrom R C, Goehring T B, Johnson D D. The relationship of hair color to selenium content of hair and selenosis in swine[J]. Nutrition Reports International, 1984, 29(1): 143-147.

[126] Judson G J, Mcfarlane J D. Mineral disorders in grazing livestock and the usefulness of soil and plant analysis in the assessment of these disorders[J]. Australian Journal of Experimental Agriculture, 1998, 38(7): 707-723.

[127] Suttle N F. Predicting the risk of mineral deficiencies in grazing animals[J]. South African Journal of Animal Science, 1988, 18(1): 15-22.

[128] Masters D G, White C L. Detection and treatment of mineral nutrition problems in grazing sheep [M]. Canberra: ACIAR, 1996.

[129] Juniper D T, Phipps R H, Givens D I, et al. Tolerance of ruminant animals to high dose in-feed administration of a selenium-enriched yeast[J]. Journal of Animal Science, 2008, 86(1): 197-204.

[130] Awadeh F T, Kincaid R L, Johnson K A. Effect of level and source of dietary selenium on concentrations of thyroid hormones and immunoglobulins in beef cows and calves[J]. Journal of Animal Science, 1998, 76(4): 1204-1215.

[131] Thomson C D. Assessment of requirements for selenium and adequacy of selenium status: a review[J]. European Journal of Clinical Nutrition, 2004, 58(3): 391-402.

[132] Pavlata L, Misurova L, Pechova A, et al. Direct and indirect assessment of selenium status in sheep-a comparison[J]. Veterinarni Medicina-UZEI (Czech Republic), 2012, 57(5): 219-223.

[133] Ludvikova E, Pavlata L, Vyskocil M, et al. Selenium Status of Horses in the Czech Republic[J]. Acta Veterinaria Brno, 2005, 74(3): 369-375.

[134] Misurova L, Pavlata L, Pechova A, et al. Selenium metabolism in goats-maternal transfer of selenium to newborn kids[J]. Veterinarni Medicina, 2009, 54(3): 125-130.

[135] Constable P D, Hinchcliff K W, Done S H. Veterinary Medicine: A Textbook of the Diseases of Cattle, Horses, Sheep, Pigs, and Goats [M]. 11th ed. Amsterdam: Elsevier Ltd, 2017.

[136] Pavlata L, Misurova L, Pechova A, et al. The effect of inorganic and organically bound forms of selenium on glutathione peroxidase activity in the blood of goats[J]. Veterinarni Medicina, 2011, 56(2): 75-81.

[137] kesson B, Huang W, Persson-Moschos M, et al. Glutathione Peroxidase, Selenoprotein P and Selenium in Serum of Elderly Subjects in

Relation to Other Biomarkers of Nutritional Status and Food Intake[J]. Journal of Nutritional Biochemistry, 1997, 8(9): 508-517.

[138] Huang W, Akesson B, Svensson B G, et al. Selenoprotein P and glutathione peroxidase (EC 1 • 11 • 1 • 9) in plasma as indices of selenium status in relation to the intake of fish[J]. British Journal of Nutrition, 1995, 73(3): 455-461.

[139] Perssonmoschos M, Huang W, Srikumar T S, et al. Selenoprotein P in serum as a biochemical marker of selenium status[J]. Analyst, 1995, 120(3): 833-836.

[140] Xia Y, Hill K E, Burk R F. Biochemical Studies of a Selenium-Deficient Population in China: Measurement of Selenium, Glutathione Peroxidase and Other Oxidant Defense Indices in Blood[J]. Journal of Nutrition, 1989, 119(9): 1318-1326.

[141] Xia Y, Hill K E, Byrne D W, et al. Effectiveness of selenium supplements in a low-selenium area of China[J]. The American Journal of Clinical Nutrition, 2005, 81(4): 829-834.

[142] Neve J. New Approaches to Assess Selenium Status and Requirement[J]. Nutrition Reviews, 2009, 58(12): 363-369.

[143] Duffield A J, Thomson C D, Hill K E, et al. An estimation of selenium requirements for New Zealanders[J]. The American Journal of Clinical Nutrition, 1999, 70(5): 896-903.

[144] Salman S, Kholparisini A, Schafft H, et al. The role of dietary selenium in bovine mammary gland health and immune function[J]. Animal Health Research Reviews, 2009, 10(1): 21-34.

[145] Rainard P, Riollet C. Innate immunity of the bovine mammary gland[J]. Veterinary Research, 2006, 37(3): 369-400.

[146] Malbe M, Klaassen E, Kaartinen L, et al. Effects of oral selenium supplementation on mastitis markers and pathogens in Estonian cows[J]. Veterinary Therapeutics Research in Applied Veterinary Medicine, 2003, 4(2): 145-154.

第二章　硒蛋白的生物合成及其生物学功能

绝大多数金属微量元素在生物体内是以辅因子的形式与蛋白质相互作用发挥其生物学功能，而硒在人和动物体内主要以共价结合的硒代半胱氨酸(SeCys)的形式存在于蛋白质中，通常把这类蛋白称为硒蛋白。在人和动物体内，硒主要以硒蛋白的形式发挥其广泛的生物学功能。本章主要阐述硒蛋白的生物合成机制以及硒蛋白的生物学功能。

第一节　硒蛋白的生物合成

在硒酶或硒蛋白中，硒总是以 Sec 的形式存在并位于硒酶的活性中心。Sec 插入蛋白质的合成过程并不像硒代蛋氨酸随机取代蛋氨酸那样随机取代半胱氨酸合成到蛋白质中，而是和其他 20 种氨基酸一样通过特定的编码方式合成到蛋白质中的。硒酶中的 Sec 若被半胱氨酸取代，硒酶的催化活性会明显降低甚至消失，因为在生理 pH 条件下，硒醇组较硫醇组更易去质子化。因此，Sec 被认为是生物合成和掺入到蛋白质分子中的第 21 种氨基酸。众所周知，所有 64 个核苷酸码已被解译为 20 个已知的氨基酸码和基因转录的终止码，似乎不可能有另一个特殊的编码对应于这个 Sec。那么，究竟 Sec 是如何掺入到蛋白质中的呢？事实上，Sec 掺入蛋白质有一套特殊的分子机制，其特殊性主要表现在三个方面：首先，传统的终止密码子 UGA 是 Sec 的遗传密码，且这种解码机制依赖于硒蛋白 mRNA 中的顺式作用元件；其次，Sec 是在一种特殊的能够负荷 Sec 的 tRNA 上合成的；最后，负荷有 Sec 的特殊 tRNA 需通过特定的反式作用蛋白被引渡到核糖体。

一、UGA 是 Sec 的遗传密码

1986 年，Chambers 等[1]采用克隆技术获得了小鼠细胞 cGPX 的 cDNA 序列，发现与 Sec 残基(47 号氨基酸残基)相对应的密码子是 TGA。之后相继在人、牛、大鼠 GPX 的基因中以及大肠杆菌甲酸脱氢酶(一种硒蛋白)的基因中也得到了同样的结果。在已经发现并得到完整基因的硒蛋白 mRNA 分子中，其开读框(ORF)内含有一个或多个 UGA 密码子，且 UGA

在 mRNA 中的位置与硒蛋白一级结构中 Sec 的位置相对应[2]。将大肠杆菌甲酸脱氢酶的 fdhF 基因与 lacZ 基因重组在质粒 pFM54 中发现，当培养基中硒减少时，fdhF-lacZ 基因的产物减少，当培养基中不含硒时，翻译就终止在 UGA 密码子处[3]。一系列大肠杆菌甲酸脱氢酶的突变实验发现，当用 UCA（编码丝氨酸）代替 UGA 时，Sec 不能掺入到 fdhF 基因的产物中，当用 UGC 或 UGU（编码半胱氨酸）代替 UGA 时，Sec 的掺入非常少，代替 UGA 后硒缺乏并没有阻断翻译过程，但甲酸脱氢酶的活力明显下降甚至消失。上述研究结果表明，UGA 是 Sec 的遗传密码，而且更重要的一点是 UGA 作为 Sec 的遗传密码并不妨碍其在通常的蛋白质翻译过程中作为终止密码子的作用。因此，UGA 解读为 Sec 必然有其特殊的机制。

二、UGA 解读为 Sec 的信号元件

在硒蛋白 mRNA 的 UGA 密码子下游存在一个特殊的茎环结构，称为硒代半胱氨酸插入元件（SECIS），是 UGA 解读为 Sec 的信号元件。SECIS 对 UGA 重编码成 Sec 的密码子以及 Sec 掺入到蛋白质的过程必不可少且具有高度的特异性[4]。通常情况下，UGA 是蛋白质合成过程中的终止密码，而当 UGA 密码子下游存在 SECIS 时，UGA 码就成了编码 Sec 的密码子。SECIS 一级序列和二级结构的任何改变都会影响 UGA 密码子的解读。

（一）原核生物硒蛋白基因中的 SECIS

原核生物硒蛋白 mRNA 的 SECIS 位于紧邻 UGA 密码子的下游编码区内，SECIS 顶环与 UGA 码的距离通常为 20～25 个核苷酸[5]。在细菌中，转录和翻译是相耦联的，当翻译到 UGA 码时，SECIS 有可能还没有被转录，UGA 码和 SECIS 的紧邻定位方式可很好地避免这一问题的出现。但这种紧邻的定位方式也有不足之处，紧邻 UGA 码下游的核苷酸序列既要具有 SECIS 的结构特征，又要具有编码功能以及和延长因子的结合功能，因而会限制 Sec 下游氨基酸序列的多样性。

在原核生物中，研究最为清楚的是大肠杆菌（*Escherichia coli*）中编码甲酸脱氢酶 H（fdhF）的基因上的 SECIS。大肠杆菌 fdhF 中的 SECIS 包含 40 个核苷酸，是一个距 UGA 码 11 nt 的由 17 个核苷酸（15～31）组成的特殊茎环结构。这个茎环结构是 UGA 解读为 Sec 所必需的，茎环结构中 U17 必须是非配对的，而 15～19 与 28～31 间的核苷酸必须配对形成茎样结构，顶环要有保守的 GU 序列，G23 必须位于顶环上（图 2-1）。研究表

明，－7～－1 和 42～50 碱基的变化对通读性影响不大，而当 4～14 与 32～41 之间的碱基不能形成配对时，则通读性下降约 25％，U17 和 G23 参与了与延伸因子的相互结合，两者中任何一个突变都会导致 UGA 通读性完全丧失[6,7]。UGA 码与 SECIS 顶环的距离对通读性的影响也非常显著，当 UGA 左移 1 个密码子时，通读性下降 24％，左移 2 个密码子时，通读性下降 80％，左移 3 个密码子时，通读性下降 60％[7]。

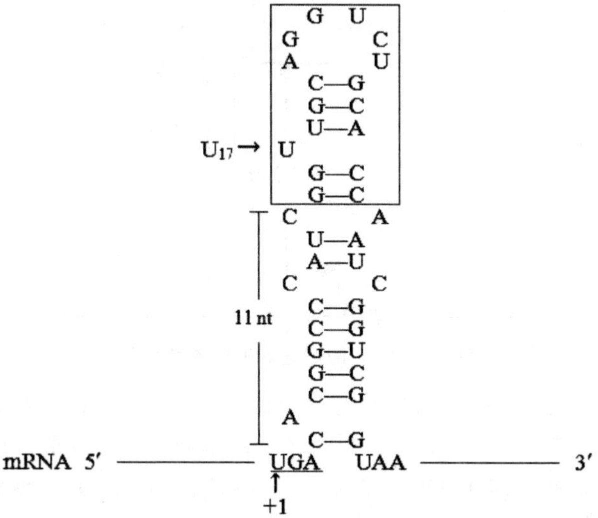

图 2-1　大肠杆菌(*Escherichia coli*)fdhF 中的 SECIS[8]

尽管已经很好地阐明了 *Escherichia coli* fdhF 中 SECIS 的结构特征，但由于原核生物 SECIS 缺乏非常保守的一级序列和二级结构特征，此结构并不一定适用于所有细菌的硒蛋白基因，而且在所有细菌的硒蛋白基因组中是否都存在 SECIS 还有待进一步研究确定。Zhang 等[9]通过分析已知细菌硒蛋白基因中各种 SECIS 的组成和结构特征，提出了统一的细菌 SECIS 结构模型(图 2-2)。该模型的核心部分为一个含 3～14 个核苷酸的顶环和一个邻近的含 4～16 个 bp 的茎，顶环的前两个核苷酸至少有一个是 G，其后常跟着一个 U，UGA 码与 SECIS 顶环的距离在 16～37 个核苷酸之间。

(二)真核生物硒蛋白基因中的 SECIS

真核生物硒蛋白 mRNA 的 SECIS 位于远离 UGA 密码子的下游 3′-非翻译区(3′-UTR)内，各种硒蛋白 SECIS 与 UGA 码的距离存在很大差异。SECIS 定位于 3′-UTR 可以使其与 SECIS 结合蛋白牢固结合，从而提高 Sec 的插入效率。SECIS 定位于 3′-UTR 还可以实现用一个茎环结构指导多个 Sec 残基的插入。但与原核生物相比，真核生物硒蛋白合成中 Sec 的插入速率可能会因 SECIS 与 UGA 码的距离较远而变慢。此外，有些硒蛋白 mRNA 的 3′-UTR 含有的特殊茎环结构不止一个，如 SelP mRNA 的 3′-UTR 就含有两个 SECIS，SelP 的氨基酸序列中含有 10 个 Sec 残基，因此，不止一个的 SECIS 可能与其能够支持多个 UGA 密码子的解读有关。

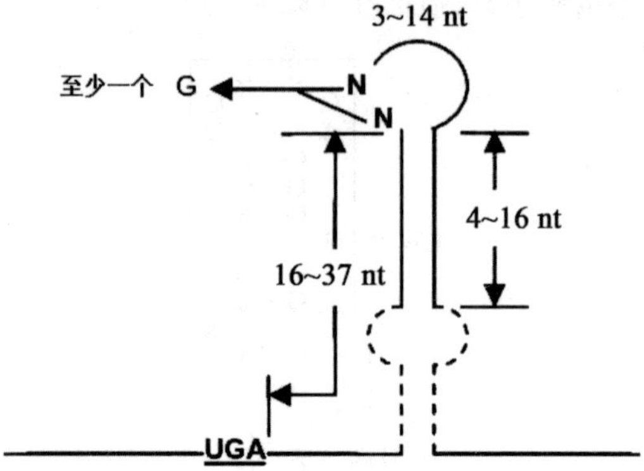

图 2-2 统一的细菌 SECIS 结构模型

真核生物 SECIS 的结构比较大，约有 80～150 个核苷酸，UGA 码与 SECIS 之间的距离不定，从几百到几千个核苷酸不等，但至少应相距 51～111 个核苷酸才有利于 UGA 密码子的解读，有些真核生物则需要相距 2 000 个以上的核苷酸才有利于 UGA 密码子的解读。在真核生物硒蛋白基因中，SECIS 结构按构型不同分为Ⅰ型和Ⅱ型两种(图 2-3)[10,11]。Ⅰ型 SECIS 结构主要由螺旋Ⅰ、内环、四联体或 SECIS 核心、螺旋Ⅱ和顶环五个部分构成。Ⅰ型 SECIS 最显著的特点是螺旋Ⅱ中包含一个非 Watson-Crick 四联体碱基对(UGAN……NGAN)，G、A/A、G 串联于中部，称为四联体或 SECIS 核心，该核心前是一个未成对的 A/G，绝大多数为 A，顶环上包含一个未成对的 AA 基序，该基序与 SECIS 核心的距离为 11～12 个核苷酸。Ⅱ型 SECIS 与Ⅰ型 SECIS 的主要区别在于其顶环上还附加有一个茎环结构。通常，在已知的 SECIS 中，约有 66%为Ⅱ型，而且往往当顶环较大时出现Ⅱ型结构，通过突变实验发现，Ⅰ型和Ⅱ型 SECIS 在功能和活性上无明显区别[11]。以上两个方面暗示，Ⅱ型 SECIS 中附加的茎环结构可能主要起稳定 SECIS 结构的功能。

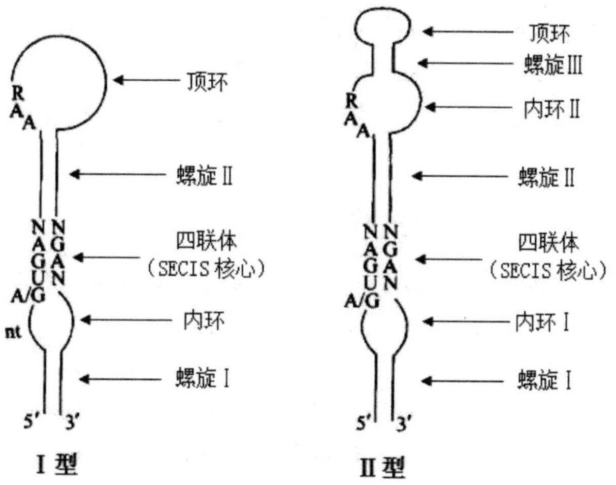

图 2-3　真核生物硒蛋白基因的 SECIS 结构示意图

三、tRNA^{Sec}（SelC 基因的产物）

众所周知，tRNA 是指具有携带并转运氨基酸功能的一类小分子核糖核酸，大多数 tRNA 由七十几至九十几个核苷酸折叠形成的三叶草形短链组成，相对分子质量为 25 000～30 000，沉降常数约为 4S。tRNA 的主要作用是携带氨基酸进入核糖体，以 mRNA 为模板，通过密码子与反密码子的相互作用，将 mRNA 中具有密码意义的核苷酸顺序翻译成蛋白质中的氨基酸顺序。在肽链生成过程中，第一个进入核糖体与 mRNA 起始密码子结合的 tRNA 称为起始 tRNA，其余 tRNA 参与肽链延伸，称为延伸 tRNA，按照 mRNA 上密码的排列，携带特定氨基酸的 tRNA 依次进入核糖体。由此推断，Sec 要想掺入到硒蛋白，必须要有一个能够携带 Sec 的特殊 tRNA，而且这个特殊的 tRNA 必须具有能够识别 mRNA 中 UGA 密码子的反密码子。

1988 年，Bock 等在大肠杆菌中发现了这种特殊的 tRNA，它是一种称为 SelC 基因的产物[12]，为了区别于其他延伸 tRNA，Lee 等将其指定为 tRNA^{Sec}[13]。对 tRNA^{Sec} 的系统研究发现，它能够携带 Sec 并具有能够识别 UGA 码的反密码子 UCA，带有这一基因片段的大肠杆菌甲酸脱氢酶突变株能够合成含 Sec 的蛋白质，反之，Sec 合成到蛋白质的过程被阻断[12]。这一发现引起科学家们的极大关注，对一些熟知的哺乳动物抑制性 tRNA^{Ser} 的重新分析发现，这些磷酸丝氨酰 tRNA 能够携带 Sec[13,14]。目前，这样的能够携带 Sec 的 tRNA 在细菌[15-17]、原生生物[18]、真菌[19]以及整个动物界[20]中均有发现,而在酵母以及高等植物中并未证实存在 $tRNA^{Sec}$。与典型的延伸 tRNA 相比,这种能够负荷 Sec 的 tRNA 有其独一无二的结构特征,具有一个较长的可变臂、一个延长的接受臂和一个固定长度(6 bp)的 D 茎。

(一)细菌 $tRNA^{Sec}$的结构

基于对三个细菌基因组的分析,典型的延伸 tRNA 的核苷酸序列中包括 22 个不变或半不变核苷酸:U8、A14、R15、G18-G19、A21、U33、R37、Y48、R52-G-U-A-C-R-A58、Y60-C-Y62 和 C74-C-A[21],多数位于 tRNA 结构的肘部,参与稳定其三级结构。二级结构由一个 7 bp 长的接受臂、一个 3~4 bp 长的 D 茎、一个 5 bp 长的反密码子茎和一个 5 bp 长的 TΨC 茎组成,反密码子环和 T 环均包括 7 个核苷酸,而 D 环通常由 8~10 个核苷酸组成。由于具有 7 bp 长的接受臂和 5 bp 长的 TΨC 茎,故称为 7/5 模式。依据可变环长度的不同,一般把典型的延伸 tRNA 分为Ⅰ型和Ⅱ型两类,Ⅰ型 tRNA 的可变序列较短,由 4 个或 5 个核苷酸组成(图 2-4 A),而Ⅱ型 tRNA 的可变序列较长,由 10~24 个核苷酸组成(图 2-4 B)。

与典型的延伸 tRNA 相比,细菌 $tRNA^{Sec}$的一级序列有几处发生了改变,如位置 8 和位置 14 的保守核苷不再出现。最显著的区别在于二级结构,$tRNA^{Sec}$的接受臂多出一对碱基,D 茎延长到 6 个 bp,同时 D 环减少到仅 4 个核苷酸(图 2-4 C)。由于接受臂多出一对碱基而 TΨC 茎长度不变,故称为 8/5 模式。此外,细菌 $tRNA^{Sec}$的可变环较长,甚至要比Ⅱ型 tRNA 的还要长。

(二)真核生物 $tRNA^{Sec}$的结构

对真核生物 $tRNA^{Sec}$结构的认识远不及对细菌 $tRNA^{Sec}$结构的认识。高等真核生物的延伸 tRNA 具有典型的三叶草型结构,包含 26 个不变或半不变核苷酸:U9、Y11、G18-G19、Pu24、Py25、Py32、Pu37、G53、U55、C56、A58 和 C74-C-A[21]。当牛肝 $tRNA^{Sec}$被第一次测序时,由于其一级结构与典型的延伸 tRNA 的差异很大而且可能形成了延长的 D 茎,因此其二级结构并不清楚[14]。随着对其他真核生物 $tRNA^{Sec}$的测序以及对 *Xenopus laevis* $tRNA^{Sec}$化学特征的探讨,逐步认识了真核生物 $tRNA^{Sec}$的三叶草保守结构[23],包含一个 9 bp 长的接受臂,比典型的延伸 tRNA 多出两个 bp,由于接受臂的延长,导致 TΨC 茎的长度缩短,仅为 4 个 bp(9/4 模式),同样与细菌 $tRNA^{Sec}$相似,其 D 茎延长到 6 个 bp,同时 D 环减少到仅 4 个核苷酸(图 2-5 A)。然而,依据典型延伸 tRNA 三叶草型结构的作用特点,TΨC 茎需要有 5～6 层堆叠的核苷酸才能保证 D/T 环的正常相互作用[24]。如果满足这一规则,则真核生物可供选择的另一种三叶草型结构应该含有通常的 7 bp 的接受臂和 5 bp 的 TΨC 茎,即 7/5 模式,这样,接受臂和 D 茎之间有 4 个核苷酸连接而不是 2 个,而且在 TΨC 茎上凸出了 1 个核苷酸(图 2-5 B)[25]。酶和化学探针试验的结果更支持真核生物 tRNA^{Sec}的二级结构为 9/4 模式[26],古细菌 *Methanococcus jannaschii* tRNA^{Sec}的二级结构也为 9/4 模式[27]。

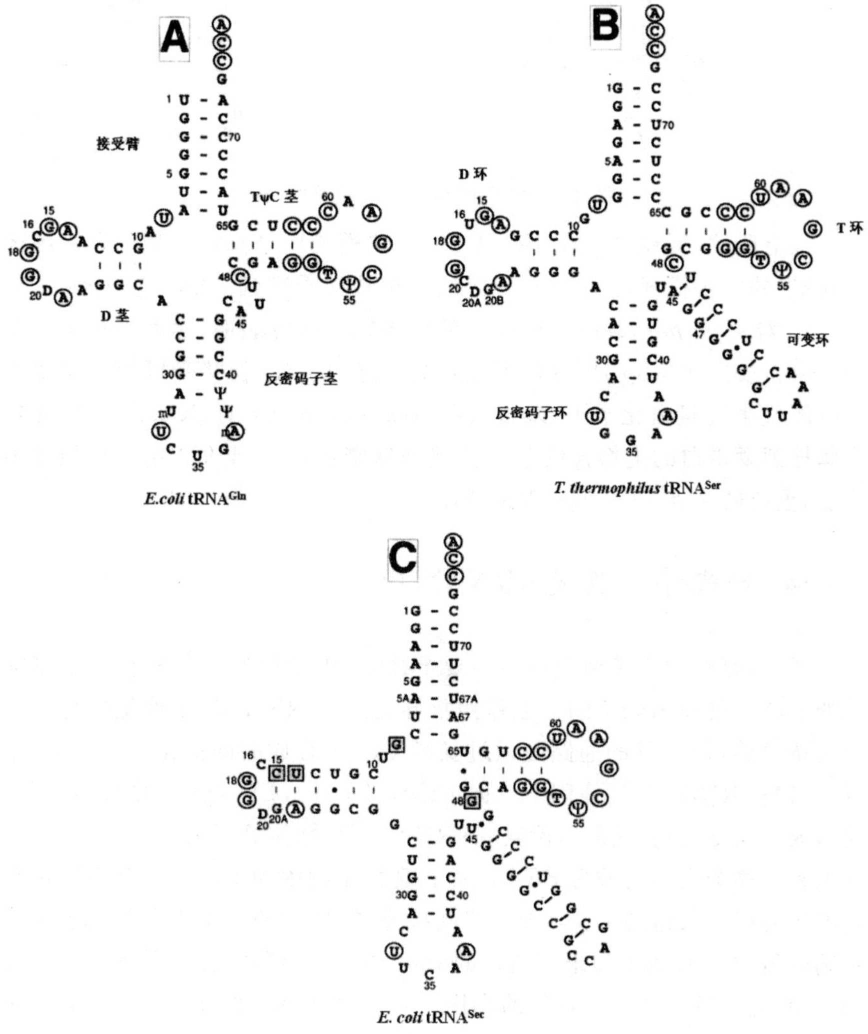

图 2-4　细菌典型延伸 tRNA 以及 tRNA^{Sec}的二级结构[22]

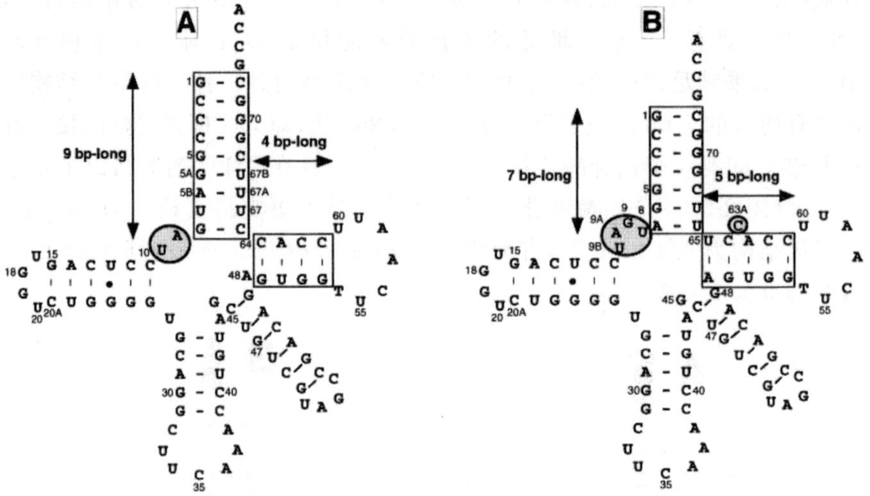

图 2-5　牛肝 tRNA^{Sec}的二级结构[22]

综上所述，无论是细菌 tRNA^{Sec}的 8/5 模式还是真核生物以及古细菌 tRNA^{Sec}的 9/4 模式，所有 tRNA^{Sec}的结构从接受臂到 TΨC 茎均为 13 bp 长[28]。对 *Xenopus laevis* tRNA^{Sec}的体外转录研究表明，然而当这个长度由 13 bp 减少到 8 bp 时，硒代半胱氨酰化能力丧失，但接受臂上点突变不影响硒代半胱氨酰化水平，来自人和 *Xenopus laevis* 的 tRNA^{Sec}基因支持大肠杆菌硒蛋白的生物合成[17]。这些证据暗示，8/5 或 9/4 结构可能并不重要，重要的是其 13 个碱基对的长度。

四、硒代半胱氨酰-tRNA^{Sec}的合成

通常，氨基酸合成到蛋白质需要 mRNA 中的密码子与核糖体中氨酰化的 tRNA 的反密码子的相互作用指导完成。tRNAs 的正确氨酰化一般有两条通路，最常见的通路是通过氨酰-tRNA 合成酶的催化一步完成，生成的氨酰-tRNAs 与延伸因子（elongation factors，EF）结合，并与 GTP 一起形成三元复合物，从而引渡氨酰-tRNAs 到核糖体 A 位点（图 2-6 A）[29]。在细菌中这个延伸因子为 EF-Tu，而在真核生物中为 EF-1α。tRNAs 正确氨酰化的第二条通路至少需要两步间接完成，增加的一步主要是为了区分正确氨酰化的 tRNA 和未氨酰化的前体 tRNA。例如，在各种细菌以及古细菌中，通过特定的转酰氨酶的作用，Glu-tRNA^{Gln}转变为 Gln-tRNA^{Gln}，在古细菌中，Asp-tRNA^{Asn}转变为 Asn-tRNA^{Asn}（图 2-6 B）[30]。

早在 1975 年，Young 等[31]就从大肠杆菌中发现了含硒 tRNA 的存在。

1982 年，Hawkes 等[32]从鼠肝切片中同样发现了含硒 tRNA，并且证实 Sec 可以键合在 $tRNA^{Sec}$ 上，进一步的研究发现，用 Se^{75} 标记的亚硒酸钠能够进入硒代半胱氨酰-$tRNA^{Sec}$（Sec-$tRNA^{Sec}$），而且在非细胞体系中，这种带有 Se^{75} 的 Sec-$tRNA^{Sec}$ 是 GPX 合成的最好底物，但增加 Sec 的浓度反而使 Se^{75} 进入 GPX 的量减少。Sunde 等[33]的研究发现，硒的无机形式与 Sec 相比更易进入 GPX，用标记的氨基酸证明，GPX 中 Sec 的碳架来自丝氨酸而不是 Sec，Leinfelder 等[12]的研究发现，$tRNA^{Sec}$ 不能直接携带培养基中供给的游离 Sec，结合 Sunde 等的发现，Leinfelder 等推测蛋白质中的 Sec 残基可能是由丝氨酸上的羟基氧被硒取代而来。Mizutani 等[34]的研究发现，在标准条件下，$tRNA^{Sec}$ 不能与 Sec 结合，只有当 H_2Se 和一些酶共存时，微量的磷酸丝氨酰-tRNA 才能转变为 Sec-$tRNA^{Sec}$，当酪蛋白、H_2Se 和一些酶共同培养时，大量的磷酸丝氨酸残基能够转变为硒代半胱氨酸残基，这一研究结果很好地证明了 Leinfelder 等的推测。

Sec 的合成是在 $tRNA^{Sec}$ 上完成的，因此 Sec 的合成亦即 Sec-$tRNA^{Sec}$ 的合成。现已知道，在各种细菌中，Sec-$tRNA^{Sec}$ 是 Ser-$tRNA^{Sec}$ 在硒代半胱氨酸合成酶的催化下合成的，需要两步间接完成。首先，$tRNA^{Sec}$ 在丝氨酰-$tRNA^{Sec}$ 合成酶的作用下消耗 1 分子 ATP 与丝氨酸反应生成 Ser-$tRNA^{Sec}$，然后 Ser-$tRNA^{Sec}$ 在硒代半胱氨酸合成酶的催化下以硒代磷酸为供硒体生成 Sec-$tRNA^{Sec}$，生成的 Sec-$tRNA^{Sec}$ 与硒代半胱氨酸延伸因子以及 GTP 形成三元复合物，从而引渡 Sec-$tRNA^{Sec}$ 到核糖体 A 位点（图 2-6 C）[35]。

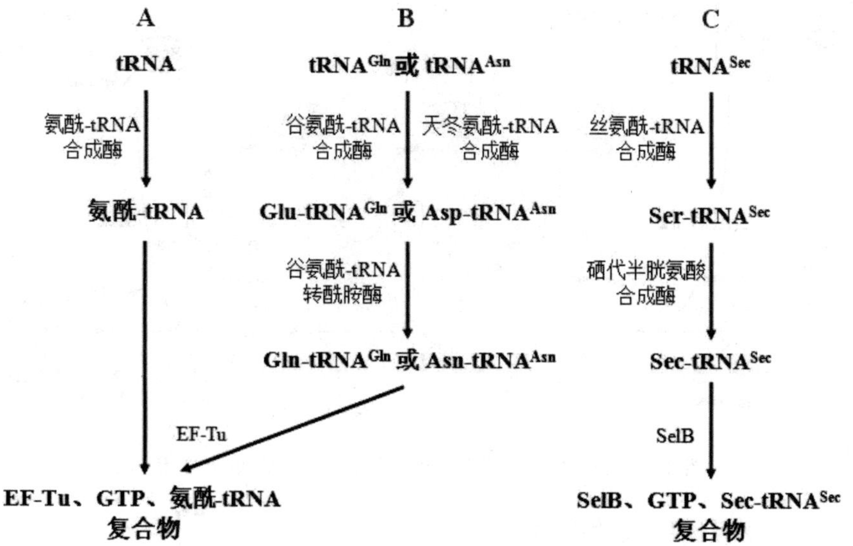

图 2-6 细菌氨酰-tRNA 的合成通路

(一)硒代半胱氨酸合成酶的催化机理

在大肠杆菌中,硒代半胱氨酸合成酶是 SelA 基因编码的产物,分子量为 500 kDa[36],该酶蛋白上的一个氨基可以和吡哆醛-5-磷酸键合形成吡哆醛磷酸盐形式,当与 Ser-tRNA^{Sec}反应时,吡哆醛-5-磷酸的甲酰基与丝氨酸残基的 α-氨基共价结合,进而发生 2,3-消去反应,脱去一分子水,形成氨基丙烯酰-tRNA^{Sec}复合物,然后,作为供硒体的硒代磷酸与该复合物中氨基丙烯酰基团的双键发生亲核加成反应,生成 Sec-tRNA^{Sec}(图 2-7)[37]。大肠杆菌硒代半胱氨酸合成酶是迄今为止发现的唯一一种能与核苷酸结合的 PLP 酶。

尽管各种实验证据表明上述反应在真核生物系统中也存在,但真核生物硒代半胱氨酸合成酶还未被鉴定[38]。早在 1970 年,在鸡肝中就发现一种激酶能够使 Ser-tRNA 磷酸化[39]。对古细菌基因组的研究也发现一个类似于哺乳动物的磷酸化 Ser-tRNA 的激酶基因。在哺乳动物中,Ser-tRNA 的磷酸化具有高度的特异性[40],目前对磷酸化的 Ser-tRNA 的生物学功能还不清楚,最初认为是一种 Ser-tRNA 的存储形式[41],后来认为它可能参与硒蛋白的生物合成,即 Ser-tRNA^{Sec}转变为 Sec-tRNA^{Sec}需经过一个 Ser-tRNA 磷酸化的中间过程[42]。但体外硒代半胱氨酸的合成不需要磷酸化 Ser-tRNA 激酶的参与[43]。

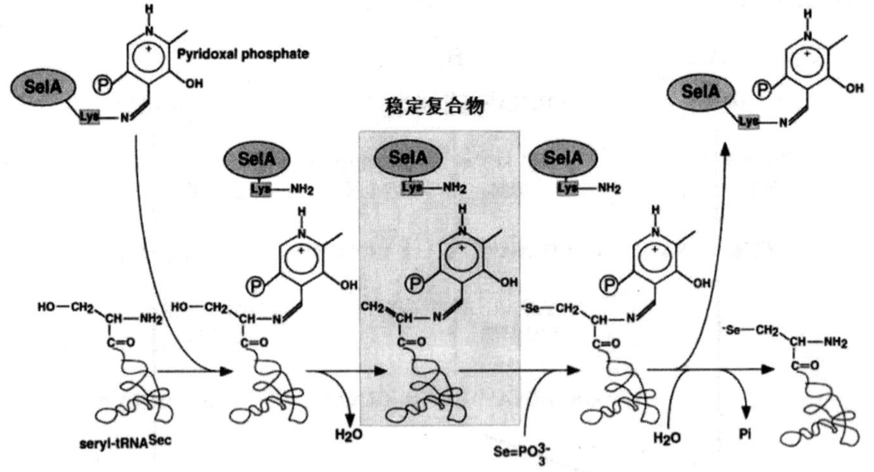

图 2-7 硒代半胱氨酸合成酶催化 Sec-tRNA^{Sec}的合成[22]

(二)硒代磷酸的合成

如上所述,Sec-tRNA^{Sec}的合成需要硒代磷酸作为供硒体,硒代磷酸是在硒代磷酸合成酶(SPS)的催化下合成的。在大肠杆菌中,SPS 是 SelD 基因编码的产物,分子量为 37 kDa,含有 7 个半胱氨酸残基[44],ATP 是 SPS 的特异性底物,Mg^{2+} 是 SPS 活性所必需。在硒代磷酸分子中,硒原子与磷直接键合,P-Se 键相对较弱,适合于在 Sec 的加成以及 tRNA 转录后的取代反应(Se 取代 S)中起作用[45]。SPS 催化硒代磷酸合成的反应如下:

$$\text{ATP}+\text{硒化物}+H_2O \xrightarrow{\text{SPS}} H_3PO_3Se+\text{Pi}+\text{AMP}$$

真核生物硒蛋白的合成也需要硒代磷酸作为供硒体,在哺乳动物体内广泛存在 SPS,而且还发现一种与大肠杆菌 SPS 不同的 SPS,其酶蛋白中含有一个 Sec。为了区别,将不含 Sec 的 SPS 定义为 SPS1,将含有 Sec 的定义为 SPS2[46]。与 SPS1 一样,SPS2 的专一性底物也是 ATP,Mg^{2+} 也是 SPS2 活性所必需。SPS2 合成硒代磷酸的途径可能与 SPS1 不同,因为细菌 SPS 主要利用硒蛋白的降解产物合成硒代磷酸,而真核生物可以利用亚硒酸钠代谢产物合成硒代磷酸,但具体的合成机制还有待进一步研究证实。

五、Sec-tRNA^{Sec}向核糖体 A 位点的引渡

在大肠杆菌通常的蛋白质合成过程中,氨酰-tRNA 需要与 GTP 以及 EF-Tu 键合形成三元复合物,当特定的密码子与反密码子相互作用发生后,GTP 被水解,EF-Tu 构象改变,氨酰-tRNA 被引渡到核糖体 A 位点,然后 EF-Tu/GTP 从核糖体上解离下来,GDP 被 GTP 交换因子代替。

在大肠杆菌中,EF-Tu 并不能识别 Sec-tRNA^{Sec}。体外研究表明,EF-Tu 与 Sec-tRNA^{Sec}的结合能力要比与常规氨酰-tRNAs 的结合能力弱 100 倍[47],这一结果暗示可能有一种特异的延伸因子参与硒蛋白的合成。现已证明,这种特异性延伸因子是 SelB 基因的产物,SelB 不与 Ser-tRNA 反应,但特异地且唯一与 Sec-tRNA 结合。对四种细菌 SelB 基因产物的序列分析表明,SelB 的 N-末端区域(结构域Ⅰ、Ⅱ和Ⅲ)与 EF-Tu 相应的二级结构相一致。SelB 的鸟嘌呤核苷酸结合位点与 EF-Tu 的相似,与 GTP 的亲和力远高于 GDP,因而不需要 GTP 交换因子。SelB 与 Sec-tRNA^{Sec}的结合区域,即氨酰基结合口袋更加开放,包括一个精氨酸残基[48]。SelB 的 C-末端存在一个分子量为 24 kDa 的扩展区,称为结构域Ⅳ,而在 EF-Tu 上不具有相应的结构[49]。结构域Ⅳ具有 SECIS 的结合位点,能够键合 SECIS 元件。RNA 足迹法研究表明,SelB 结合在 SECIS 元件的顶环区域[7],而且 SelB 与 Sec-tRNA^{Sec}的特异性结合会诱导增强其与 SECIS 的亲和力[49]。在大肠

杆菌中，SelB 需要与 GTP、Sec-tRNA^{Sec}和 SECIS 形成四元复合物，才能引渡 Sec-tRNA^{Sec}到核糖体 A 位点（图 2-8）[50]。

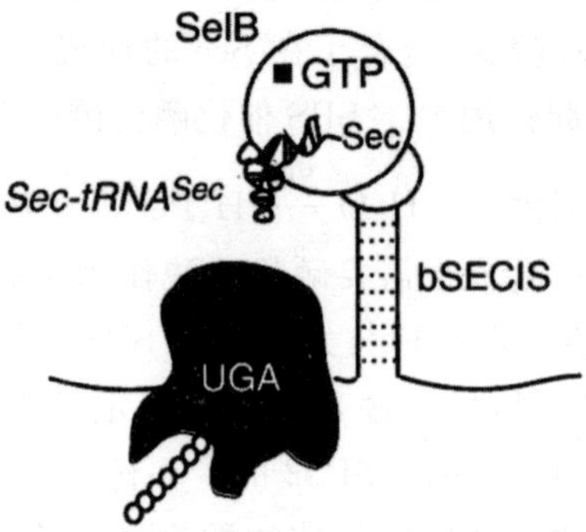

图 2-8　细菌 Sec-tRNA^{Sec}向核糖体 A 位点的引渡

在真核生物中，也存在 Sec 特异性延伸因子，定义为 mSelB 或 EFsec，与 SelB 相似，在体内和体外环境下，EFsec 也特异性地与负荷有 Sec 的 tRNA^{Sec}结合，且与 GTP 的亲和力要远高于 GDP[51,52]。但体外研究表明，EFsec 不具有结合 SECIS 的功能，而是通过与 SECIS 键合蛋白 2（SECIS-binding protein 2，SBP2，一个分子量为 120 kDa 的适配蛋白）或其他未鉴定的因子结合而识别 SECIS，进而引渡 Sec-tRNA^{Sec}到核糖体 A 位点（图 2-9）[52]。SBP2 拥有一个独一无二的 N 末端区域和一个 C 末端区域，尽管不能和 Sec-tRNA^{Sec}结合，但具有结合 SECIS 和核糖体的功能以及体外 Sec 参入的功能[53]。其保守甘氨酸残基（G669）的突变导致 SECIS 结合和 Sec 参入功能丧失，但不影响核糖体结合功能[54]。

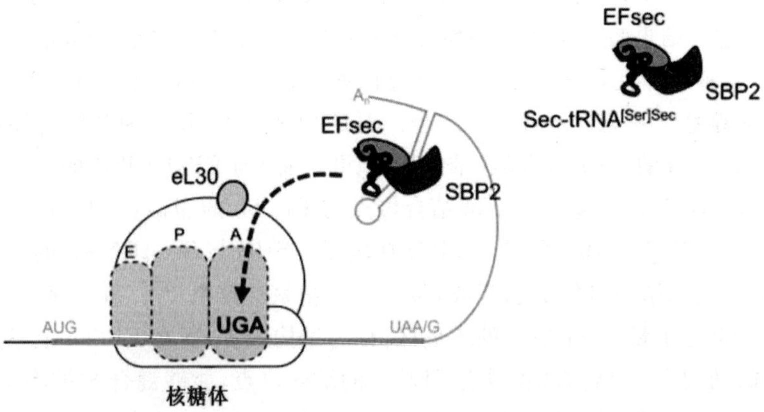

图 2-9　真核生物 Sec-tRNA^{Sec}向核糖体 A 位点的引渡

第二节　硒蛋白及其生物学功能

不同生物体内被鉴定的硒蛋白的数量差异很大，少则1个，多则50多个，有些生物体如高等植物体内还未发现有硒蛋白存在。现今为止发现的硒蛋白有25种存在于人和动物体内(表2-1)，包括GPX家族、TrxR家族、脱碘酶(DIO)家族、SPS2、15-kDa硒蛋白(SEP15，SelF)、硒蛋白H(SelH)、硒蛋白I(SelI)、硒蛋白K(SelK)、硒蛋白M(SelM)、硒蛋白N(SelN)、硒蛋白O(SelO)、SelP、蛋氨酸硫氧化物还原酶B1(MsrB1，SelR)、硒蛋白S(SelS)、硒蛋白T(SelT)、硒蛋白V(SelV)和硒蛋白W(SelW)[55,56]。

多数硒蛋白具有抗氧化功能，如GPX、TrxR和SelP具有直接清除过氧化物的作用。其他特定的生理生化过程如DNA的合成、氧化蛋白的还原、转录因子的氧化还原调节、细胞凋亡的调节、免疫调节、甲状腺激素的活化和灭活、硒的转运和贮存、硒代磷酸的合成以及内质网中蛋白质的折叠和未折叠蛋白质的降解，均与硒蛋白有关。还有一些硒蛋白，其生物学功能仍然是未知的。

除了SelP以外，每种硒蛋白序列中都只含有一个Sec残基，而且该残基常常出现在蛋白质的活性位点中心。依据Sec残基在硒蛋白序列中的位置，可将硒蛋白分为两组，一组由TrxR家族、SelK，SelS、SelR、SelO和SelI组成，Sec残基位于硒蛋白序列的C-末端区域，其余硒蛋白的Sec残基位于N-末端区域。

表2-1　25种硒蛋白的组织分布、细胞定位和分子量

硒蛋白	蛋白质长度	SeCys位置	分子量(kDa)	组织分布	细胞定位
GPX1	201	47	87	普遍存在	胞质、细胞核、线粒体
GPX2	190	40	93	肝脏、胃肠道上皮	胞质、细胞核
GPX3	226	73	93	血浆	胞外
GPX4	197	74	22	睾丸	胞质、线粒体、细胞核
GPX6	—	—	23	嗅上皮、胚胎组织	胞外
TrxR1	499	498	60～108	普遍存在	胞质、细胞核
TrxR2	523	522	60～106	普遍存在	线粒体

续表

硒蛋白	蛋白质长度	SeCys位置	分子量(kDa)	组织分布	细胞定位
TrxR3	656	655	75	睾丸	胞质、细胞核、内质网、微粒体
DIO1	249	126	4～29	肝脏、肾脏、甲状腺、脑垂体、卵巢	内质网、血浆、膜
DIO2	265	133	30、34	心脏、肾脏、脑、脊髓、胰腺、甲状腺、骨骼肌、胎盘	内质网膜
DIO3	278	144	31	胎盘、胎组织、皮肤	细胞膜、核内体膜
SPS2	448	60	47	肝脏	胞质
Sep15	162	93	13、15	肾脏、肺、脑、前列腺、甲状腺	内质网腔
SelH	122	44	13	各种组织(以胚胎和肿瘤组织为主)	细胞核
SelI	397	387	45	各种组织(脑组织最多)	内质网膜
SelK	94	92	10	各种组织(心脏最多)	内质网膜
SelM	145	48	14	肾脏、肺脏、脑组织	内质网腔、高尔基体
SelN	556	428	61～62	普遍存在	内质网膜
SelO	669	667	73	各种组织	未知
SelP	381	59	45～57	肝脏、肾脏、心脏、脑组织	胞外
SelR	116	95	5～14	心脏、肝脏、肌肉、肾脏	胞质、细胞核
SelS	189	188	21	血浆、各种组织	内质网膜
SelT	182	36	20	普遍存在	内质网、高尔基体
SelV	—	—	17	睾丸	未知
SelW	87	13	9	各种组织(肌肉最多)	胞质

一、GPX家族及其生物学功能

在生物体内，GPX是一个非常大的家族，其中大多数GPX中活性位点Sec被Cys取代，主要存在于植物、古细菌、细菌和原生生物体内。迄今为止在哺乳动物体内发现8种形式的GPX，其中前5种分别为GPX1、胃肠道谷胱甘肽过氧化物酶（GI-GPX，GPX2）、GPX3、磷脂氢谷胱甘肽过氧化物酶（PH-GPX，GPX4）和嗅上皮谷胱甘肽过氧化物酶（GPX6）。GPX1-4在所有哺乳动物体内均为硒蛋白，GPX6在多数哺乳动物体内为硒蛋白，但在一些物种体内GPX6是一种含Cys的蛋白质。

哺乳动物GPX1-4的共同特征是都包括一个保守的由Sec、Gln、Asn和Trp组成的催化四元组，主要的区别在于组织分布和底物特异性不同。

GPX的主要功能是以硫醇（RSH）作为还原辅因子催化过氧化氢（H_2O_2）和氢过氧化物（ROOH）的还原[57]，因而在保护细胞免受氧化损伤方面起重要作用。由于H_2O_2和ROOH对GPX活性中心Sec的攻击，导致其氧化为亚硒酸，亚硒酸再被RSH还原为活性形式。在哺乳动物体内，GPX的还原剂主要以GSH为主。这一反应过程被称为乒乓机制（图2-10），具体的反应过程如下：①H_2O_2或ROOH攻击GPX中以硒醇盐形式存在的硒醇（-Se^-），在自身还原为H_2O和相应的醇的同时，GPX中的-Se^-被氧化为亚硒酸（-SeOH）；②-SeOH首先与一分子RSH反应生成硒硫化物；③另一分子RSH攻击硒硫化合物，通过硫醇-二硫化物交换，使酶活性中心的Se恢复为-Se^-形式。

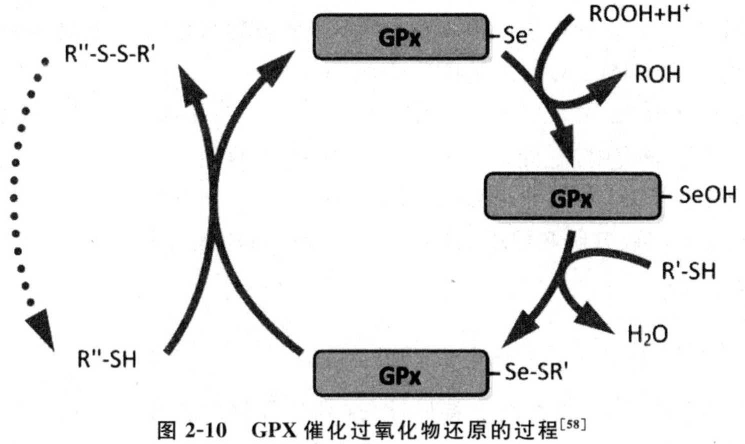

图2-10　GPX催化过氧化物还原的过程[58]

事实上，从化学反应的角度讲，RSH可直接与氢过氧化物反应（反应1、2、3），但在生物体内，由于RSH活性相对较低，氢过氧化物并不能直接攻击RSH，而是先由酶中高活性的硫醇或硒醇与氢过氧化物反应，然后在硫醇的作用下，使酶再恢复到有活性的还原状态，这样可以使反应速率提高几个数量级。这一过程中RSH被氧化为RSSR，在NADPH存在下，RSSR被谷胱甘肽还原酶（GR）还原为RSH（反应4）。

$$H_2O_2+2RSH \xrightarrow{GPX} RSSR+2H_2O \tag{1}$$

$$ROOH+2RSH \xrightarrow{GPX} RSSR+ROH+H_2O \tag{2}$$

$$H_2O_2+GSH+RSH \xrightarrow{GPX} GSSR+2H_2O \tag{3}$$

$$RSSG+NADPH+H^+ \xrightarrow{GR} 2RSH+NADP^+ \tag{4}$$

需要注意的是，上述反应只能还原处于游离状态的过氧化物，对于一些处于酯化状态的脂肪酸过氧化物（细胞膜上）则需要 GPX 家族的另一个成员 GPX4 来还原，或者先通过一些酶系统（如磷脂酶）将其转变为游离状态，然后再通过上述反应来处理。

GPX 活性的维持或提高依赖于组织中的硒营养状况，日粮硒添加可以提高包括人、鼠、鸡、猪、马、牛、羊、鹿、鱼等各种动物组织或血液中的 GPX 活性。

（一）GPX1

1. GPX1 的发现、细胞定位和组织分布

GPX1 是 Mills 在 1957 年发现的，能保护 H_2O_2 诱导的红细胞溶血。16 年后，GPX1 被证实是一种硒酶，是第一个被发现的高等哺乳动物硒蛋白，俗称经典 GPX。GPX1 是由 4 个相同亚基构成的四聚体蛋白，每个亚基的分子量为 22～23 kDa。GPX1 是一种胞内酶，主要存在于胞质、细胞核和线粒体基质中，是哺乳动物体内普遍存在的硒蛋白。当硒供应充足时，所有组织细胞中均可观察到 GPX1 的表达，过氧化物生成较多的组织如红细胞、肝脏、肾脏和肺中 GPX1 的水平较高，而睾丸组织中 GPX1 的水平相对较低。

2. GPX1 的生物学功能

GPX1 对 GSH 具有高度特异性，这种特异性归因于该酶的活性位点 Sec 周围存在 4 个 Arg 残基，GSH 的羧基与 4 个 Arg 的胍基结合相继引导 2 个 GSH 分子进入能够与 Sec 反应的位点[59]。该酶专一地利用 GSH 作为底物还原 H_2O_2 和 ROOH，包括叔丁基过氧化氢、异丙苯过氧化氢、游离的长链脂肪酸氢过氧化物和胆固醇氢过氧化物[57]。研究表明，在正常生理条件下，GPX1 基因敲除的小鼠能够健康生长且能繁殖后代，与野生型小鼠相比并没有表现出更多的氧化损伤和敏感性[60]。这让那些认为 GPX1 是一种“神奇的酶”的人感到非常失望。但是后来的研究发现，在氧化应激条

件下，转GPX1基因小鼠与野生型小鼠相比更能抵抗百草枯和 H_2O_2 诱导的氧化应激[61,62]，而GPX1基因敲除小鼠对百草枯、H_2O_2、LPS以及巨噬细胞活化诱导的氧化应激更为敏感[61-63]。而且当机体处于缺硒状态时，这些效应表现的更为明显[64]。这说明，GPX1对ROS所致的氧化应激的保护作用依赖于硒的状况和氧化应激的严重程度[65]，在硒缺乏或应激状态下，即使非常少量的GPX1对抗氧化防御系统也是非常关键的，GPX1被认为是抵御急性氧化应激的主要保护因子[66]。

GPX1还与抵抗病毒感染有关，当GPX1基因敲除小鼠暴露于非致命的柯萨奇病毒株时，病毒株突变为致命形式，和硒缺乏小鼠一样导致心肌病发生，症状类似于克山病[67]。然而，GPX1基因敲除或硒缺乏小鼠暴露于流感病毒时，比硒充足小鼠有更高的生存率[68]。GPX1基因过表达降低了肿瘤坏死因子α(TNF-α)诱导的核因子κB(NF-κB)活化[69]、增强了HIV感染组织的病毒扩散[70]、刺激了对乙酰氨基酚中毒[71]、提高了DMBA/TPA诱导的癌变[72]、触发了肥胖和胰岛素耐受[73]。

尽管GPX1可清除活性氮(RNS)类物质过氧亚硝基阴离子($ONOO^-$)，具有抑制这些物质对蛋白质的硝酰化[74]，但GPX1基因敲除小鼠的肝细胞与野生型肝细胞相比更能抵抗过氧亚硝基诱导的细胞凋亡、细胞色素C的释放以及caspase的激活[75]。上述这些不一致甚至矛盾的研究结果提示，GPX1抗氧化功能的发挥取决于其在细胞或组织中的水平，维持在适宜或平衡的生理水平是前提，过高或缺失均不利于其最优抗氧化功能的发挥，而且仅仅把GPX1看作是一种抵抗氧化应激的酶有些过于片面或简单。

(二)GPX2

1.GPX2的细胞定位和组织分布

Chen等人于1993年首次描述了GPX2的特征。GPX2与GPX1具有相似的底物特异性，也可以催化还原 H_2O_2、叔丁基过氧化氢和异苯丙过氧化氢。GPX2主要分布于胃肠道上皮，从食道一直到结肠末梢均可检测到GPX2的存在。人的肝脏中也可检测到GPX2。在一些上皮细胞源性肿瘤中，如结肠腺瘤[76]、Barrett食管瘤[77]、鳞状细胞瘤[78]以及吸烟者的肺腺瘤[79]中，GPX2的基因表达上调。GPX2在小肠中的分布有其独有的特征，在潘氏细胞中可见GPX2存在，在小肠上皮隐窝处尤其丰富，从隐窝基底到肠绒毛顶端GPX2的分布逐渐降低[80]。GPX2是一个由4个分子量为22 kDa的亚基构成的分泌型四聚体酶蛋白，主要存在于胞质中。GPX2中

65%的氨基酸序列和60%的核苷酸序列与GPX1相似，而且在Sec活性位点周围存在3个保守的Arg残基，因而GPX2与GPX1具有相似的底物特异性。

2. GPX2的生物学功能

GPX2的表达主要被限制在胃肠道，其主要功能是保护肠道上皮免受氧化应激的影响并保证黏膜的内稳态。隐窝基底GPX2水平高主要是为了保证上皮细胞持续增殖以及保护细胞免受氧化损伤，而肠绒毛向肠腔延伸过程中GPX2水平逐渐降低主要是为了保证上皮细胞的生理性凋亡，这种作用对于维持小肠上皮细胞增殖和凋亡的平衡非常重要。GPX2基因敲除小鼠小肠隐窝细胞的自发性凋亡显著增加，这种效应在硒供应受限时表现更为明显[81]，而GPX1/GPX2双基因敲除小鼠会自发地发展为结肠炎和肠癌[82]。GPX1和GPX2的这种抗炎和抗癌的作用归功于它们的抗氧化作用，同时这一现象也暗示GPX1和GPX2在功能上可以相互补偿，这与二者具有相似的底物特异性是一致的，但由于二者组织分布的差异，这种补偿作用是有限的。

与GPX1相比，GPX2更能耐受日粮硒缺乏的影响，在硒供应受限时，GPX2的mRNA表达可保持在稳定水平，而GPX1的mRNA水平则显著降低。细胞在低硒环境下生长，其多种GPX的活性和免疫化学反应均较低，一旦补充硒后，GPX2蛋白快速表达并达到峰值，而其他硒蛋白的表达提高则相对滞后。GPX2对硒缺乏的高耐受性、对补硒的高反应性以及其组织分布特征说明，这种硒蛋白是机体抵御由日粮消化产生的过氧化物和肠道微生物诱导产生的氧化应激的第一道防线。

（三）GPX3

1. GPX3的细胞定位和组织分布

GPX3由Takahashi等人于1987年纯化，是由4个亚基构成的分子量为93kDa的糖基化的分泌型四聚体蛋白。GPX3由肾脏近曲小管上皮细胞合成并分泌到血浆中，眼房水和羊水中也可见GPX3存在[83]。其他组织如心脏、甲状腺、脂肪、肺脏、小肠、附睾也可表达GPX3。

组织细胞合成的GPX3仅有部分进入到血液循环系统，如近曲小管上皮细胞合成的GPX3仅有15%～20%分泌到血浆，其余大部分结合于近曲小管上皮细胞的基质膜上[83]，在小肠、肺脏和附睾中也均观察到这种膜结合特性[84]。

2. GPX3 的生物学功能

和 GPX1 相似，GPX3 也能以 GSH 作为底物催化还原 H_2O_2 和游离脂肪酸氢过氧化物，但其 Sec 活性位点周围仅存在 2 个保守的 Arg 残基，因而对 GSH 的特异性较低，对 GSH 表现出饱和动力学特征。GPX3 对氢过氧化物的催化还原活性仅为 GPX1 的 1/10[85]。GPX3 是一种胞外酶，因而工作环境在胞外，胞外环境中 GSH（血浆中 GSH 的浓度不足 0.5 μM）和还原型硫氧还蛋白（Trx）水平很低而且缺乏这些还原型物质的再生系统，GPX3 对抗胞外氧化应激的能力在几个催化循环周期中即被耗竭。因此，GPX3 的胞外抗氧化能力是有限的，但这种低能力可能可以满足防止低密度脂蛋白（LDL）的氧化因而预防动脉粥样硬化。这种有限的能力也可以通过降低胞外过氧化物水平而下调脂氧合酶（LOX）和环氧合酶（COX）的活性因而防止白三烯和前列腺素的过量生成。此外，氧化了的 GPX3 可能作为一种信号分子“感知”局部过氧化物的释放，并把这种信号传递给周围对氧化还原敏感的受体。

上文提到 GPX3 可以与一些组织细胞的基底膜结合，这种结合特性可以使其处于相对较高的 GSH 环境，因而可能增强其抗氧化作用。

研究发现，在许多类型的癌细胞中均可见 GPX3 基因表达水平的降低，而且在 Barrett 食道癌和前列腺癌病人中同时观察到 GPX3 基因启动子的高甲基化[86,87]，而在前列腺肿瘤细胞中过量表达 GPX3 可抑制细胞增殖，降低肿瘤体积、肿瘤转移和动物死亡率[87]。这些研究暗示 GPX3 可能具有抑癌作用。

（四）GPX4

1. GPX4 的细胞定位和组织分布

GPX4 由 Ursini 等人于 1985 年首次报道。与前三种 GPX 不同，GPX4 是一种单体蛋白，分子量为 22 kDa。GPX4 是一种胞内蛋白，其 mRNA 和蛋白在所有组织中均有表达，但在睾丸组织中的基因表达水平和酶活性最高。GPX4 有三种亚型，分别存在于胞浆、线粒体和核膜中，前两种分别被命名为胞浆型 GPX4（cGPX4）和线粒体型 GPX4（mGPX4）。第三种亚型是在大鼠精子细胞核膜上发现的，最初被命名为精子核特异性 GPX4，但由于不能排除该蛋白在其他组织核膜上的表达，目前称之为核型 GPX4（nGPX4）。

2. GPX4 的生物学功能

GPX4 的独特作用是能够催化还原酯化的脂质过氧化物（磷脂氢过氧化物、胆固醇酯氢过氧化物）。研究表明，豚鼠 104C1 细胞过表达 GPX4 可保护细胞免于卵磷脂氢过氧化物诱导的氧化损伤[88]，乳腺癌上皮细胞 COH-BR1 过表达线粒体型 GPX4 能保护细胞免于胆固醇氢过氧化物诱导的细胞死亡[89]，人乳腺癌 MCF-7 细胞过表达线粒体型 GPX4 可保护细胞免于单线态氧诱导的脂质过氧化和细胞死亡[90]，$GPX4^{+/-}$ 基因小鼠的细胞对 γ 射线、百草枯、叔丁基过氧化氢以及 H_2O_2 诱导的氧化应激更为敏感[91]。这些研究结果充分说明，GPX4 在防止细胞膜脂质过氧化中起着重要的作用。除此之外，GPX4 还能还原胸腺嘧啶氢过氧化物[92]。

GPX4 的 Sec 活性位点周围不存在任何保守的 Arg 残基，因而对还原性底物的要求不像 GPX1 那样专一。当 GSH 存在时，GPX4 优先被 GSH 还原，主要通过结合 GSH 到 Lys 残基实现。但当 GSH 水平较低或受限时，其他巯基化合物如二硫苏糖醇和二硫赤藓糖醇均可代替 GSH 作为 GPX4 的还原性底物，半胱氨酸残基丰富的精子线粒体相关蛋白也可作为 GPX4 的还原性底物[93]，甚至 GPX4 可以通过聚合作用以自身作为其还原性底物。

GPX4 在胚胎形成及发育过程中起着至关重要的作用。GPX4 全基因敲除小鼠导致在怀孕中期胚胎死亡[94]，而选择性敲除 nGPX4[95] 和 mGPX4[96] 基因的小鼠能够正常繁殖后代，暗示 cGPX4 对保证胚胎的正常发育更为重要。cGPX4 保证胚胎正常发育的原因可能与其抗细胞凋亡有关。在细胞分化过程中，12/15-LOX 参与生物膜的氧化破坏，并通过由凋亡诱导因子 AIF 调节的特定通路触发细胞凋亡，GPX4 能阻止 12/15-LOX 催化的膜脂质过氧化[97]。神经特异性 GPX4 的耗竭导致神经细胞退化，而 AIF 基因沉默或抑制 12/15-LOX 活性能有效防止这一现象的发生[97]。这种 GPX4 与 12/15-LOX 的对抗作用能高效维持氧化还原平衡，而氧化还原平衡的维持对于胚胎的形成和发育可能非常重要。

GPX4 参与精子的生成和成熟。动物试验表明，硒缺乏与雄性低生育能力甚至不育有关，而且 GPX4 在睾丸中的基因表达水平最高，暗示 GPX4 在精子生成以及成熟中起重要作用。在精子发生的早期阶段，GPX4 大量合成而且以可溶性的还原活性形式存在；而在成熟的精子细胞中，GPX4 形成一种不具有酶活性的蛋白质聚合体，在成熟精子中段，这种聚合体作为一种结构蛋白在嵌入线粒体螺旋的内囊物质中占 50%以上[93]。无论何种原因导致的 GPX4 缺乏均会引起精子中段结构的不稳定和精子形态的改变，直到失去精子尾部进而失去受精能力[98-100]。在精子发生的早期阶段，GPX4 以 GSH 为还原性底物发挥其抗氧化功能。而在精子发生晚期以及成熟阶段，大量 GSH 被氧化为 GSSG 且 GSSG 在胞外被 γ-谷氨酰转肽酶降解，在 GSH 缺乏的情况下，GPX4 与其自身发生反应形成线性 GPX 聚合物[101]，利用半胱氨酸残基丰富的特别是具有 PCCP 基序的蛋白质作为替代底物，最终将自身通过硒硫键和二硫键交联形成一个三维的蛋白质聚合体[102]。精子发生晚期 GSH 的缺失是由于过氧化物的爆发式产生所致，但对于过氧化物爆发式产生的原因还知之甚少。值得强调的是，这一过程中 GPX4 由一种清除过氧化物的过氧化物酶转变为一种硫醇/硒醇氧化了的蛋白，并利用过氧化物的大量产生建立了一种生物学结构，而这种结构对精子功能的正常发挥是必不可少的。但是，由于 GPX4 变成了几乎完全被氧化的蛋白，其抗细胞凋亡的作用转变为促细胞凋亡，因而成熟精子细胞易于氧化凋亡，通常在几天内就会死亡。尽管核型 GPX4 基因敲除小鼠可以存活并能繁殖后代，但核型 GPX4 作为一种蛋白质硫醇过氧化物酶对精子染色质的压缩起重要作用[95]。从核型 GPX4 基因敲除小鼠分离出来的精子其 DNA 压缩能力明显降低，且对热、酸、变性剂处理等物理损伤的抵抗力下降[95]。

GPX4 的另一个重要作用是抑制炎症反应。在胞外，GPX4 与 GPX3 一样可通过清除 H_2O_2 抑制其对 LOX 和 COX 的激活，进而抑制促炎脂质代谢物如前列腺素和白三烯的生成[103]。然而在胞内，由于膜脂质过氧化物相比 H_2O_2 是更好的 LOX 和 COX 的激活剂，因此 GPX4 相比 GPX1 对炎症反应的抑制作用可能更加明显[92,104]。与 GPX1 相比，GPX4 对 15-LOX、12-LOX 和 COX-2 的抑制作用更强[105]。细胞内 NF-κB 信号通路的活化可上调促炎基因的表达。研究表明，GPX4 过表达抑制了紫外线诱导的皮肤成纤维细胞 NF-κB 的活化[106]以及白介素(IL)-1 诱导的动脉平滑肌细胞 NF-κB 的活化，而 GPX1 过表达不具有这种抑制效应[106]。

GPX4 参与细胞的生存和死亡。GPX4 具有促进细胞生存，抑制细胞死亡的功能。研究表明，GPX4 过表达通过降低截断磷脂的形成抑制了 Jurkat T 淋巴细胞的死亡[107]，通过降低心磷脂氧化进而抑制促凋亡因子的生成保护了敌草快诱导的肝损伤[108]。近年来的研究表明，GPX4 参与一种称为“铁死亡”的新的细胞死亡程序，膜脂质过氧化是导致细胞铁死亡的关键事件[109,110]。

(五)GPX6

与其他 GPX 相比，GPX6 的发现最晚。GPX6 在大鼠和小鼠体内是一种含有 Cys 残基的蛋白，因而不是硒蛋白，但在猪及人体中该蛋白含有 Sec 残基，是一个同型二聚体硒蛋白[55]。GPX6 与 GPX3 的同源性最接近，该酶在人体中仅表达于胚胎组织和嗅上皮细胞[55]，其具体的功能尚不清楚。

二、TrxR 家族及其生物学功能

TrxR 最初是在大肠杆菌中发现的，以 NADPH 为辅因子催化氧化型硫氧还蛋白(Trx)还原为还原型 Trx。还原型 Trx 可作为许多酶如核糖核苷酸还原酶和蛋白质二硫化物还原酶以及许多信号分子的电子供体，因而在调节细胞内氧化还原平衡以及信号传导方面起着重要作用。

现已知道，哺乳动物和高等真核细胞中普遍存在 TrxR，参与巯基依赖的细胞内氧化还原过程。TrxR 是吡啶核苷酸二硫化物氧化还原酶的黄素蛋白家族成员，除 TrxR 外，这一蛋白家族还包括谷胱甘肽还原酶、硫辛酰胺脱氢酶、汞离子还原酶和 NADPH 过氧化物酶。

哺乳动物 TrxR 包括三种类型：胞质 TrxR(TrxR1)、线粒体 TrxR(TrxR2)和睾丸特异性硫氧还蛋白谷胱甘肽还原酶(TGR，TrxR3)。TrxR 具有广泛的底物特异性，除氧化型 Trx 外，还能催化还原亚硒酸盐、H_2O_2、脂质氢过氧化物、脱氢抗坏血酸、抗坏血酸自由基、硫辛酸、细胞毒素肽 NK-lysin、钙结合蛋白、谷氧还蛋白 2 和肿瘤抑制蛋白 p53。

(一)哺乳动物 TrxR 的蛋白质结构及特点

哺乳动物 TrxR 为同型二聚体硒酶，每个亚基的分子量为 55～67 kDa，较大肠杆菌 TrxR 大(亚基分子量 35 kDa)。哺乳动物 TrxR1 和 TrxR2 与谷胱甘肽还原酶具有同源相似性，由两个头尾相邻的亚基组成，每个亚基由 N-端 FAD 结合域、NADP(H)结合域和 C-端交界域三个结构组成(图 2-11)，这种结构组成与中国哲学中描述的“道”极其相似，两个亚基分别表示“阴”和“阳”。哺乳动物 TrxR3 的每个亚基除了上述三个结构域外，在 N-末端还包括一个谷氧还蛋白(Grx)域。

以大鼠 TrxR1 为例，FAD 结合域中存在一个由两个邻近的 Cys(Cys_{59} 和 Cys_{64})形成的分子内氧化还原活性二硫化物，与谷胱甘肽还原酶结构不同的是，TrxR 每个亚基的 C-末端含一个具有 Cys-Sec(Cys_{497} 和 Sec_{498})的 16 个氨基酸的延伸端，其中含有特征性的-Gly-Cys-Sec-Gly 保守序列，为酶的活性中心，内含酶催化所必需的 Sec 残基，位于 C-末端的倒数第二位(图 2-12)。当酶处于氧化型时，催化位点的 Cys_{497} 和 Sec_{498} 以硒代二硫化物形式存在，而当酶处于还原型时，催化位点的 Cys_{497} 和 Sec_{498} 以硒代二硫醇形式存在。

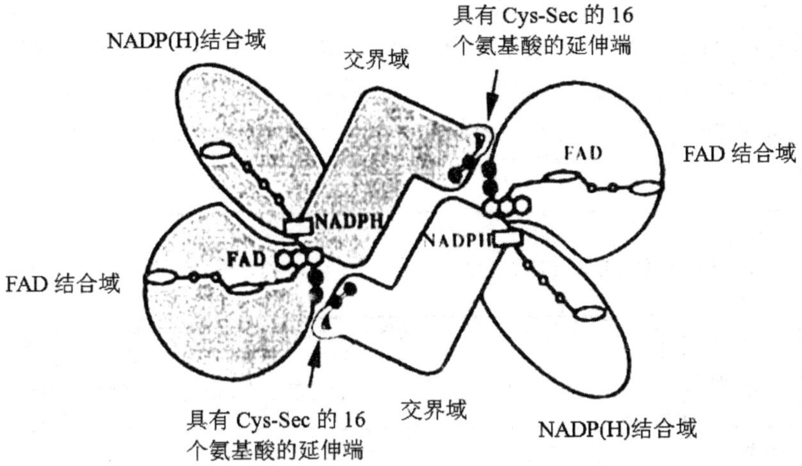

图 2-11　哺乳动物 TrxR 结构模型[111]

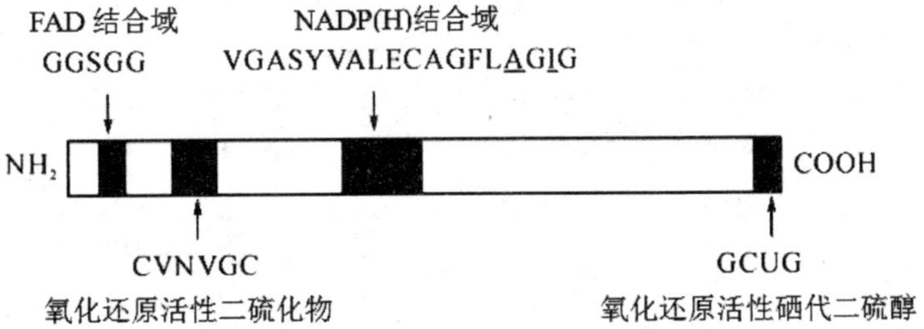

图 2-12　哺乳动物 TrxR 功能结构域[112]

(二)哺乳动物 TrxR 的催化机制

与大肠杆菌相比,哺乳动物 TrxR 的催化机制尚未完全阐明,推测的催化步骤可归纳为两大步。

第一步是 NADPH 结合引起的 TrxR 内部的电子传递:NADPH 结合引起构象变化,使二氢吡啶环处于接近辅因子 FAD 的位置,电子由 NADPH 传递给 FAD 使 FAD 还原,还原的 FAD 使 N-端的氧化还原活性二硫化物(-CVNVGC-)还原,其中的一个 Cys(Cys_{64})以硫醇阴离子(S^-)形式存在并与 FAD 形成加合物,另一个 Cys(Cys_{59})也形成硫醇阴离子并攻击 C-末端的氧化还原活性硒代二硫化物,使 Cys_{497} 和 Sec_{498} 分别转变为硫醇和硒醇形式(图 2-13B)。需要注意的是,电子从 N-端的氧化还原活性二硫化物向 C-端的 Cys-Sec 的传递发生在不同亚基间而非同一亚基内(图 2-13A)。TrxR 的每个亚基至少需要 4 个电子以完全还原氧化型 TrxR 中的 FAD,在巯基-二硫键交换后,保守催化位点被氧化,而 C-末端的-Cys-Sec-催化位点被还原，由硒代二硫化物转变为硒代二硫醇，且在生理 pH 条件下，硒醇以硒醇阴离子（Se^-）形式存在。还原型 TrxR 的 C-末端为表面暴露构型，易被羧肽酶 Y 或胰蛋白酶降解导致酶失活，而氧化型 TrxR 的 C-末端的最后氨基酸残基埋藏于活性位点，不易为蛋白酶所接近或被化学修饰[111,113]。

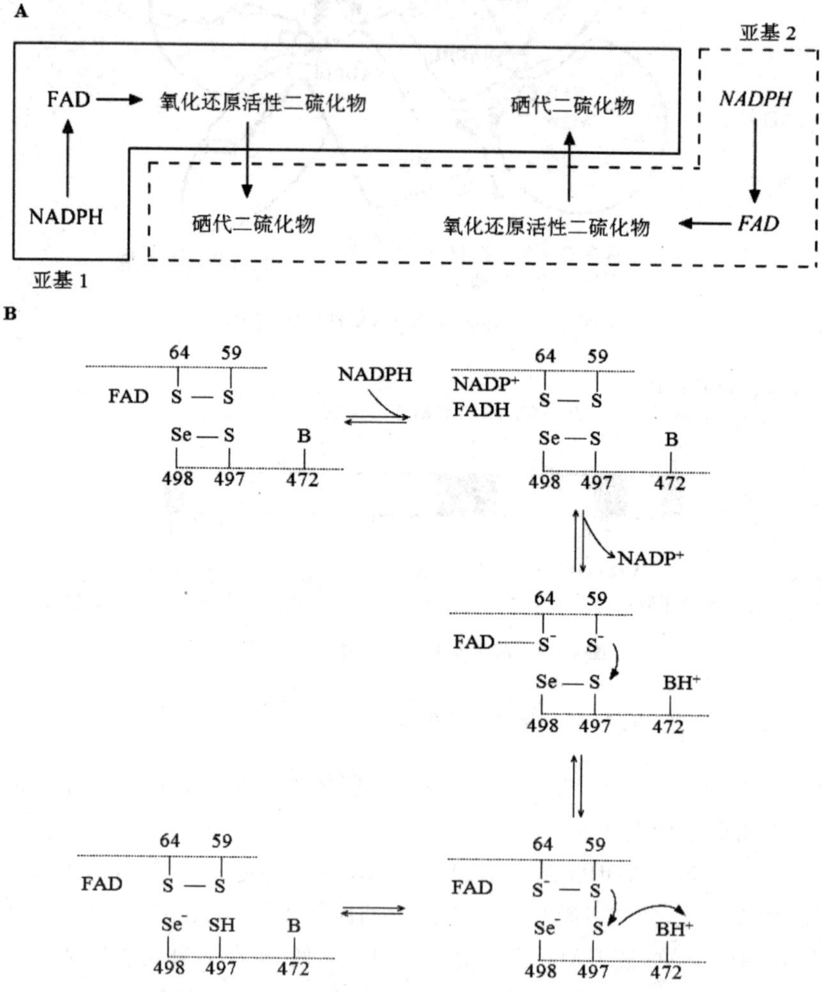

图 2-13　哺乳动物 TrxR 电子传递途径[112,114]

第二步是还原型 TrxR 催化的底物还原：由于哺乳动物 TrxR 具有两个活性位点，因而具有广泛的底物特异性，可催化蛋白质如 Trx、蛋白质二硫化物异构酶（PDI）和 Grx 以及小分子物质如硒酸盐、亚硒酸盐、硫辛酸和 H_2O_2 的还原。下文仅以还原 Trx 和 H_2O_2 为例说明其催化机制。

还原型 TrxR C-末端的硒醇阴离子（Se^-）攻击氧化型 Trx 的二硫键，使 TrxR 与 Trx 之间形成硒代二硫化物，随后 TrxR C-末端与 Sec 邻近的

Cys 巯基攻击该二硫化物从而使 Trx 还原并将其释放，同时 TrxR C-末端的-Cys-Sec-形成硒代二硫键，还原型 TrxR 转变为氧化型 TrxR，当氧化型 TrxR 被 NADPH 还原时即进入下一个催化反应循环(图 2-14A)。

TrxR 催化的 H_2O_2 还原如图 2-14B 所示，当 TrxR 被 NADPH 通过 FAD 和 N-端氧化还原活性二硫化物还原后，其 C-末端的硒醇阴离子(Se^-)可直接与 H_2O_2 反应生成亚硒酸(-SeOH)，随后很可能是 TrxR C-末端与 Sec 邻近的 Cys 巯基与亚硒酸(-SeOH)反应产生一分子水，同时 TrxR C-末端的-Cys-Sec-形成硒代二硫键，还原型 TrxR 转变为氧化型 TrxR。由于 TrxR 作用于 H_2O_2 的 K_m 为 2.5 mM[115]，TrxR 催化的 H_2O_2 的还原可能仅在 H_2O_2 浓度很高时起作用。

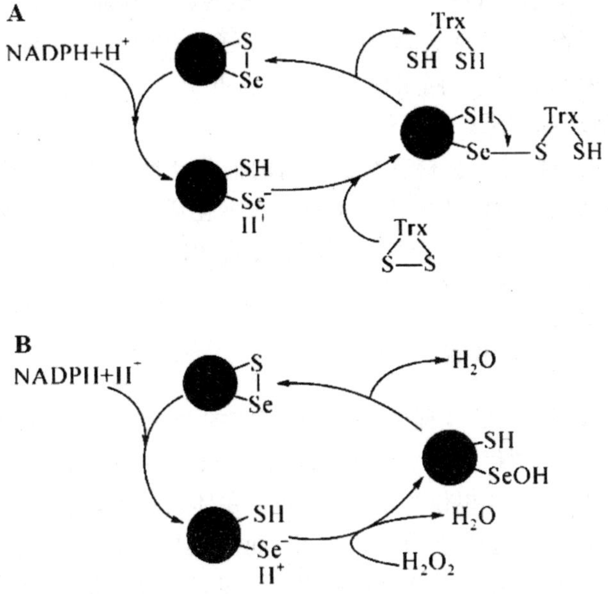

图 2-14　哺乳动物 TrxR 催化的氧化型 Trx 和 H_2O_2 的还原[112]

(三) TrxR 的生物学功能

如前所述，TrxR 可催化氧化型 Trx 的还原，TrxR 和 NADPH 以及 Trx 共同组成 Trx 系统，是一个广泛分布的 NADPH 依赖性二硫化物还原酶系统。在这个系统中，TrxR 通过 NADPH 还原 Trx，Trx 具有高度保守的 Cys-Gly-Pro-Cys 活性位点序列，当 Trx 处于还原状态时，活性位点上的两个 Cys 残基以硫醇形式存在，当处于氧化状态时，两个 Cys 残基以分子内二硫键形式存在。在哺乳动物细胞中存在两种形式的 Trx，一种为 Trx1，存在于胞质，某种条件下可易位到细胞核或分泌到胞外，另一种为

Trx2，存在于线粒体。

TrxR 介导的 Trx 还原可为多种底物提供氢，TrxR 的很多生物学功能是通过 Trx 系统实现的，如非特殊说明，下文提到的 Trx 均为 Trx1。

此外，由于哺乳动物 TrxR 本身具有易接近且活性高的含 Sec 的催化活性位点，除了能还原 Trx 外，还可还原其他蛋白质以及非蛋白质底物，TrxR 广泛的底物特异性赋予其广泛的生物学功能。

1. 促进 DNA 合成

TrxR 对 DNA 合成的促进作用主要是通过 Trx 系统实现的。TrxR 通过 NADPH 催化产生的还原型 Trx 可为核糖核苷酸还原酶提供还原力，核糖核苷酸还原酶为 DNA 的合成提供脱氧核糖核苷酸[116]。

Grx 也可作为核糖核苷酸还原酶的电子供体[117]，Grx 与 NADPH、谷胱甘肽还原酶和 GSH 共同组成 Grx 系统，电子从 NADPH 传递到谷胱甘肽还原酶，然后到 GSH，最后传递到 Grx。

Grx 对大肠杆菌核糖核苷酸还原酶的催化活性与 Trx 相比更加有效，尽管二者对哺乳动物核糖核苷酸还原酶的催化活性无显著差异，但由于在哺乳动物细胞中 Trx 与 GSH 相比浓度较低，因而 Grx 还原核糖核苷酸还原酶的作用更加显著。

2. 参与硒化合物代谢

在硒掺入硒蛋白的过程中，Sec-tRNA^{Sec}的合成需要硒代磷酸作为供硒体，硒代磷酸是在硒代磷酸合成酶的催化下合成的，催化的底物为 H_2Se。无论何种形式的硒，要想作为供硒体，必须在体内代谢转化为 H_2Se。还原型 Trx 可催化硒酸盐还原为亚硒酸盐，同时也可催化亚硒酸盐还原生成 H_2Se[118,119]。因此，哺乳动物 TrxR 不仅本身是一种硒蛋白，而且可通过代谢低分子量硒化合物来促进其他硒蛋白的生物合成，在联系硒和细胞抗氧化方面至关重要。

亚硒酸盐还原生成 H_2Se 还存在另外一条途径，即先生成中间产物硒代二谷胱甘肽（GS-Se-SG），然后在谷胱甘肽还原酶的催化下还原生成 H_2Se 和 GSH[120]。

3. 抗氧化防御

当细胞暴露于高 ROS 如超氧化物（O_2^-）、H_2O_2 和有机氢过氧化物水平时，会导致 DNA 损伤、脂质过氧化、铁硫簇解聚、二硫键形成、蛋白质糖基化以及其他潜在的致命影响。本身作为一种抗氧化酶，TrxR 既可通过直接清除 ROS 实现其抗氧化防御，也可通过 Trx 系统实现其抗氧化防御。Trx 系统生成的还原型 Trx 具有二硫化物还原酶活性，在保护抗氧化酶活性、维持抗氧化剂再生从而保护细胞免受氧化损伤方面起着自关重要的作用。Trx 系统与 Grx 系统一起并称生物体内两大氧化还原系统。下文就 Trx 系统对一些关键抗氧化酶的保护作用以及对一些非酶抗氧化剂的再生作用进行阐述。

(1)硫氧还蛋白过氧化物酶

硫氧还蛋白过氧化物酶(TPX)是过氧化物氧还蛋白家族(Prx)家族的一员。1988 年，第一个酵母 Prx 蛋白被纯化，是一个由两个相同单体通过分子内二硫键形成的反平行二聚体。Prx 能够防止谷氨酰胺合成酶被氧化，暗示其为抗氧化蛋白[121]。在有氧条件下，Prx 基因敲除酵母相比野生型酵母生长率明显降低，而在无氧条件下生长率无明显差异，该研究结果进一步证实 Prx 是一种抗氧化蛋白。Prx 的功能与 GPX 类似，可清除机体细胞产生的 H_2O_2、脂质氢过氧化物和过氧亚硝酸盐。如图 2-15 所示，H_2O_2 氧化还原态 Prx N-端 Cys 的巯基，使其转变为次磺酸，次磺酸再于 Prx 另一个单体 C-端 Cys 的巯基反应形成分子内二硫键，Prx 由还原态转变为氧化态，还原型 Trx 可还原氧化了的 Prx，使其重新获得催化活性[122]。

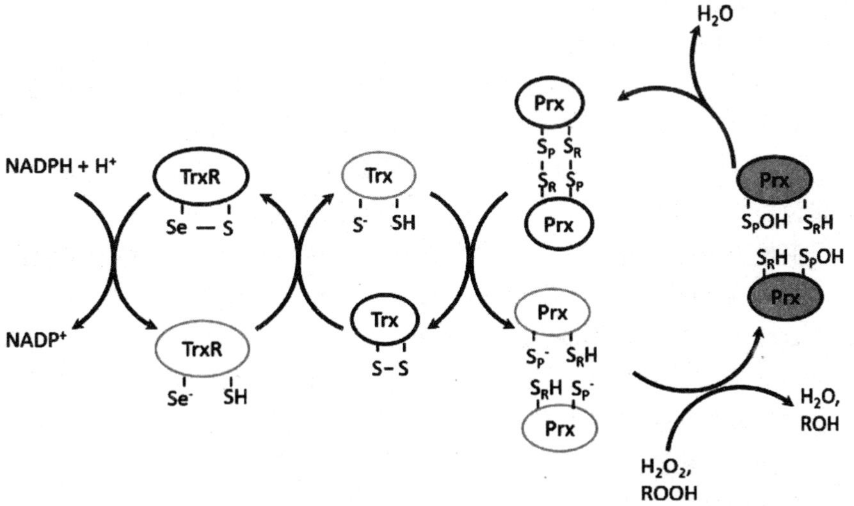

图 2-15　Trx 系统介导的 TPX 催化氢过氧化物的还原[123]

(2)血红素加氧酶 1

血红素加氧酶 1(HO-1)催化血红素转变为胆绿素、铁和 CO。HO-1 在许多细胞类型中均可表达，促氧化剂如血红素、重金属、氧化了的低密度脂蛋白、促炎细胞因子和 ROS 均可诱导 HO-1 的表达[124－126]，HO-1 基因表达水平的提高被认为是细胞抗氧化应激的一种保护反应。

HO-1 的诱导与硒蛋白的活性存在关系，对小鼠肝细胞的研究表明，硒缺乏引起的硒蛋白酶活性降低导致非硒依赖性抗氧化酶 HO-1 表达水平的补偿性提高，用以对抗硒缺乏引起的氧化应激[127]。进一步研究发现，硒蛋白中 TrxR 可能是参与调节 HO-1 表达的主要硒蛋白，然而关于 TrxR 影响 HO-1 表达的研究结果并不一致。对肝细胞的研究发现，TrxR 活性提高降低了 HO-1 的基因表达水平，原因可能与 TrxR 降低 ROS 水平从而解除其对 HO-1 表达的诱导有关[127]。对巨噬细胞的研究发现，TrxR 活性增强提高了 HO-1 的基因表达水平，对这一结果的解释是提高的 TrxR 活性使还原性 Trx 生成增加，还原性 Trx 通过调节 HO-1 上游氧化还原敏感的信号蛋白导致 HO-1 表达水平提高[128]。对血管内皮细胞的研究发现，促氧化剂 15-羟基过氧二十碳四烯酸（15-HPETE）刺激上调了硒充足细胞 HO-1 的基因表达，而且 HO-1 的基因表达与细胞凋亡呈负相关，而当细胞处于硒缺乏状态时，HO-1 的基础表达水平较高，当用 15-HPETE 刺激时，细胞凋亡增加，但并没有诱导 HO-1 基因表达的进一步提高[129]，结合用化学抑制剂抑制 TrxR 活性以及用 SiRNA 使 TrxR 基因沉默的研究结果，作者认为 TrxR 活性降低导致还原型 Trx 水平降低可能是 HO-1 这种抗氧化酶的基因表达未能上调的主要原因。

（3）蛋氨酸硫氧化物还原酶

蛋氨酸硫氧化物还原酶（Msrs）是一类硫醇依赖性抗氧化酶，能够催化蛋氨酸硫氧化物还原为蛋氨酸。Msrs 可分为三类，分别为 MsrA、MsrB 和 fRMsr，后者仅存在于单细胞生物[130]。在氧化应激条件下，游离蛋氨酸以及蛋白质蛋氨酸被氧化为蛋氨酸硫氧化物，因而使蛋白质功能受到影响。氧化形成的蛋氨酸硫氧化物一种为蛋氨酸-S-硫氧化物，另一种为蛋氨酸-R-硫氧化物。哺乳动物 MsrA 是已知唯一一种能够还原游离和蛋白基础的蛋氨酸-S-硫氧化物的酶[131]。MsrB 有 MsrB1、MsrB2 和 MsrB3 三种亚型，MsrB 能游离和还原蛋白基础的蛋氨酸-R-硫氧化物，但主要是还原蛋白基础的蛋氨酸-R-硫氧化物，而且 MsrB1 本身就是一种硒蛋白[132]。

作为抗氧化酶，Msrs 的主要功能是还原氧化损伤了的蛋白质[133]，其催化功能的再生与 Trx 有关，Msrs 从 Trx 系统得到电子，是一种 Trx 依赖性抗氧化酶。Msrs 与许多疾病的发生有关，而且还参与衰老退化过程[132]。随着大鼠年龄的增加，MsrA 和 MrsB 的活性降低，而且还伴有胞质 Trx 和 TrxR 活性的降低[134]。MsrA 基因敲除导致小鼠寿命缩短，氧化应激水平提高[135]，而且当这种小鼠处于硒缺乏状态时，TrxR 活性降低，氧化应激程度加剧[136]。以上研究结果暗示，Trx 和 TrxR 可通过调节蛋氨酸硫氧化物还原酶发挥其抗氧化防御功能。

(4)其他抗氧化酶

如前所述,Grx 与 NADPH、谷胱甘肽还原酶以及 GSH 共同组成 Grx 系统,是维持细胞内氧化还原平衡的重要系统。对线粒体 Grx2 的研究发现,Grx2 具有刺激细胞生长、防止细胞死亡的作用,既可以被 TrxR 还原,同时也可以被 GSH 还原,当 GSH 水平较低时,Grx2 主要靠 TrxR 还原[137]。TrxR 可还原的另一个蛋白质底物为蛋白质二硫化物异构酶(Protein disulfide isomerase,PDI),PDI 是一种内质网蛋白,和其他内质网蛋白如钙结合蛋白一样具有 Trx 域,能够被 TrxR 还原[138]。Trx 样蛋白 1(Trx-like-1)是 Trx 家族成员,在胞质中广泛表达,参与葡萄糖代谢[139]和胞吞作用[140],TrxR1 可直接还原该蛋白[139]。

(5)非酶抗氧化剂

一些低分子量化合物,如抗坏血酸(维生素 C)、α-生育酚(维生素 E)、硫辛酸、类黄酮和辅酶 Q,是动物体内重要的抗氧化剂。Trx 系统利用其广泛的底物特异性,在维持这些抗氧化剂的再生从而保证这些物质抗氧化功能的发挥方面起重要作用。

抗坏血酸是一种多羟基化合物,是人血浆中主要的抗氧化剂,可还原 α-生育酚自由基、过氧化物以及活性氧自由基。红细胞中脱氢抗坏血酸(氧化型抗坏血酸)的还原通常是由 GSH 完成的,然而当 GSH 耗竭时,细胞仍能循环产生抗坏血酸,这主要是通过 Trx 系统对脱氢抗坏血酸的还原实现的[141]。Trx 系统不仅可以还原脱氢抗坏血酸,而且可还原抗坏血酸自由基。

α-生育酚是一种高亲脂性化合物,为生物膜上的重要组成成分,所含羟基能与未配对电子反应而还原过氧自由基,因而也是细胞膜上的主要抗氧化剂,在防止细胞膜磷脂的多不饱和脂肪酸发生过氧化反应方面起重要作用。众所周知,硒与维生素 E 在抗氧化方面具有相互协同的作用,这种联系主要是通过硒蛋白 TrxR 建立起来的,Trx 系统可使脱氢抗坏血酸和抗坏血酸自由基还原为抗坏血酸,而抗坏血酸又可将 α-生育酚自由基还原,从而维持 α-生育酚的再循环[142]。这一机制对于防止细胞膜发生脂质过氧化链式反应具有重要作用。

硫辛酸属于 B 族维生素中的一类化合物,是一种存在于线粒体的辅酶,在多种酶系统中起辅酶作用,能消除自由基,具有抗氧化功能。硫辛酸在体内经肠道吸收后进入细胞,再由细胞内的硫辛酰胺脱氢酶和谷胱甘肽还原酶还原为二氢硫辛酸,二氢硫辛酸具有强效还原能力。在体内或体外,TrxR1 也能有效地催化还原硫辛酸,其催化活性大于谷胱甘肽还原酶,小于硫辛酰胺脱氢酶[143]。在线粒体中，TrxR 并不能催化还原硫辛酸，说明 TrxR 是胞内硫辛酸还原的重要物质之一[144]。

辅酶 Q 的主要功能是在线粒体中作为电子传递链的一部分而发挥作用，在血浆和细胞膜中，辅酶 Q 的浓度一般较低，但可作为抗氧化剂防止血浆脂蛋白和细胞膜脂质发生过氧化反应。辅酶 Q 的再生由硫辛酰胺脱氢酶催化产生，TrxR 同样也能催化该反应[145]。

4. 调控氧化还原敏感的细胞信号通路

氧气是哺乳动物代谢必不可少的底物，因而在有氧呼吸过程中 ROS 的产生也是不可避免的。ROS 过量生成会导致氧化应激，对 DNA、蛋白质、膜脂质等生物大分子造成破坏，相反 ROS 水平过低会导致还原应激，对细胞正常功能的发挥也是不利的，氧化还原应激是多种疾病的诱因。因此，为了很好地适应这种环境，生物体进化出一套对氧化还原特别敏感的信号系统，通过在不同水平以及利用不同机制对这种信号进行调控，来维持细胞的正常功能、活动和稳态。

由于细胞内的强还原环境，大部分蛋白质处于还原状态，具有强抗氧化缓冲能力而不介入氧化还原信号传导，只有具备高巯基/二硫化物氧化电势且巯基易接近的蛋白质是氧化还原敏感的，才可能成为信号通路的组成部分。Trx 在氧化还原信号传导中具有枢纽作用，是信号传导的控制节点和分子开关，通过与其靶蛋白和转录因子（氧化还原敏感的信号分子）相互作用实现信号传导，进而调节生长、分化、增殖、迁移、凋亡、代谢和炎症等细胞进程（图 2-16）。下文就 Trx 对与信号通路相关的靶蛋白和转录因子的调节进行阐述。

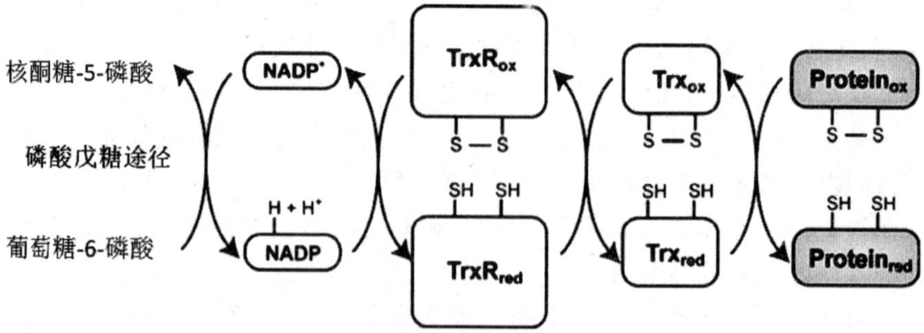

图 2-16　Trx 系统的氧化还原级联反应[146]

（1）凋亡信号调节激酶 1

在真核细胞中，丝裂原活化蛋白（MAP）激酶信号通路进化上相对保守，包括 MAP 激酶激酶激酶（MAPKKK）、MAP 激酶激酶（MAPKK）和

MAP 激酶(MAPK)三种顺序激活的激酶,多种 MAPK 信号级联可以同时激活,也可依次激活。MAPK 家族包括胞外信号调节激酶(ERK)、JNK 和 p38 MAPK。凋亡信号调节激酶 1(ASK1)属于 MAPKKK 家族成员。胞外刺激或应激,如生长因子、ROS、紫外线辐射、LPS、炎症细胞因子,首先被 MAPKKK 监测到,随后 MAPKK 和 MAPK 依次激活,MAPK 调节下游靶标进而调控应激反应、先天免疫、细胞周期阻滞、细胞生存、增殖、分化和凋亡等活动(图 2-17)。

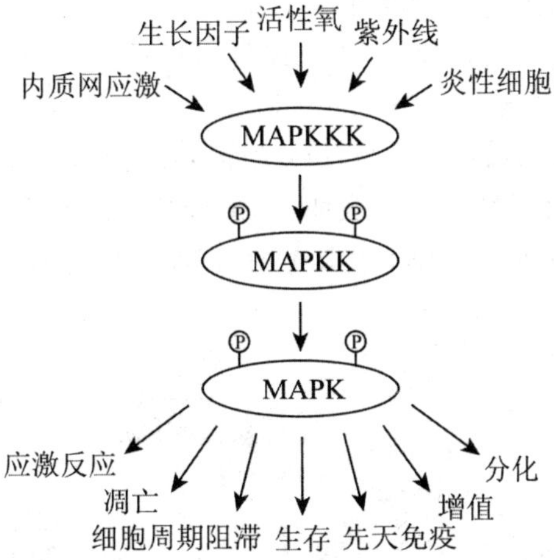

图 2-17 MAPK 信号级联[146]

ERK 家族包括两大类,一类是广泛研究的 ERK1 和 ERK2,另一类是分子量较大的 ERK3、ERK5、ERK7 和 ERK8。细胞生存和分化刺激可使 ERK1/2 激活从而促进细胞的分化、增殖和生存[147]。氧化应激和紫外线可通过磷酸化和活化多种生长因子受体引起 ERK1/2 活化[148,149]。氧化损伤条件下,ERK1/2 活化促进了细胞的生存,刺激了角质细胞的角化增殖[150,151]。尽管 ERK1/2 通常被认为是一种促生存标志物,但 ERK1/2 也可激活促凋亡通路,在 HeLa 细胞中,ERK1/2 活化促进了铅化合物诱导的细胞凋亡[152]。与 ERK1/2 的作用相反,JNK 和 p38 主要刺激应激条件下的促炎和促凋亡通路。当 ASK1 监测到 ROS、紫外辐射、LPS 和炎症细胞因子等环境应激时,ASK1 激活,进而激活 JNK 和 p38 MAPK 信号通路。研究发现,H_2O_2 刺激 HEK293 细胞导致 ASK1 水平提高,细胞凋亡增加[147],未折叠或错误折叠蛋白增加导致的内质网应激也可激活 ASK1,刺激细胞死亡[153]。然而对于一些角质细胞和神经细胞,ASK1 的活化刺激了细胞的增殖和分化[154,155]。

ASK1 信号蛋白的活动受到翻译后调节作用的调控。翻译后调节过程如磷酸化使其激活，泛素化使其降解，S-硝酰化使其活性抑制。

ASK1 信号蛋白的活动还受到蛋白-蛋白相互作用的调控。还原型 Trx、Grx、硫氧还蛋白互作蛋白(Txnip)和丝氨酸/苏氨酸蛋白磷酸酯酶 5 可与 ASK1 相互作用而调节其活性。

细胞在非应激状态下，ASK1 与还原型 Trx 结合形成一个低聚复合物，该复合物被称作信号小体(Signalosome)，呈无活性的静默状态。当细胞处于氧化应激状态时，还原型 Trx 被氧化而从信号小体上解离，一旦解离，一些 ASK1 激活剂如肿瘤坏死因子受体相关因子(TRAF)即被 ASK1 招募而使其恢复活性(图 2-18)。研究表明，TRAF2 和 TRAF6 的缺失抑制了 TNF-a 和 H_2O_2 诱导的 ASK1 活化以及其下游 JNK 和 p38 的活化[156]。

Grx 对 ASK1 活性的调节与 Trx 相似，被 GSH 还原的 Grx 也可与 ASK1 结合而抑制其活性。研究表明，葡萄糖缺乏同时阻止了 Trx 和 Grx 与 ASK1 的结合，激活了 ASK1 及其下游的 MAPK 信号通路，刺激了细胞的凋亡[157,158]。

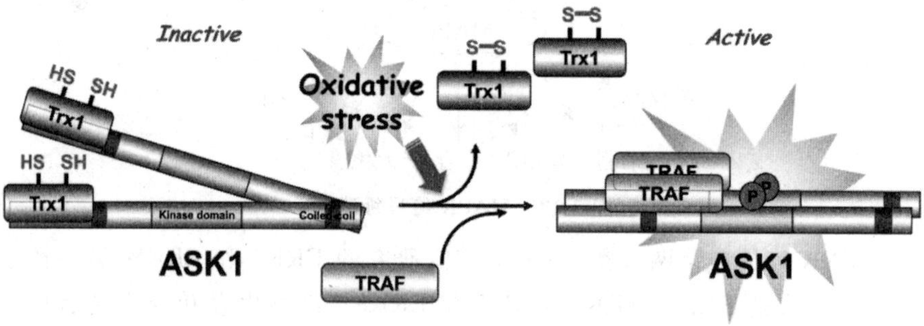

图 2-18 氧化应激诱导的 ASK1 活化[159]

Txnip 也被称作硫氧还蛋白结合蛋白 2 和维生素 D_3 上调蛋白 1，Txnip 对 ASK1 活性的调节是间接的。Txnip 能与还原型 Trx 结合，该结合竞争性地抑制了 ASK1 与还原型 Trx 的结合，从而间接激活了 ASK1。当细胞处于氧化应激状态时，Txnip 的表达显著提高，Txnip 过表达显著抑制了 Trx 与 ASK1 的相互作用。Txnip 不仅可以调节氧化还原信号通路，还可以调节代谢通路和细胞周期。Txnip 在介导硝化应激方面也扮演着重要角色，内源生成的一氧化氮(NO)使蛋白质的酪氨酸发生 S-硝酰化，是一种重要的蛋白质翻译后修饰过程，但过量生成的 NO 及其衍生的 RNS 会导致硝化应激。还原型 Trx 可使蛋白质的酪氨酸去硝酰化，因而具有抗硝化应激的作用，Txnip 可通过与 Trx 结合抑制其去硝酰化作用而介导硝化应激。但 Txnip 的表达又可被过量生成的 NO 抑制，因此，Txnip 对蛋白质酪氨酸的 S-硝酰化和硝化应激具有反馈调节作用。

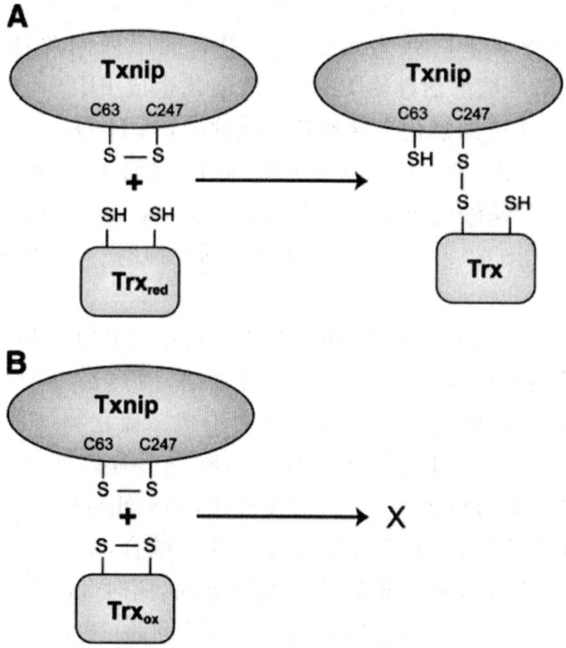

图 2-19 Txnip 与 Trx 的相互作用[146]

(2)磷酸酶张力蛋白同源物

磷脂酰肌醇 3-激酶(PI3Ks)是一类细胞内脂质激酶,根据底物特异性和结构的不同可分为三类,其中研究最为广泛的为Ⅰ类 PI3K。根据使其激活的受体的不同,Ⅰ类 PI3K 又可进一步分为 1A 类 PI3K 和 1B 类 PI3K 两种亚型,酪氨酸激酶耦联受体可激活 1A 类 PI3K,G 蛋白耦联受体可激活 1B 类 PI3K[160,161]。PI3K 激活的结果使质膜上的磷脂酰肌醇-4,5-二磷酸(PIP2)磷酸化为磷脂酰肌醇-3,4,5-三磷酸(PIP3),PIP3 是一种脂质第二信使,可招募细胞内含有 PH(pleckstrin homology)结构域的信号蛋白并与之结合。

Akt 是 PIP3 下游主要的效应分子,是一种丝氨酸/苏氨酸蛋白激酶,也称蛋白激酶 B。Akt 以非活性构象存在于细胞质中,细胞被刺激后,Akt 被 PIP3 招募至膜并与之结合,3-磷酸肌醇依赖性蛋白激酶 1(PDK1)通过磷酸化 Akt 蛋白的 Thr308 而激活 Akt。mTOR 是哺乳动物雷帕霉素靶蛋白(mTOR),也是一种丝氨酸/苏氨酸蛋白激酶,细胞内存在 mTORC1 和 mTORC2 两种不同的复合体,mTORC2 通过磷酸化 Akt 蛋白的 Ser473 而激活与 PIP3 结合的 Akt。Akt 信号通路对调控细胞的生长、增殖、迁移、存活以及糖代谢起到十分重要的作用,大量证据表明,PIP3 诱导的 Akt 信号通路的活化上调了促生存因子的表达,因而具有保护细胞免于死亡的作用。

磷酸酯酶张力蛋白同源物(PTEN)是一种肿瘤抑制剂[162],与酪氨酸磷酸酯酶和细胞骨架蛋白张力蛋白具有很高的序列同源性[163]。活化的PTEN通过其磷酸酯酶活性使PIP3去磷酸化生成PIP2而抑制Akt信号通路[164,165]。PTEN的氧化会导致其活性抑制,当 H_2O_2 存在时,PTEN因氧化形成二硫化物(C71-C124)而失去活性,因而不能抑制Akt信号通路[166]。PTEN的磷酸化也会导致其活性丧失,酪蛋白激酶Ⅱ可通过磷酸化PTEN C-端的S380、T382和T383使其活性抑制[167]。胞质PTEN也可被泛素化而降解。

Trx可调控PTEN的活性,还原型Trx可还原PTEN因氧化形成的二硫化物使其活性恢复,从而抑制Akt信号通路。Trx除了直接还原PTEN使其活性恢复外,还可通过与PTEN结合形成二硫化物($C32_{Trx}$-$C212_{PTEN}$)而使PTEN灭活,这在Trx过表达导致Akt信号通路活化以及细胞死亡抑制的研究结果中得以证实[168]。如前所述,Trx可与Txnip结合,当Trx与Txnip发生作用时,Trx不再能够与PTEN结合,解离下来的Trx形成还原型,可激活PTEN从而抑制Akt信号通路,同时Trx与Txnip的结合也会使Trx从ASK1上解离,因而激活ASK1通路(图2-20)。

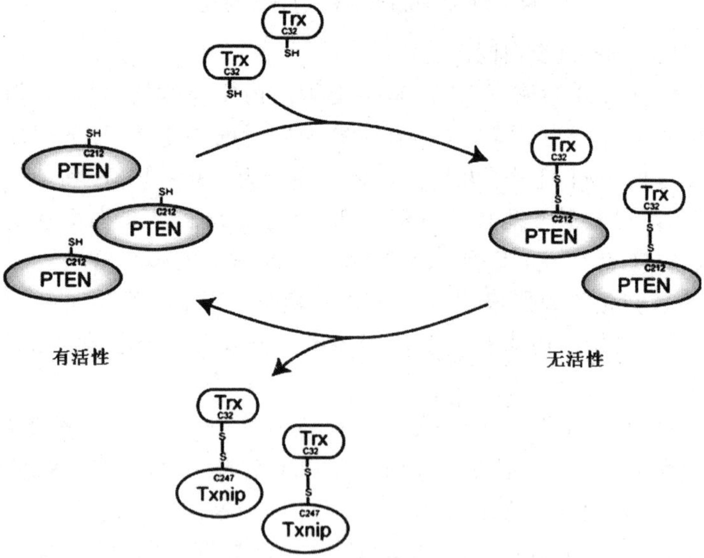

图2-20 Trx和Txnip对PTEN活性的调节[146]

(3)NF-κB

NF-κB是一种细胞内核转录因子,其家族成员主要包括RelA(p65)、c-Rel、RelB、NF-κB1(p50蛋白及其前体p105)和NF-κB2(p52蛋白及其前体p100),各成员间可形成同源或异源二聚体。在哺乳动物细胞中最常见的是 NF-κB p65 与 p50 结合形成的 p65/p50 二聚体。NF-κB 抑制因子(IκB)是一类 NF-κB 抑制蛋白,包括 IκBα、IκBβ、IκBλ、IκBε、IκBNS、Bcl-3、IκBζ 等亚基。细胞在静息状态下,IκB 通过与 NF-κB 结合掩蔽了 NF-κB 二聚体的核定位信号,同时也干扰了 NF-κB 与 DNA 的结合,从而阻抑了 NF-κB 的转录功能。多种刺激如生长因子、细胞因子、淋巴因子、紫外线、氧化应激等可激活 NF-κB 信号通路。在外来信号的作用下,NF-κB 与某些配体以及细胞表面受体结合,招募接头蛋白继而招募 IκB 激酶(IKK)复合物并导致 IKK 复合物激活,IKK 复合物活化的结果使 IκB 磷酸化,继而导致 IκB 的泛素化和降解,IκB 的降解使 NF-κB 二聚体的核定位信号暴露,NF-κB 活化进而易位到细胞核中,与 DNA 结合诱导相关基因的表达。

TrxR 以及 Trx 可调控 ROS 诱导的 NF-κB 信号通路的活化。在细胞质中,ROS 可通过激活蛋白激酶 A 导致 IκB 从 NF-κB 复合体上释放[169],而 Trx 过表达可显著抑制 ROS 诱导的 NF-κB 活化[170]。然而在细胞核中,Trx 可通过还原 NF-κB 中氧化了的 Cys 残基增强其与 DNA 的结合[171](图 2-21)。还原型 Trx 主要存在于胞质中,而在氧化刺激下,Trx 可易位到细胞核中[172]。硒依赖的抗氧化酶 GPX 和 TrxR 可通过直接清除刺激因子 ROS 而抑制 NF-κB 的活化,也可通过调节接头蛋白的氧化还原状况影响 NF-κB 信号通路(图 2-21)。

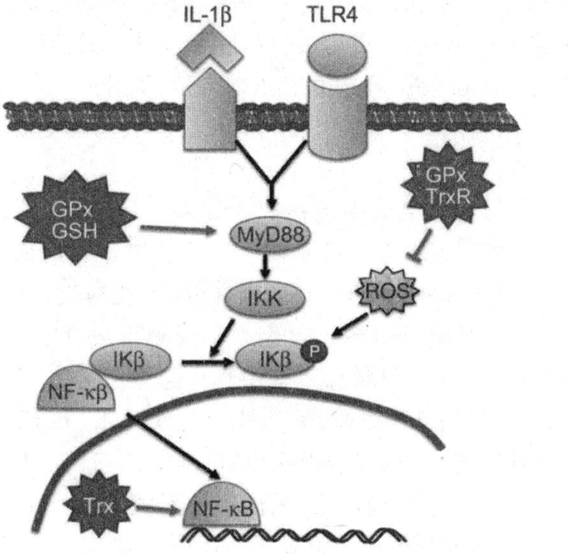

图 2-21 GPX、TrxR 和 Trx 对 NF-κB 信号通路的调节[173]

(4)Nrf2

核因子 E2 相关因子 2(Nrf2)是一种转录因子,能够大范围激活细胞内抗氧化系统[174,175]。Nrf2在癌细胞中高度活化，用以对抗癌细胞中高水平的氧化应激[176]。然而在正常细胞中，如免疫细胞也需要Nrf2激活从而发挥其抗肿瘤作用[177]。在其他正常细胞中，Nrf2活化可提高细胞的抗氧化能力，阻止生物大分子如DNA的氧化损伤，因而可能具有防止正常细胞癌变的可能。许多文献报道了通过抑制TrxR1活性阻止癌症发展，然而许多具有抑制TrxR1活性作用的抗癌药物能够激活Nrf2，而Nrf2的活化有利于癌细胞生存。

三、DIO家族及其生物学功能

在大多数脊椎动物体内发现的DIO是硒蛋白，存在于细胞质膜上，具有亲脂性，包括1型、2型和3型三种类型，主要功能是催化甲状腺激素的脱碘代谢。

（一）甲状腺激素

甲状腺激素是碘化了的甲腺原氨酸，主要由具有甲状腺激素活性的3，5，3′，5′-四碘甲腺原氨酸（T_4）和3，5，3′-三碘甲腺原氨酸（T_3）组成，此外生物体内还存在不具有甲状腺激素活性的3，3′，5′-三碘甲腺原氨酸（简称反式T_3，rT_3）和二碘甲腺原氨酸。甲状腺最初合成的是完全碘化了的T_4，T_4全部在甲状腺中合成，合成后由甲状腺释放进入血液循环并被运送到机体的各种组织中。血液循环中存在的T_3和rT_3除少量在甲状腺中生成外，大部分在甲状腺以外的组织中由T_4通过脱碘代谢生成，进一步脱碘则生成二碘和一碘化合物。

甲状腺激素的化学结构中包括两个苯环，可分别接纳两分子碘，因此完全碘化的甲状腺激素含有4分子碘。甲状腺激素在脱碘代谢过程中，当外侧苯环脱去1分子碘时称为外环脱碘（ORD），相应地，当内侧苯环脱去1分子碘时称为内环脱碘（IRD）。T_4经外环脱碘形成T_3，T_3的生物活性约是T_4的5～8倍，因此在甲状腺激素中真正起作用的是T_3。T_4经内环脱碘形成rT_3，rT_3不具有甲状腺激素活性。生成的T_3和rT_3进一步分别经内环和外环脱碘生成3，3′-二碘甲腺原氨酸（3，3′-T_2）（图2-22）。T_2还包括3，5-T_2和3′，5′-T_2两种形式，无论哪种形式的T_2均不具有甲状腺激素活性，属于甲腺原氨酸的清除形式[178]。

甲状腺激素在调节机体中枢神经系统发育、生长发育、能量代谢以及三大营养物质代谢方面起重要作用。

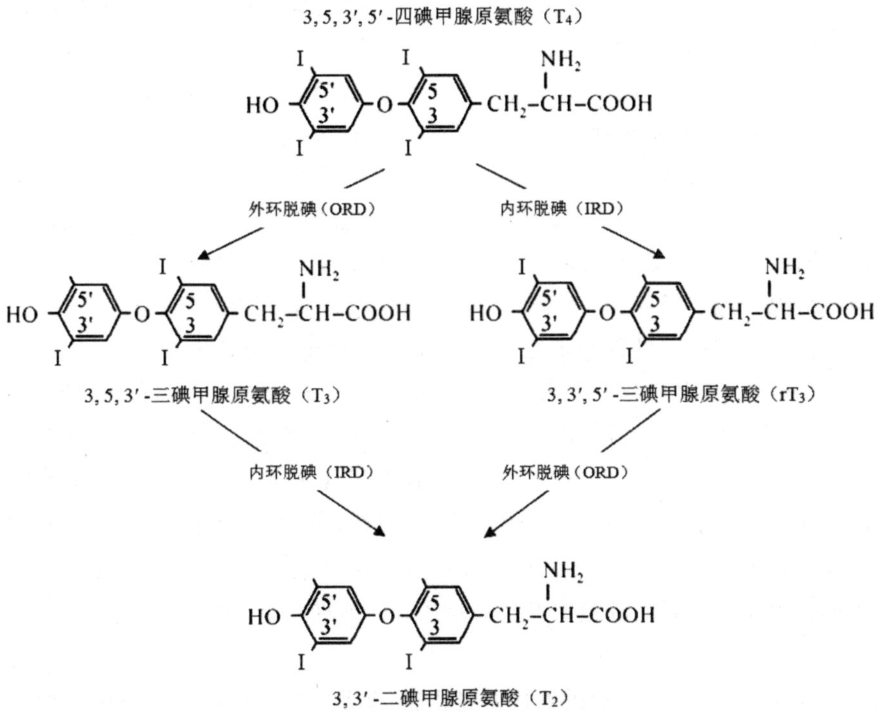

图 2-22　甲状腺激素内环和外环脱碘代谢

（二）DIO

DIO1 是第一个被发现的能够催化甲状腺激素脱碘的硒酶[179]。所有脊椎动物体内均发现有 DIO1 存在，不同种属的 DIO1 具有高度同源性，其全酶由两个分子量为 27 kDa 的亚基单位组装而成。DIO1 主要存在于浆膜上，是一个整合膜蛋白，其单体 N-端的 12 个氨基酸残基伸展进入内质网腔内，C-端的酶催化部位处于细胞质中[180]。DIO1 的 mRNA 主要在肝脏、肾脏、甲状腺和垂体中表达。DIO1 是三种脱碘酶中唯一一个即能催化外环脱碘，又能催化内环脱碘的脱碘酶。在辅因子二硫苏糖醇参与下，DIO1 通过外环脱碘作用可将 T_4 降解为更高活性的 T_3，也可将 rT_3 进一步降解为 3，3′-T_2；通过内环脱碘作用可将 T_4 降解为无生物活性的 rT_3，也可将高生物活性的 T_3 进一步降解为 3，3′-T_2。与外环脱碘作用相比，DIO1 的内环脱碘活性相对较弱，因此哺乳动物 DIO1 的主要作用包括两个方面：①将甲状腺中合成的 T_4 脱碘为血液循环中的 T_3 并提供给外周组织，是血液循环以及外周组织中 T_3 的主要来源；②降解 rT_3，是机体清除 rT_3 的主要途径。研究表明，缺硒引起的亚急性克山病患者血清 T_4 水平明显提高；用病区粮食饲喂大鼠也得到类似的结果；缺硒动物肝脏中 T_4 水平明显提高，T_3 水平以及 DIO1 活性显著降低[181,182]。

DIO2 是在对甲状腺功能低下小鼠脑垂体的研究中发现的。除蜥蜴可能是个例外，其他所有种类的脊椎动物体内均发现有 DIO2 存在。类似于 DIO1，DIO2 也需要由两个亚基单位二聚化后才能形成具有催化活性的全酶。DIO2 主要存在于内质网膜上，是一个整合内质网膜蛋白，其 N-端伸入内质网腔内，C-端处于胞浆中[183]。其 mRNA 主要在中枢神经系统、棕色脂肪组织、胎盘、甲状腺、皮肤、松果体、乳腺、肾脏、心脏、脾脏和前列腺中表达。哺乳动物的肝脏不表达 DIO2，而肺鱼和多数硬骨鱼类肝脏中 DIO2 的活性最高。DIO2 仅具有外环脱碘活性，可将 T_4 转变为更高活性的 T_3，也可将 rT_3 降解为 3，3′-T_2。与 DIO1 不同的是，在细胞水平上由 DIO2 催化产生的 T_3 并不进入血液循环，可能起着优先为局部组织细胞的细胞核提供 T_3 的作用[184]。在冷应激条件下，脂肪组织中 T_3 水平提高，而 DIO2 基因敲除小鼠在冷应激条件下无法调节体温且表现为骨骼发育异常[185]。硒缺乏大鼠脑、垂体、棕色脂肪组织中 DIO2 活性降低，脑和垂体中 T_3 水平下降[111]。

DIO3 是甲状腺激素脱碘代谢过程中涉及的第三种硒酶。在所有种类的脊椎动物体内均发现有 DIO3 存在。DIO3 主要存在于细胞膜上，也是一个整合膜蛋白，大部分蛋白质分子位于细胞外，催化部位也位于细胞外[183]。DIO3 的 mRNA 主要在子宫、胎盘和新生儿组织中表达。DIO3 仅具有内环脱碘活性，可催化 T_4 转变为 rT_3 以及 T_3 转变为 3，3′-T_2。鉴于 DIO3 的组织分布和亚细胞定位的特征，DIO3 是使循环系统中 T_4 和 T_3 迅速失活、防止甲状腺以外组织 T_4 和 T_3 积累的主要脱碘酶，在维持组织中甲状腺激素的内稳态、防止甲状腺以外组织过早暴露于成年动物甲状腺激素水平中、避免甲状腺激素对胎儿发育器官的过度刺激以及保证胎儿的正常生长发育方面起重要作用。

四、SelP 及其生物学功能

SelP 是 20 世纪 70 年代首次在大鼠血浆中发现的一种硒蛋白，因主要存在于血浆中，因而以血浆的第一个字母 P 命名为硒蛋白 P。目前已知血浆中仅存在两种硒蛋白，即 SelP 和 GPX3，在个体硒摄入正常的情况下，血浆中 50％的硒以 SelP 形式存在[186]。SelP 是继 GPX 之后被发现的第二种硒蛋白，是一种高度糖基化的分泌型硒蛋白。

(一)SelP 的结构

人、大鼠和牛的 SelP 的 N-末端都含有一段 19 个氨基酸的信号肽,这是分泌型蛋白的典型结构。SelP 与其他已发现硒蛋白最大的区别在于其多肽链中含有不止一个 Sec 残基,根据 SelP 中 Sec 的分布,可将其分为两个结构域:一个较长的 N-末端结构域,从 N-末端开始到第二个 Sec 残基之前(1-244),占整个氨基酸序列的 2/3,包含 1 个含有第一个 Sec 残基的 U-X-X-C 基序(U 表示 Sec,类似于 Trx 的 C-X-X-C 基序)、3 个 N-糖基化位点(64、155、169)、1 个肝素结合域(80-86)和 2 个富组氨酸区(图 2-23A);一个较短的 C-末端结构域,从第二个 Sec 残基开始到 C-末端(245-366),共包含 9 个 Sec 残基(大鼠、小鼠、人)以及 1 个 O-糖基化位点(图 2-23B)。较长的 N-末端结构域可能起着酶促结构域的作用,而较短的 C-末端结构域起着将高活性的 Se 原子安全运送到机体全身的作用。从大鼠血浆中纯化出的 SelP 存在 4 种亚型,除了以上描述的完全型 SelP 外,还存在 3 种缩短型的 SelP,分别终止于第二、第三和第七个 Sec 残基之前(图 2-23)。

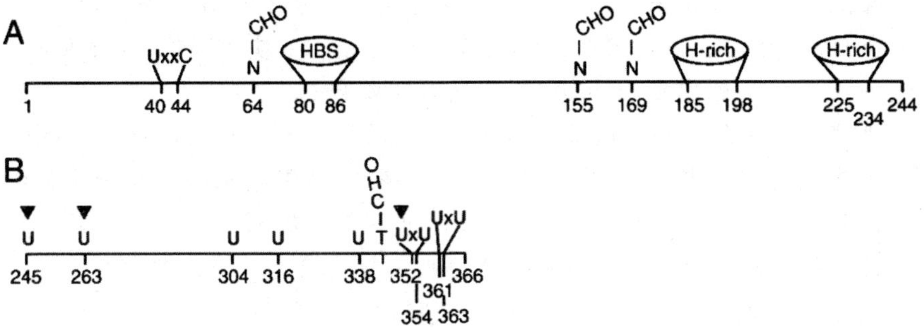

图 2-23 大鼠 SelP 结构域[187]

(二)SelP 的生物学功能

1. 转运和分配硒的功能

SelP 在肝脏合成,肝细胞所有的硒蛋白 mRNA 中,编码 SelP 的占到 25%,肝脏病变可导致血清硒和 SelP 的含量显著降低。肝脏从多种来源获得硒,并在合成硒蛋白和生成排泄性硒代谢物之间进行分配。SelP 基因敲除小鼠尿液硒排泄增加以及机体硒存留降低的现象很好地证实了肝脏的这种作用。肝脏合成的硒蛋白包括两部分,一部分为其自身所需的硒蛋白,另一部分即为 SelP。肝脏合成的硒蛋白(包括 SelP)中的硒注定要被机体保留,而未被合成硒蛋白的硒则进入硒排泄通路。如果肝脏对硒蛋白合成和硒排泄具有主动调控能力,那么这对机体硒稳态的维持具有极其重要的意义,然而到底能否主动调控还有待进一步研究确定。

1982 年,Motsenbocker 和 Tappel 给大鼠腹腔注射^{75}Se-亚硒酸钠发现,1 h 后有 50%的^{75}Se 存在于肝组织中,3 h 后血浆^{75}Se 有 25%存在于 SelP 中,而 25 h 后血浆中几乎不存在^{75}Se,大部分已累积于肝外组织[188]。基于该研究结果,Motsenbocker 和 Tappel 提出了"SelP 运输硒到肝外组织供组织细胞利用"的假说。SelP 中的硒是以共价结合的方式存在于蛋白质中的,在细胞利用硒之前,蛋白质必须降解以释放出细胞可利用的硒,但该蛋白不能像转铁蛋白负载和卸载铁那样加载和卸载硒[189]。

1993 年的研究发现,SelP 与细胞膜具有亲和性,这为 SelP 具有向肝外组织转运硒的功能提供了有力的证据[190]。给正常和缺硒大鼠分别注射^{75}Se-SelP 后 24 h,发现不同组织间^{75}Se 的摄取差异很大,最值得关注的是,注射^{75}Se-SelP 后 2 h,硒缺乏动物脑组织中^{75}Se 的水平显著提高,是对照组的 5 倍,而其他组织并未表现出这样大的差异[191]。该试验提示 SelP 可能转运硒给脑组织。进一步研究发现,硒缺乏大鼠静脉注射^{75}Se-亚硒酸钠后,在血浆 GPX3 中未检测到^{75}Se,血浆中的^{75}Se 主要存在于 SelP 和其他小分子成分中,随着时间的延续,血浆中小分子^{75}Se 快速降低,而血浆中^{75}Se-SelP 在 1 h 后出现,4 h 最高,之后开始降低(图 2-24A)。图 2-24B 显示,肝脏中^{75}Se 的提高与血浆中小分子^{75}Se 的下降密切相关,而脑组织中^{75}Se 的提高是在血浆 SelP 出现之后。这一研究结果暗示脑组织能够利用 SelP 形式的^{75}Se 而不能利用小分子^{75}Se,进一步说明 SelP 能够转运硒给脑组织。从牛血清中纯化出来的 SelP 与亚硒酸钠相比对神经细胞的生存更加有效,这也说明了脑组织更加倾向于利用 SelP 形式的硒[192]。

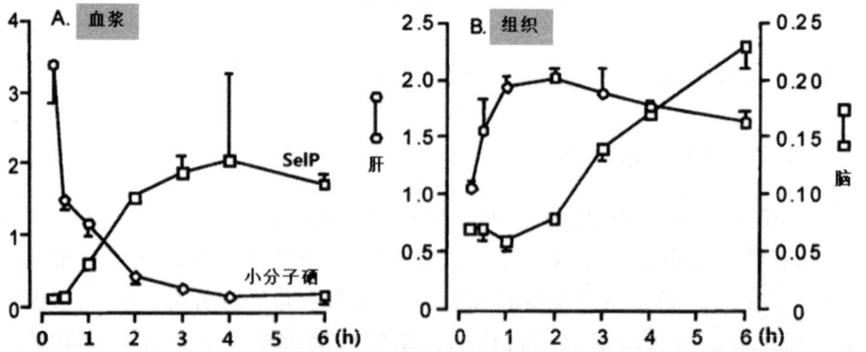

图 2-24　硒缺乏大鼠注射^{75}Se-亚硒酸钠后血浆以及组织中^{75}Se 的分布及变化[189]

硒为大鼠、小鼠精子发生所必需。对缺硒大鼠的研究表明,睾丸组织对 SelP 具有很强的亲和力[193]。通过免疫组化技术发现小鼠睾丸支持细胞中存在含有 SelP 的囊泡，这与受体介导的蛋白质内吞是一致的，进一步采用放射受体分析技术证实在小鼠睾丸组织的细胞膜上存在阿朴脂蛋白 E 受体 2(ApoER2)，ApoER2 基因敲除小鼠睾丸组织中硒含量极低[194]。进一步研究发现，ApoER2 基因敲除小鼠脑组织中硒含量降低 56%，而机体总硒水平无显著变化[195]。这些研究表明，睾丸和脑组织对 SelP 的摄入依赖受体 ApoER2。

SelP 基因敲除可导致肾脏硒水平降低。由于肾脏本身也可以合成 SelP，因此通过灭活肝细胞特异性 tRNASec 后发现，血浆中 SelP 水平明显降低，同时肾脏硒含量也显著降低[196]。对近曲小管上皮细胞的研究发现，细胞中也存在含有 SelP 的囊泡[197]，然而 ApoER2 基因敲除小鼠肾脏和机体总硒水平无显著改变[195]，暗示肾脏 SelP 摄入不依赖该受体。进一步研究发现，近曲小管上皮细胞的刷状缘存在另外一种被称为巨蛋白(Megalin)的脂蛋白受体[198]，与野生型相比，Megalin 基因敲除导致近曲小管上皮细胞内含 SelP 的囊泡消失[197]，说明肾脏摄入 SelP 依赖受体 Megalin。近曲小管上皮细胞可合成 GPX3，因此，这种依赖 Megalin 受体的对肾小球滤过液中 SelP 的重吸收具有循环利用硒的作用，对于防止硒从尿液中的流失具有重要意义。

综上所述，肝脏合成的 SelP 具有将硒运送到肝外组织的功能，但仅有一些特殊的组织如脑组织、睾丸组织和肾脏享有优先利用 SelP 的权利。

2. 抗氧化功能

硒蛋白因其活性中心含有 SeCys 而具有氧化还原酶活性。SelP 的一级结构中含有许多 SeCys 残基，因而具有许多潜在的氧化还原活性中心。研究发现，在 GSH 存在情况下，SelP 不能还原 H_2O_2，因而不具有 cGPX 样催化活性[199]。然而在同样的条件下，当以合成的卵磷脂氢过氧化物作为底物时，SelP 能够还原该底物[200]。在反应体系中加入非离子型表面活性剂 Triton X-100 和脱氧胆酸盐大大提高了 SelP 的还原能力。Triton X-100 可使脂质双分子层转变为透明的混合微胶团溶液，暗示 SelP 的催化反应发生在脂-水界面[201]。在 Triton X-100 存在条件下，脱氧胆酸盐可刺激 PH-GPX 活性因而可能也刺激了 SelP 活性。体外条件下，SelP 还能够将脂质氢过氧化物 15-HPETE 还原为其相应的羟基二十碳四烯酸[202]。这些研究结果表明，SelP 具有 PH-GPX 样催化活性，其 N-末端的 U-X-X-C 结构域可能是酶的活性中心。和 PH-GPX 类似，SelP 除了以 GSH 作为还原性底物外，巯基乙醇、半胱氨酸、DTT 以及 Trx 均可作为 SelP 的还原剂，而且以 DTT 和 Trx 作为还原性底物时其催化活性比以 GSH 作为还原性底物时的催化活性更高[85]。

体内研究表明，缺硒大鼠发生的肝坏死与脂质过氧化有关（血浆 F2 异前列腺素水平提高）[203]，给缺硒大鼠补硒后血浆 SelP 水平的提高与敌草快诱导的脂质过氧化程度的降低密切相关[204]，SelP 可保护缺硒大鼠 GSH 耗竭引起的氧化损伤[188]。结合体外研究结果推测，SelP 在生物体内应具有防止脂质过氧化的作用。一方面，SelP 本身作为一种抗氧化剂，和血浆中的 GPX3 一样具有直接防止 LDL 和其他血浆蛋白发生过氧化的作用[205]，一些研究还发现，SelP 还具有捕获过氧亚硝基阴离子的作用[206]；另一方面，SelP 可通过其转运和供应硒的功能提高其他抗氧化硒蛋白如 cGPX、PH-GPX 和 TrxR 的基因或蛋白表达以及活性，从而间接发挥其抗氧化作用[207,208]。如前所述，SelP 结构中具有 2 个富组氨酸区，因而具有金属结合特性。过渡金属如 Fe^{3+} 可通过加剧 H_2O_2 向羟自由基的转变而诱导氧化应激[209]。在细胞培养中，金属螯合剂可减缓氧化应激诱导的细胞死亡[210]。因此，SelP 还可能通过其金属螯合特性间接发挥其抗氧化作用。SelP 与血液中重金属离子的结合还可起到消除重金属毒性的作用。

五、内质网硒蛋白及其生物学功能

内质网是哺乳动物细胞中一个极为重要的细胞器，既是蛋白质合成与翻译后修饰，多肽链正确折叠与装配的重要场所，也是细胞第二信使 Ca^{2+} 的贮存器，还与糖类、脂类和甾体类激素的合成、血糖浓度的调节、解毒、转运、同化等有关。在 25 中哺乳动物硒蛋白中有 7 种定位于内质网上，分别是 15kDa 硒蛋白、DIO2、SelK、SelM、SelN、SelS 和 SelT。除 DIO2 外，对大部分内质网硒蛋白的功能知之甚少或尚不清楚。近年来的研究发现，多数内质网驻留蛋白在氧化还原平衡调节、蛋白质折叠质量控制、内质网相关蛋白降解、Ca^{2+} 稳态调节、内质网应激调节以及炎症调节等过程中发挥作用。

（一）15kDa 硒蛋白及其生物学功能

15kDa 硒蛋白（Sep15），也被称为 SelF，1998 年在人的组织中首次被鉴定并表征[211]。与其他多数硒蛋白一样，SEP15 的表达受到机体硒营养状况的调节，随着机体硒水平的降低而降低。SEP15 蛋白由 162 个氨基酸组成，包含 1 个内质网信号肽、1 个 UGGT 结合域和 1 个硫氧还蛋白样折叠结构域（图 2-25）。内质网信号肽位于 N-末端，在蛋白进入内质网后被裂解。UGGT 结合域富含 Cys，能够和 UDP-葡糖糖：糖蛋白葡糖糖基转移酶(UGGT)形成 1∶1 紧密复合体，UGGT 位于内质网，作为分子伴侣参与内质网中蛋白质折叠的质量控制。硫氧还蛋白样折叠域位于 C-末端，含有具有氧化还原活性的 CxU(U 表示 Sec)基序。哺乳动物 Sep15 具有广泛的组织分布，肝脏、肾脏、睾丸、脑、甲状腺和前列腺中的基因表达水平很高。

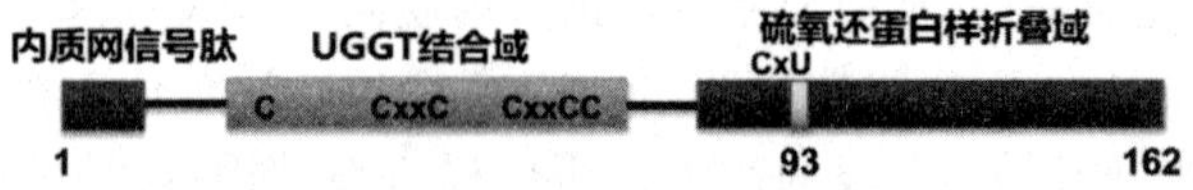

图 2-25 人 15kDa 硒蛋白的结构特征[212]

Sep15 存在于内质网腔，参与调节蛋白质折叠。SEP15 基因敲除小鼠可导致白内障发生，原因可能与晶状体中未折叠或错误折叠蛋白的积累有关[213]。果蝇重组 Sep15 的氧化还原电势为－225mV，在已知的位于内质网协助蛋白质折叠的巯基依赖性氧化还原酶中，其氧化还原电势最低[214]。目前为止还不知道什么体系能够还原 Sep15 中 Cys 和 Sec 间形成的硒-硫键。

内质网腔 Ca^{2+} 耗竭、内质网蛋白质糖基化抑制、二硫键错配、内质网蛋白质向高尔基体转运减少等可导致未折叠或错误折叠蛋白质在内质网腔蓄积从而使内质网功能发生改变，统称为内质网应激。适度的内质网应激可促进内质网对网腔中未折叠或错误折叠蛋白质的处理，从而有利于恢复细胞内环境稳态和维持细胞存活，但过度的持续的内质网应激则可导致细胞凋亡。Sep15 与内质网应激有关。衣霉素和蓝菌素诱导的适度内质网应激引起 Sep15 基因表达上调，而由二硫苏糖醇和毒胡萝卜素诱导的过度内质网应激则导致 Sep15 被蛋白水解酶快速降解[215]。相反，Sep15 表达水平的降低并不能够诱导内质网应激，提示机体存在其他对 SEP15 功能的补偿机制[215]。近年来的研究发现，SEP15 参与氧化还原稳态调节，SEP15 基因敲除小鼠肝脏的氧化应激水平明显提高[216]。

Sep15 与癌症的发生有关，人 Sep15 基因位于染色体 lp31，在癌症患者中该染色体经常发生缺失或突变。早期研究发现，癌变的人肺、乳腺、前列腺癌和肝脏组织中 Sep15 的基因表达水平明显降低[217,218]，提示 Sep15 具有防癌作用。然而近年来的研究表明，与正常的人肝细胞相比，HepG2 和 Huh7 肝癌细胞中 SEP15 的基因表达水平明显提高[219]，化学诱导结肠癌的研究发现，SEP15 基因敲除小鼠与野生型小鼠相比异常隐窝病灶的形成明显降低[220]，提示 Sep15 具有促癌作用。因此，关于 Sep15 与癌症的关系尚需进一步研究。

(二)SelM 及其生物学功能

SelM 是 2002 年采用生物信息学方法对人类基因组进行分析预测并

实验验证的一种硒蛋白[221]。人类 SelM 基因全长 3 kb，位于 22 号染色体，由 5 个外显子和 4 个内含子组成，编码 145 个氨基酸残基，Sec 位于氨基酸序列的第 48 位（图 2-26）。

SelM 在其 N-末端存在一段由 25 个氨基酸组成的信号肽，该信号肽对蛋白质易位是必需的，被剪切后蛋白进入内质网[221]。C-末端存在一段由 4 个氨基酸组成的 ER 驻留信号肽，是 SelM 特异性地定位于内质网的关键所在[222]。中间较长的一段为硫氧还蛋白样折叠结构域，是巯基-二硫化物氧化还原酶最普遍的特征，主要由 3 个 α 螺旋（α1-α3）和 4 个反平行 β 折叠组成（β1-β4），在 α1-螺旋结构的上游存在一个 CxxU 基序[222]（图 2-26）。CxxC 或 CxxU 基序通常是氧化还原中心，如硫氧还蛋白、蛋白二硫化物异构酶、谷氧还蛋白以及二硫化物氧化酶均含有 CxxC 基序，定位于 α-螺旋结构的上游且作为蛋白的活性中心。因此，SelM 中的 CxxU 基序也被认为是其氧化还原活性中心，SelM 可能具有巯基-二硫化物氧化还原酶活性[223]。

图 2-26　人 SelM 的结构特征

SelM 在多数组织中均可表达，在大脑中的表达水平最高[223]。

SelM 可能具有抗氧化、调节细胞内氧化还原平衡的作用。对转染 SelM 基因的大鼠的研究发现，过表达人 SelM 可降低大鼠血清 H_2O_2 水平，提高脑、肺、肝和肠组织中 GPX 活性，增强大脑皮层、海马和肠组织中 SOD 活性[224]。对小鼠 HT22 海马细胞、C8-D1A 小脑细胞和原代培养的皮层神经元的研究发现，SelM 过表达可降低由 H_2O_2 诱导引起的超氧化物的产生和凋亡性细胞死亡，SelM 基因敲除则加剧细胞的氧化损伤，在无 H_2O_2 诱导情况下，SelM 基因敲除也会导致细胞存活率下降及 ROS 水平提高[225]。

SelM 可能参与胞内钙稳态的调节。阿尔茨海默症（AD）是一种与年龄密切相关的神经退行性疾病，目前已证明氧化应激在 AD 的发生发展中起关键性作用，氧化应激可诱导细胞内 Ca^{2+} 水平增加，神经元钙稳态失调在 AD 发病机制中具有重要地位。对小鼠 HT22 和 C8-D1A 细胞的研究发现，过表达 SelM 明显降低了 H_2O_2 诱导引起的胞质中 Ca^{2+} 水平的增加，而胞内总钙水平没有受到显著影响[225]。对原代皮层细胞的研究发现，SelM 基因敲除会大大增加胞质中 Ca^{2+} 的本底水平[225]。以上研究说明 SelM 参与胞内钙稳态的调节。进一步研究发现，细胞经毒胡萝卜素（内质网 Ca^{2+}-

ATP 酶的选择性抑制剂，能不可逆地使内质网钙池排空，进而使胞内 Ca^{2+} 水平居高不下）作用后，H_2O_2 无法诱导胞质中 Ca^{2+} 水平的增加；而细胞经 EGTA（螯合胞质中的本底钙）作用后，H_2O_2 诱导引起的胞质 Ca^{2+} 水平持续增加。说明 SelM 参与内质网钙池中钙的内稳态调节，可影响细胞从钙池释放钙的过程。

SelM 可能参与金属离子内稳态的调节。SelM 结构中的 CxxU 基序能够和 Zn^{2+}、Cd^{2+}、Hg^{2+} 等金属离子发生竞争性配位结合以维持生物体内必需微量金属离子的内稳态和拮抗有毒金属离子。研究发现，将 SelM 中的 Sec_{48} 突变为 Cys_{48} 获得的 SelM 重组蛋白可与 Zn^{2+} 以摩尔比 2∶1 的比例结合，从而抑制 Zn^{2+} 介导的 β-淀粉样蛋白（$A\beta_{42}$）聚集以及 Zn^{2+}-$A\beta_{42}$ 诱导的神经细胞毒性和胞内 ROS 水平升高[226]。

SelM 可能参与能量代谢调节。SelM 基因敲除小鼠没有认知障碍，然而表现为体重增加、白色脂肪沉积提高及血清瘦素水平提高，但葡萄糖耐受和肝脏胰岛素信号与对照相比无明显差异。进一步研究发现，SelM 存在于小丘脑表达瘦素受体的神经元中，SelM 基因敲除降低了下丘脑由于瘦素表达引起的 STAT3 磷酸化，表现为瘦素耐受[227]。内质网应激是导致肥胖和瘦素耐受的主要原因[228,229]。因此，SelM 基因敲除可能导致内质网应激水平提高和下丘脑对瘦素的敏感性降低。瘦素可刺激下丘脑促肾上腺皮质激素释放因子和脑垂体促肾上腺皮质激素的释放[230]，SelM 基因敲除在提高血清瘦素水平的同时降低了早晨皮质酮的基础水平的事实说明 SelM 基因敲除破坏了瘦素信号与下丘脑—脑垂体—肾上腺轴的耦联。褐色脂肪组织活化可通过自适应产热（燃烧脂肪酸和葡糖糖）刺激能量消耗和降低脂肪沉积[231]。自适应产热受到下丘脑的调控，在许多啮齿类动物肥胖模型中均观察到自适应产热功能的缺陷[232]。下丘脑多个区域可刺激褐色脂肪组织活化，主要分布在背内侧核[233]和室旁核[234]。SelM 基因敲除小鼠与野生型相比白色脂肪沉积增加，褐色脂肪组织减少，而且褐色脂肪组织表现为颜色发白[227]，说明 SelM 在调节体重及能量代谢方面起着重要作用。

（三）SelK 及其生物学功能

SelK 是 2003 年由 Kryukov 等鉴定出的一种新型硒蛋白[235]。人 SelK 分子量大小为 11 kDa，含有 94 个氨基酸，是一种单跨膜整合蛋白，包含一个短的 N-端序列，一个跨膜区域和一个 C-端序列。N-端序列约含 20 个氨基酸，在内质网腔内；C-端序列在细胞质中，含有保守的 M（A/G）GGUGR 模序，其中 92 位氨基酸为 Sec[236]。SelK 含有多个可与信号蛋白结合的模序：一个 Src 同源性 3(Src-homology 3，SH3)结合域，一个次级非典型 SH3 结构域和一个假定的磷酸化位点(Ser 51)[237]。

SelK 在多数组织中均可表达。用 RNA 印迹技术检测人体组织中 SelK 的表达，表明该蛋白在心脏、脑、胎盘、肺、肝脏、骨骼肌、肾和胰腺中均有表达，其中在心脏、骨骼肌和胰腺中的表达水平特别高[238]。用实时荧光定量 PCR 技术测定小鼠组织中 SelK 的表达，显示该蛋白在肝脏、脑、心脏、肺、睾丸、小肠和肾脏中均有表达，其中在睾丸中表达水平特别高[239]。

作为一种内质网跨膜蛋白，SelK 在免疫细胞中高度表达，其表达水平对日粮以及细胞培养液中的硒水平非常敏感[240]。SelK 参与调节 ER 应激，作为伴侣蛋白协助 ER 相关的未折叠蛋白的降解[241,242]。在免疫细胞中，Ca^{2+} 主要贮存在内质网中，在白细胞表面特定受体的作用下，内质网贮存的 Ca^{2+} 释放进入胞浆，这触发胞外空间大量 Ca^{2+} 进入胞浆[243]。Ca^{2+} 内流对于免疫细胞的活化、增殖、迁移以及吞噬活动至关重要。SelK 基因敲除导致小鼠 T 细胞、中性粒细胞和巨噬细胞受体调节的 Ca^{2+} 内流破坏。SelK 调节的 Ca^{2+} 内流对于 ERK 的激活以及巨噬细胞通过产生 NO 吞噬 IgG 调理化的细菌颗粒非常关键。

SelK 基因敲除破坏了巨噬细胞对氧化的低密度脂蛋白(oxLDL)的摄入，导致泡沫细胞形成降低和动脉粥样硬化[244]。CD36 是单核细胞、巨噬细胞和内皮细胞表面的跨膜蛋白。巨噬细胞稳定转染 CD36 刺激了泡沫细胞的形成，CD36 是巨噬细胞表面表达的捕获 oxLDL 的主要受体[245]。棕榈酰化是指在蛋白质的 Cys 残基上通过硫酯键可逆地添加一个十六碳的脂肪酸，是一种重要的蛋白质翻译后修饰过程，这种修饰可以促进胞质蛋白的膜结合或跨膜蛋白的稳定表达。SelK 缺乏降低了巨噬细胞 CD36(摄取 oxLDL 的受体)的棕榈酰化[246]，因而降低了巨噬细胞表面 CD36 的表达，导致对 oxLDL 的捕获降低，泡沫细胞的形成降低。

上文提到，Ca^{2+} 内流对于免疫细胞的活化、增殖、迁移以及吞噬活动至关重要。内质网膜上四聚体 Ca^{2+} 通道蛋白的装配需要内质网膜跨膜蛋白 IP3 受体的棕榈酰化，而 IP3 受体的棕榈酰化需要 DHHC6(一种催化蛋白质棕榈酰化的酶，属于 DHHC 家族成员)催化完成。DHHC6 包含一个推定的 SH3 结构域，而 SelK 包含一个 SH3 结合域，SelK 作为 DHHC6 的辅因子与 DHHC6 通过 SH3/SH3 结合域结合。SelK 基因敲除 T 细胞和巨噬细胞 IP3 受体的棕榈酰化和基因表达水平降低，培养在低硒环境的 T 细胞 SelK 的水平以及 IP3 受体的棕榈酰化和基因表达水平降低[247]，这可能从机理上解释了为什么 SelK 缺乏导致依赖 Ca^{2+} 内流的免疫细胞功能的降低(图 2-27)。

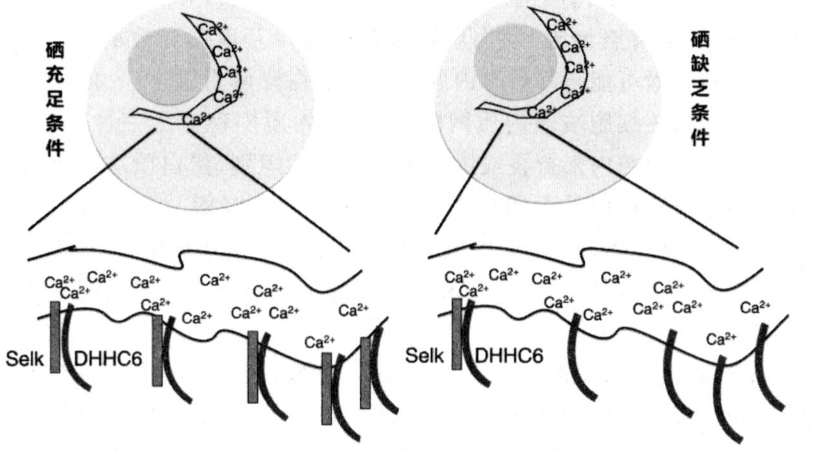

图 2-27　硒充足和缺乏条件下 DHHC6 对蛋白质的棕榈酰化[247]

（四）SelS 及其生物学功能

SelS 是由 Walder 等于 2002 年在典型的Ⅱ型糖尿病模型大鼠肝脏中首次发现的，最初被命名为 Tanis[248]。Kryukov 等于 2003 年在人类基因组中发现了一种新的硒蛋白 SelS，和大鼠 Tanis 是同一个蛋白[235]。人 SelS 分子量为 21 kDa，含 189 个氨基酸，是一个单跨膜蛋白，包含一段短的 N-端序列，一段跨膜区和一段 C-端序列。N-端序列由 1～25 位氨基酸组成，在内质网腔内；内质网跨膜区由 26～48 位氨基酸组成；C-端序列在细胞质内，由 49～189 位氨基酸组成，52～122 位氨基酸残基形成两段伸展的 α-螺旋，称卷曲螺旋结构域，SelS 可能通过该区域与 p97 ATP 酶结合或形成二聚体。SelS 螺旋结构域下游的 C-端序列（123～189 号氨基酸）富含甘氨酸、脯氨酸和精氨酸，无明显的二级结构，第 188 位氨基酸为 Sec，这种非结构化的、带正电荷多的结构区可能易于结合表面带负电荷的蛋白质[236]。

SelS 由于含有 Sec 残基，可能具有抗氧化功能。研究表明，SelS 基因过表达能降低 H_2O_2 诱导的胰岛 B 细胞系 Min6 细胞的凋亡[249]和人脐静脉内皮细胞的损伤[250]，而用 RNA 干扰技术使 SelS 基因表达降低加剧了 H_2O_2 诱导的血管平滑肌细胞的氧化应激程度，表现为 ROS 水平和脂质过氧化程度进一步提高，GPX1 活性进一步降低，氧化应激诱导的细胞凋亡程度进一步加剧[251]。

SelS 可能参与调节内质网应激。内质网应激可促进 SelS 基因的表达，暗示 SelS 是一种受内质网应激调节的蛋白，可作为细胞发生内质网应激的生物标志物[249]。体外细胞实验表明，SelS 过表达可保护巨噬细胞[252]和星状胶质细胞[253]免受内质网应激导致的损伤，SelS 表达降低可加剧肝癌细胞[254]和血管平滑肌细胞[251]内质网应激程度以及由此导致的凋亡。SelS 调节内质网应激可能与其参与内质网相关蛋白降解有关[255]。内质网相关蛋白降解是真核细胞减弱内质网应激、恢复内质网稳态的一种重要途径，主要包括内质网腔内未折叠或错误折叠蛋白的识别、蛋白质从内质网腔向细胞质的逆向转运以及蛋白质在细胞质中泛素化降解三个过程。在内质网中，SelS 与 p97 ATP 酶、Derlin 伴侣蛋白、E3 泛素连接酶和 SelK 共同组成内质网相关蛋白复合体，促进未折叠或错误折叠蛋白从内质网腔逆向转运至细胞质而被降解[256]。

SelS 可能具有调节炎症反应的作用。酵母双杂交分析表明，SelS 能与血清淀粉样蛋白 A(serum amyloid A，SAA)相互作用，SelS 可能是 SAA1β 的受体[248]。巨噬细胞 SelS 表达抑制导致 TNF-α、IL-1β 和 IL-6 的水平显著增加[257]，人肝癌细胞 SelS 表达降低加剧了 LPS 诱导的诱导型一氧化氮合酶(inducible nitric oxide synthase，iNOS)及 SAA 的表达增加[258]。而星状胶质细胞 SelS 过表达抑制了 LPS 诱导的 IL-1β 和 IL-6 的表达，血管内皮细胞 SelS 过表达抑制了 TNF-α 诱导的 IL-1β、IL-6、IL-8 的表达以及 NF-κB 的活化[259]。

(五)SelT 及其生物学功能

SelT 是由 Kryukov 及其团队于 1999 年发现的[260]。分子克隆技术揭示，SelT cDNA 序列包含 970 个核苷酸，编码 195 个氨基酸，分子量为 22.3 kDa[261]。生物信息学分析预测该蛋白含有一段由 1～19 位氨基酸组成的信号肽，两段分别由 87～102 位和 125～143 位氨基酸组成的疏水序列，疏水序列或许代表着跨膜区，对于 SelT 靶向于内质网具有重要作用[261](图 2-28)。Sec 残基位于 N-端第 49 位，与其上游的 Cys 残基间隔 2 个氨基酸，因而构成一个 CVSU 基序[262]，该基序可能是 SelT 的氧化还原活性中心。而且该基序存在于 β 折叠和 α 螺旋之间，具有 Trx 样折叠的结构特点[263]。

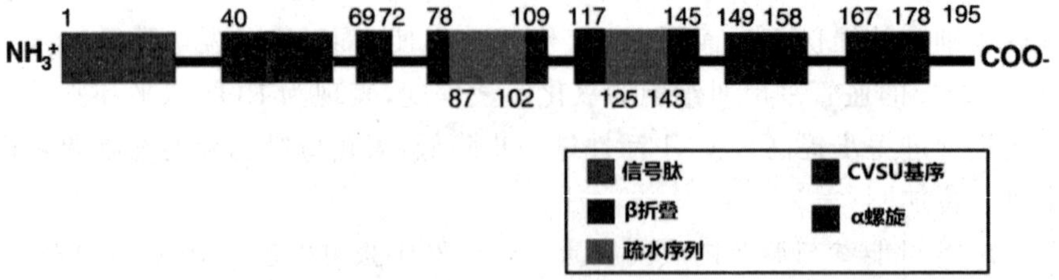

图 2-28　SelT 结构示意图[263]

亚细胞定位以及重组蛋白强有力的氧化还原酶活性暗示 SelT 发挥着关键的氧化还原功能，在控制内质网中蛋白质加工，应对细胞氧化应激，保证内质网稳态方面起重要作用。与内质网其他硒蛋白（Sep15、SelK、SelS）类似，SelT 可能通过 TrxR 样活动影响硫醇氧化还原循环，进而影响包括分化和生存等各种细胞进程。在进化过程中，SelT 基因相当保守，该基因的敲除会导致胚胎早期死亡[264]。

垂体腺苷酸环化酶激活肽（pituitary adenylate cyclase activating polypeptide，PACAP）是一种神经营养因子，具有促进神经细胞分化和生存的作用。在神经细胞分化过程中，Ca^{2+} 作为胞内信号参与神经营养因子 PACAP 的促分化和促生存效应[265]。研究表明，SelT 过表达提高了胞内 Ca^{2+} 水平，而 SelT 基因沉默破坏了 PACAP 诱导的 Ca^{2+} 从内质网和胞外向胞质的动员[266]。结合其他内质网硒蛋白（SelN、SelK、SelM）对胞内钙稳态以及 Ca^{2+} 通道蛋白的调节作用，SelT 被认为具有调节胞内钙稳态从而促进细胞分化和生存的效应。SelT 利用其氧化还原活性，直接或间接与 Ca^{2+} 通道蛋白的巯基作用，调节蛋白质的功能活性，这可能是 SelT 调节胞内钙稳态的机制所在。

在小鼠纤维瘤细胞中，SelT 基因敲除改变了细胞的黏附以及与细胞形态和氧化应激相关的基因表达，暗示 SelT 参与细胞黏附，具有控制氧化还原平衡的功能[267]。氧化应激是导致多巴胺能神经元死亡的主要原因，因此患有帕金森病（PD）的小鼠是研究 SelT 与氧化应激关系的理想模型[264]。在多巴胺能 SH-SY5Y 细胞模型中，SelT 促进了细胞生存，抑制了神经毒素 MPP^+ 诱导的氧化应激。在野生型小鼠中，神经毒素 MPTP 或鱼藤酮处理诱导了多巴胺能神经细胞以及黑质纹状体通路中星状胶质细胞 SelT 基因表达的上调[264]。而在 SelT 基因敲除小鼠中，相同的处理导致快速且严重的帕金森样疾病发生，而且与野生型小鼠相比表现出明显的氧化应激和多巴胺能神经变性[264]。

SelT 在胚胎期的脑组织中表达水平高，但在出生后几乎检测不到。与此不同，SelT 在出生后的内分泌器官（胰腺、睾丸、甲状腺）中仍然持续高表达[268]。胰岛 B 细胞 SelT 基因特异性敲除提高了血浆葡萄糖与胰岛素的比率，给 SelT 缺乏小鼠注射胰岛素不影响血浆葡萄糖水平[269]，这些结果说明，SelT 参与调节胰岛素的合成和分泌，从而控制葡萄糖的稳态。

在正常的肝脏中几乎检测不到 SelT 的表达，但在肝再生的急性期，SelT 的基因表达被强烈诱导，这与 SelT 在处于分化和增殖期细胞中的高表达是一致的[268]。SelT 在肝再生过程中的作用尚不清楚，但这种诱导性的高表达主要表现在枯否细胞，暗示 SelT 可能具有促进肝脏巨噬细胞合成细胞因子 TNF-α 和 IL-6，从而促进肝细胞增殖的作用[270]。

六、SelW 及其生物学功能

在第一章中提到，硒缺乏可导致羔羊和犊牛患白肌病，这种疾病的典型特征是由于磷酸钙（羟基磷灰石）的异常沉积而引起骨骼肌和心肌钙化，最终导致肌肉无力和心肌变性而死亡。硒缺乏可导致人患克山病，也是一种类似于白肌病的肌肉疾病。研究发现，当动物患白肌病或人患克山病时，一种小分子量的含硒蛋白的水平非常低，因而后来将这种硒蛋白命名为 SelW。

SelW 发现于 1969 年，分子量约为 10 kDa，是已知最小的硒蛋白[271]。从哺乳动物到细菌均可检测到 SelW 的存在。与多数硒蛋白相似，哺乳动物 SelW 包含 1 个 Sec 残基，与其上游的 Cys 残基间隔 2 个氨基酸，构成一个 CxxU 基序。在蛙类以及许多细菌的 SelW 中，该位置仍为 Cys 而非 Sec，同样与其上游间隔 2 个氨基酸的 Cys 构成 CxxC 基序。SelW 表达于几乎所有的组织，在骨骼肌、心肌和脑组织中的表达水平特别高。SelW 主要存在于细胞质中，部分在细胞膜上[272]。

由于含有 CxxU 或 CxxC 基序，而且还具有 Trx 样折叠结构，因而推测 SelW 具有氧化还原调节功能。大鼠神经胶质细胞 SelW 过表达显著增强了对氧化应激的抵抗力[273]。当 GSH 耗竭时，SelW 的抗氧化功能丧失，说明其酶活性依赖于 GSH 的存在[274]。此外，大脑皮层神经细胞 SelW 基因敲除显著提高了对 H_2O_2 诱导的氧化应激的敏感性[275]。这些研究证据均说明 SelW 具有抗氧化功能，但其确切的机制还有待进一步研究证实。

参考文献

[1] Chambers I, Frampton J, Goldfarb P, et al. The structure of the mouse glutathione peroxidase gene: the selenocysteine in the active site is encoded by the ‘termination’ codon, TGA[J]. The EMBO Journal, 1986, 5(6): 1221-1227.

[2] Hubert N, Walczak R, Carbon P, et al. A Protein Binds the Selenocysteine Insertion Element in the 3′-UTR of Mammalian Selenoprotein mRNAs[J]. Nucleic Acids Research, 1996, 24(3): 464-469.

[3] Zinoni F, Birkmann A, Stadtman T C, et al. Nucleotide sequence

and expression of the selenocysteine-containing polypeptide of formate dehydrogenase (formate-hydrogen-lyase-linked) from Escherichia coli[J]. Proceedings of the National Academy of Sciences of the United States of America, 1986, 83(13): 4650-4654.

[4] Berry M J, Banu L, Chen Y, et al. Recognition of UGA as a selenocysteine codon in Type I deiodinase requires sequences in the 3′ untranslated region[J]. Nature, 1991, 353(6341): 273-276.

[5] Su D, Li Y, Gladyshev V N. Selenocysteine insertion directed by the 3′-UTR SECIS element in Escherichia coli[J]. Nucleic Acids Research, 2005, 33(8): 2486-2492.

[6] Chen G F T, Fang L, Inouye M. Effect of the relative position of the UGA codon to the unique secondary structure in the fdhF mRNA on its decoding by selenocysteinyl tRNA in Escherichia coli[J]. Journal of Biological Chemistry, 1993, 268(31): 23128-23131.

[7] Baron C, Heider J, Bock A. Interaction of translation factor SELB with the formate dehydrogenase H selenopolypeptide mRNA[J]. Proceedings of the National Academy of Sciences of the United States of America, 1993, 90(9): 4181-4185.

[8] Liu Z, Reches M, Groisman I, et al. The nature of the minimal 'selenocysteine insertion sequence' (SECIS) in Escherichia coli[J]. Nucleic Acids Research, 1998, 26(4): 896-902.

[9] Zhang Y, Gladyshev V N. An algorithm for identification of bacterial selenocysteine insertion sequence elements and selenoprotein genes [J]. Bioinformatics, 2005, 21(11): 2580-2589.

[10] Fagegaltier D, Lescure A, Walczak R, et al. Structural analysis of new local features in SECIS RNA hairpins[J]. Nucleic Acids Research, 2000, 28(14): 2679-2689.

[11] Grundnerculemann E, Martin G W, Harney J W, et al. Two distinct SECIS structures capable of directing selenocysteine incorporation in eukaryotes[J]. RNA, 1999, 5(5): 625-635.

[12] Leinfelder W, Zehelein E, Mandrand-Berthelot M A, et al. Gene for a novel tRNA species that accepts L-serine and cotranslationally inserts selenocysteine[J]. Nature, 1988, 331(6158): 723-725.

[13] Lee B, Rajagopalan M, Kim Y, et al. Selenocysteine tRNA [Ser]Sec gene is ubiquitous within the animal kingdom[J]. Molecular and

Cellular Biology, 1990, 10(5): 1940-1949.

[14] Hatfield D L, Smith D W E, Lee B J, et al. Structure and function of suppressor tRNAs in higher eukaryotes[J]. Critical Reviews in Biochemistry and Molecular Biology, 1990, 25(2): 71-96.

[15] Heider J, Leinfelder W, Bock A. Occurrence and functional compatibility within Enterobacteriaceae of a tRNA species which inserts selenocysteine into protein[J]. Nucleic Acids Research, 1989, 17(7): 2529-2540.

[16] Tormay P, Wilting R, Heider J, et al. Genes coding for the selenocysteine-inserting tRNA species from Desulfomicrobium baculatum and Clostridium thermoaceticum: structural and evolutionary implications [J]. Journal of Bacteriology, 1994, 176(5): 1268-1274.

[17] Deckert G, Warren P V, Gaasterland T, et al. The complete genome of the hyperthermophilic bacterium Aquifex aeolicus[J]. Nature, 1998, 392(6674): 353-358.

[18] Hatfield D L, Lee B J, Price N M, et al. Selenocysteyl-tRNA occurs in the diatom Thalassiosira and in the ciliate Tetrahymena[J]. Molecular Microbiology, 1991, 5(5): 1183-1186.

[19] Hatfield D L, Choi I S, Mischke S, et al. Selenocysteyl-tRNAs recognize UGA in Betavulgaris, a higher plant, and in Gliocladium virens, a filamentous fungus[J]. Biochemical and Biophysical Research Communications, 1992, 184(1): 254-259.

[20] Lee B J, Worland P J W, Davis J N, et al. Identification of a selenocysteyl-tRNA(Ser) in mammalian cells that recognizes the nonsense codon, UGA[J]. Journal of Biological Chemistry, 1989, 264(17): 9724-9727.

[21] Dirheimer G, Keith G, Dumas P, et al. Primary, secondary and tertiary structures of tRNAs//Soll D. tRNA: Structure, Biosynthesis, and Function[M]. Washington DC: ASM Press. 1995.

[22] Commans S, Bock A. Selenocysteine inserting tRNAs: an overview[J]. FEMS Microbiology Reviews, 1999, 23(3): 335-351.

[23] Sturchler C, Westhof E, Carbon P, et al. Unique secondary and tertiary structural features of the eucaryotic selenocysteine tRNASec[J]. Nucleic Acids Research, 1993, 21(5): 1073-1079.

[24] Steinberg S V, Leclerc F, Cedergren R. Structural rules and

conformational compensations in the tRNA L-form[J]. Journal of Molecular Biology, 1997, 266(2): 269-282.

[25] Steinberg S V, Ioudovitch A, Cedergren R. The secondary structure of eukaryotic selenocysteine tRNA: 7/5 versus 9/4[J]. RNA, 1998, 4(3): 241-245.

[26] Hubert N, Sturchler C, Westhof E, et al. The 9/4 secondary structure of eukaryotic selenocysteine tRNA: More pieces of evidence[J]. RNA, 1998, 4(9): 1029-1033.

[27] Bult C J, White O, Smith H O, et al. Complete Genome Sequence of the Methanogenic Archaeon, Methanococcus jannaschii[J]. Science, 1996, 273(5278): 1058-1073.

[28] Mizutani T, Goto C. Eukaryotic selenocysteine tRNA has the 9/4 secondary structure[J]. FEBS Letters, 2000, 466(2): 359-362.

[29] Meinnel T, Mechulam Y, Blanquet S. Aminoacyl-tRNA synthetases: occurrence, structure and function//Soll D, RajBhandary U L. tRNA: Structure, Biosynthesis, and function [M]. Washington, DC: ASM Press. 1995.

[30] Ibba M, Curnow A W, Soll D. Aminoacyl-tRNA synthesis: divergent routes to a common goal[J]. Trends in Biochemical Sciences, 1997, 22(2): 39-42.

[31] Young P A, Kaiser I I. Aminoacylation of Escherichia coli cysteine tRNA by selenocysteine[J]. Archives of Biochemistry and Biophysics, 1975, 171(2): 483-489.

[32] Hawkes W C, Tappel A L. In vitro synthesis of glutathione peroxidase from selenite. Translational incorporation of selenocysteine[J]. BBA-Gene Structure and Expression, 1983, 739(2): 225-234.

[33] Sunde R A, Hoekstra W G. Incorporation of selenium from selenite and selenocystine into glutathione peroxidase in the isolated perfused rat liver [J]. Biochemical and Biophysical Research Communications, 1980, 93(4): 1181-1188.

[34] Mizutani T, Hitaka T. The conversion of phosphoserine residues to selenocysteine residues on an opal suppressor tRNA and casein [J]. FEBS Letters, 1988, 232(1): 243-248.

[35] Baron C, Heider J, Böck A. Mutagenesis of selC, the gene for the selenocysteine-inserting tRNA-species in E. coli: effects on in vivo

function[J]. Nucleic Acids Research, 1990, 18(23): 6761-6766.

[36] Tormay P, Wilting R, Lottspeich F, et al. Bacterial selenocysteine synthase-structural and functional properties[J]. FEBS Journal, 2010, 254(3): 655-661.

[37] Forchhammer K, Boesmiller K, Böck A. The function of selenocysteine synthase and SELB in the synthesis and incorporation of selenocysteine[J]. Biochimie, 1991, 73(12): 1481-1486.

[38] Mizutani T, Kurata H, Yamada K, et al. Some properties of murine selenocysteine synthase[J]. Biochemical Journal, 1992, 284(3): 827-834.

[39] MäenpääP H, Bernfield M R. A Specific Hepatic Transfer RNA for Phosphoserine[J]. Proceedings of the National Academy of Sciences of the United States of America, 1970, 67(2): 688-695.

[40] Wu X Q, Gross H J. The length and the secondary structure of the D-stem of human selenocysteine tRNA are the major identity determinants for serine phosphorylation[J]. EMBO Journal, 1994, 13(1): 241-248.

[41] Amberg R, Wu X G H, Mizutani T. SELENOCYSTEINE SYNTHESIS IN MAMMALIA-AN IDENTITY SWITCH FROM TRNA (SER) TO TRNA(SEC)[J]. Journal of Molecular Biology, 1996, 263 (1): 8-19.

[42] Mizutani T. Some evidence of the enzymatic conversion of bovine suppressor phosphoseryl-tRNA to selenocysteyl-tRNA[J]. FEBS Letters, 1989, 250(2): 142-146.

[43] Mizutani T, Kurata H, Yamada K. Study of mammalian selenocysteyl-tRNA synthesis with [75Se]HSe[J]. FEBS Letters, 1991, 289 (1): 59-63.

[44] Stadtman T C. Selenocysteine[J]. Annual Review of Biochemistry, 1996, 65(65): 83-100.

[45] Z V, L T, Td S, et al. Synthesis of 5-methylaminomethyl-2-selenouridine in tRNAs: 31P NMR studies show the labile selenium donor synthesized by the selD gene product contains selenium bonded to phosphorus[J]. Proceedings of the National Academy of Sciences of the United States of America, 1992, 89(7): 2975-2979.

[46] Kim I Y, Guimarães M J, Zlotnik A, et al. Fetal mouse selenophosphate synthetase 2 (SPS2): Characterization of the cysteine mutant

form overproduced in a baculovirus-insect cell system[J]. Proceedings of the National Academy of Sciences of the United States of America, 1997, 94(2): 418-421.

[47] Förster C, Ott G, Forchhammer K, et al. Interaction of a selenocysteine-incorporating tRNA with elongation factor Tu from E. coli[J]. Nucleic Acids Research, 1990, 18(3): 487-491.

[48] Hilgenfeld R, Böck A, Wilting R. Structural model for the selenocysteine-specific elongation factor SelB[J]. Biochimie, 1996, 78(12): 971-978.

[49] Kromayer M, Wilting R, Tormay P, et al. Domain structure of the prokaryotic selenocysteine-specific elongation factor SelB[J]. Journal of Molecular Biology, 1996, 262(4): 413-420.

[50] Heider J, Baron C, Böck A. Coding from a distance: dissection of the mRNA determinants required for the incorporation of selenocysteine into protein[J]. EMBO Journal, 1992, 11(10): 3759-3766.

[51] Tujebajeva R M, Al E. Decoding apparatus for eukaryotic selenocysteine insertion[J]. EMBO Reports, 2000, 1(2): 158-163.

[52] Fagegaltier D, Hubert N, Yamada K, et al. Characterization of mSelB, a novel mammalian elongation factor for selenoprotein translation [J]. EMBO Journal, 2014, 19(17): 4796-4805.

[53] Copeland P R, Fletcher J E, Carlson B A, et al. A novel RNA binding protein, SBP2, is required for the translation of mammalian selenoprotein mRNAs[J]. The EMBO Journal, 2000, 19(2): 306-314.

[54] Copeland P R, Stepanik V A, Driscoll D M. Insight into mammalian selenocysteine insertion: domain structure and ribosome binding properties of Sec insertion sequence binding protein 2[J]. Molecular and Cellular Biology, 2001, 21(5): 1491-1498.

[55] Kryukov G V, Castellano S, Novoselov S V, et al. Characterization of mammalian selenoproteomes[J]. Science, 2003, 300(5624): 1439-1443.

[56] Gladyshev V N, Arnér E S, Berry M J, et al. Selenoprotein Gene Nomenclature[J]. Journal of Biological Chemistry, 2016, 291(46): 24036-24040.

[57] Gromer S, Eubel J K, Lee B L, et al. Human selenoproteins at a glance[J]. Cellular & Molecular Life Sciences Cmls, 2005, 62(21):

2414-2437.

[58] Roman M, Jitaru P, Barbante C. Selenium biochemistry and its role for human health[J]. Metallomics Integrated Biometal Science, 2013, 6(1): 25-54.

[59] Ursini F, Maiorino M, Brigeliusflohé R, et al. Diversity of glutathione peroxidases[J]. Methods in Enzymology, 1995, 252(2): 38.

[60] Ho Y S, Magnenat J L, Bronson R T, et al. Mice deficient in cellular glutathione peroxidase develop normally and show no increased sensitivity to hyperoxia[J]. Journal of Biological Chemistry, 1997, 272(26): 16644-16651.

[61] Cheng W H, Ho Y S, Valentine B A, et al. Cellular glutathione peroxidase is the mediator of body selenium to protect against paraquat lethality in transgenic mice[J]. Journal of Nutrition, 1998, 128(7): 1070-1076.

[62] De Haan J B, Bladier C, Griffiths P, et al. Mice with a homozygous null mutation for the most abundant glutathione peroxidase, Gpx1, show increased susceptibility to the oxidative stress-inducing agents paraquat and hydrogen peroxide[J]. Journal of Biological Chemistry, 1998, 273(35): 22528-22536.

[63] Lubos E, Loscalzo J, Handy D E. Glutathione Peroxidase-1 in Health and Disease: From Molecular Mechanisms to Therapeutic Opportunities[J]. Antioxidants & Redox Signaling, 2011, 15(7): 1957-1997.

[64] Cheng W H, Quimby Fwlei X G. Impacts of glutathione peroxidase-1 knockout on the protection by injected selenium against the pro-oxidant-induced liver aponecrosis and signaling in selenium-deficient mice [J]. Free Radical Biology & Medicine, 2003, 34(7): 918-927.

[65] Sunde R A, Raines A M, Barnes K M, et al. Selenium status highly regulates selenoprotein mRNA levels for only a subset of the selenoproteins in the selenoproteome[J]. Bioscience Reports, 2009, 29(5): 329-338.

[66] De Haan J B, Crack P J, Flentjar N J, et al. An imbalance in antioxidant defense affects cellular function: the pathophysiological consequences of a reduction in antioxidant defense in the glutathione peroxidase-1 (Gpx1) knockout mouse[J]. Redox Report, 2003, 8(2): 69-79.

[67] Beck M A, Shi Q, Morris V C, et al. Rapid genomic evolution

of a non-virulent Coxsackievirus B3 in selenium-deficient mice results in selection of identical virulent isolates[J]. Nature Medicine, 1995, 1(5): 433-436.

[68] Li W, Beck M A. Selenium deficiency induced an altered immune response and increased survival following influenza A/Puerto Rico/8/34 infection [J]. Experimental Biology and Medicine, 2007, 232(3): 412-419.

[69] Kretz-Remy C, Mehlen P, Mirault M E, et al. Inhibition of I kappa B-alpha phosphorylation and degradation and subsequent NF-kappa B activation by glutathione peroxidase overexpression[J]. Journal of Cell Biology, 1996, 133(5): 1083-1093.

[70] Sandstrom P A, Murray J, Folks T M, et al. Antioxidant Defenses Influence HIV-1 Replication and Associated Cytopathic Effects[J]. Free Radical Biology & Medicine, 1998, 24(9): 1485-1491.

[71] Mirochnitchenko O, Weisbrotlefkowitz M, Reuhl K, et al. Acetaminophen Toxicity[J]. Journal of Biological Chemistry, 1999, 274 (15): 10349-10355.

[72] Lu Y P, Lou Y R, Yen P, et al. Enhanced skin carcinogenesis in transgenic mice with high expression of glutathione peroxidase or both glutathione peroxidase and superoxide dismutase[J]. Cancer Research, 1997, 57(8): 1468-1474.

[73] Mcclung J P, Roneker C A, Mu W, et al. Development of insulin resistance and obesity in mice overexpressing cellular glutathione peroxidase[J]. Proceedings of the National Academy of Sciences of the United States of America, 2004, 101(24): 8852-8857.

[74] Briviba K, Kissner R, Koppenol W H, et al. Kinetic Study of the Reaction of Glutathione Peroxidase with Peroxynitrite[J]. Chemical Research in Toxicology, 1998, 11(12): 1398-1401.

[75] Fu Y, Sies H, Lei X G. Opposite roles of selenium-dependent glutathione peroxidase-1 in superoxide generator diquat-and peroxynitrite-induced apoptosis and signaling[J]. Journal of Biological Chemistry, 2001, 276(46): 43004-43009.

[76] Chiu S T, Hsieh F J, Chen S W, et al. Clinicopathologic Correlation of Up-regulated Genes Identified Using cDNA Microarray and Real-time Reverse Transcription-PCR in Human Colorectal Cancer[J]. Cancer Epidemiology, Biomarkers & Prevention, 2005, 14(2): 437-443.

[77] Mork H, Scheurlen M, Altaie O, et al. Glutathione peroxidase isoforms as part of the local antioxidative defense system in normal and Barrett's esophagus[J]. International Journal of Cancer, 2003, 105(3): 300-304.

[78] Serewko M M, Popa C, Dahler A L, et al. Alterations in Gene Expression and Activity during Squamous Cell Carcinoma Development [J]. Cancer Research, 2002, 62(13): 3759-3765.

[79] Woenckhaus M, Kleinhitpass L, Grepmeier U, et al. Smoking and cancer-related gene expression in bronchial epithelium and non-small-cell lung cancers[J]. The Journal of Pathology, 2006, 210(2): 192-204.

[80] Florian S, Wingler K, Schmehl K, et al. Cellular and subcellular localization of gastrointestinal glutathione peroxidase in normal and malignant human intestinal tissue[J]. Free Radical Research Communications, 2001, 35(6): 655-663.

[81] Florian S, Krehl S, Loewinger M, et al. Loss of GPx2 increases apoptosis, mitosis, and GPx1 expression in the intestine of mice[J]. Free Radical Biology & Medicine, 2010, 49(11): 1694-1702.

[82] Hahn M A, Hahn T, Lee D, et al. Methylation of Polycomb target genes in intestinal cancer is mediated by inflammation[J]. Cancer Research, 2008, 68(24): 10280-10289.

[83] Malinouski M, Kehr S, Finney L, et al. High-Resolution Imaging of Selenium in Kidneys: A Localized Selenium Pool Associated with Glutathione Peroxidase 3[J]. Antioxidants and Redox Signaling, 2012, 16(3): 185-192.

[84] Burk R F, Olson G E, Winfrey V P, et al. Glutathione peroxidase-3 produced by the kidney binds to a population of basement membranes in the gastrointestinal tract and in other tissues[J]. American Journal of Physiology Gastrointestinal and Liver Physiology, 2011, 301(1): 32-38.

[85] Takebe G, Yarimizu J, Saito Y, et al. A Comparative Study on the Hydroperoxide and Thiol Specificity of the Glutathione Peroxidase Family and Selenoprotein P[J]. Journal of Biological Chemistry, 2002, 277(43): 41254-41258.

[86] Lee O J, Schneiderstock R, Mcchesney P A, et al. Hypermethylation and loss of expression of glutathione peroxidase-3 in Barrett's tu-

morigenesis[J]. Neoplasia, 2005, 7(9): 854-861.

[87] Yu Y P, Yu G, Tseng G, et al. Glutathione peroxidase 3, deleted or methylated in prostate cancer, suppresses prostate cancer growth and metastasis[J]. Cancer Research, 2007, 67(17): 8043-8050.

[88] Yagi K, Komura S, Kojima H, et al. Expression of human phospholipid hydroperoxide glutathione peroxidase gene for protection of host cells from lipid hydroperoxide-mediated injury[J]. Biochemical and Biophysical Research Communications, 1996, 219(2): 486-491.

[89] Hurst R, Korytowski W, Kriska T, et al. Hyperresistance to cholesterol hydroperoxide-induced peroxidative injury and apoptotic death in a tumor cell line that overexpresses glutathione peroxidase isotype-4 [J]. Free Radical Biology & Medicine, 2001, 31(9): 1051-1065.

[90] Wang H P, Qian S Y, Schafer F Q, et al. Phospholipid hydroperoxide glutathione peroxidase protects against singlet oxygen-induced cell damage of photodynamic therapy[J]. Free Radical Biology & Medicine, 2001, 30(8): 825-835.

[91] Yant L J, Ran Q, Rao L, et al. The selenoprotein GPX4 is essential for mouse development and protects from radiation and oxidative damage insults[J]. Free Radical Biology & Medicine, 2003, 34(4): 496-502.

[92] Imai H, Nakagawa Y. Biological significance of phospholipid hydroperoxide glutathione peroxidase (PHGPx, GPx4) in mammalian cells[J]. Free Radical Biology & Medicine, 2003, 34(2): 145-169.

[93] Ursini F, Heim S, Kiess M, et al. Dual function of the selenoprotein PHGPx during sperm maturation[J]. Science, 1999, 285(5432): 1393-1396.

[94] Imai H, Hirao F, Sakamoto T, et al. Early embryonic lethality caused by targeted disruption of the mouse PHGPx gene[J]. Biochemical and Biophysical Research Communications, 2003, 305(2): 278-286.

[95] Conrad M, Moreno S G, Sinowatz F, et al. The Nuclear Form of Phospholipid Hydroperoxide Glutathione Peroxidase Is a Protein Thiol Peroxidase Contributing to Sperm Chromatin Stability[J]. Molecular and Cellular Biology, 2005, 25(17): 7637-7644.

[96] Schneider M, Förster H, Boersma A, et al. Mitochondrial glutathione peroxidase 4 disruption causes male infertility[J]. FASEB Jour-

nal, 2009, 23(9): 3233-3242.

[97] Seiler A, Schneider M, Fã¶Rster H, et al. Glutathione peroxidase 4 senses and translates oxidative stress into 12/15-lipoxygenase dependent- and AIF-mediated cell death[J]. Cell Metabolism, 2008, 8(3): 237-248.

[98] Cheng C Y. Molecular Mechanisms in Spermatogenesis [M]. New York: Springer, 2009.

[99] Flohe L. Selenium in mammalian spermiogenesis[J]. Biological Chemistry, 2007, 388(10): 987-995.

[100] Foresta C, Flohe L, Garolla A, et al. Male Fertility Is Linked to the Selenoprotein Phospholipid Hydroperoxide Glutathione Peroxidase [J]. Biology of Reproduction, 2002, 67(3): 967-971.

[101] Mauri P, Benazzi L, Flohe L, et al. Versatility of selenium catalysis in PHGPx unraveled by LC/ESI-MS/MS[J]. Biological Chemistry, 2003, 384(4): 575-588.

[102] Maiorino M, Roveri A, Benazzi L, et al. Functional interaction of phospholipid hydroperoxide glutathione peroxidase with sperm mitochondrion-associated cysteine-rich protein discloses the adjacent cysteine motif as a new substrate of the selenoperoxidase[J]. Journal of Biological Chemistry, 2005, 280(46): 38395-38402.

[103] Cook H W, Lands W E M. Mechanism for suppression of cellular biosynthesis of prostaglandins[J]. Nature, 1976, 260(5552): 630-632.

[104] Weitzel F, Wendel A. Selenoenzymes regulate the activity of leukocyte 5-lipoxygenase via the peroxide tone[J]. Journal of Biological Chemistry, 1993, 268(9): 6288-6292.

[105] Huang H S, Chen C J, Suzuki H, et al. Inhibitory effect of phospholipid hydroperoxide glutathione peroxidase on the activity of lipoxygenases and cyclooxygenases[J]. Prostaglandins and Other Lipid Mediators, 1999, 58(2-4): 65-75.

[106] Wenk J, Schüller J, Hinrichs C, et al. Overexpression of phospholipid-hydroperoxide glutathione peroxidase in human dermal fibroblasts abrogates UVA irradiation-induced expression of interstitial collagenase/matrix metalloproteinase-1 by suppression of phosphatidylcholine hydroperoxide-mediated[J]. Journal of Biological Chemistry, 2004, 279(44): 45634-45642.

[107] Latchoumycandane C, Marathe G K, Zhang R, et al. Oxidatively truncated phospholipids are required agents of tumor necrosis factor α (TNFα)-induced apoptosis[J]. Journal of Biological Chemistry, 2012, 287(21): 17693-17705.

[108] Nomura K, Imai H, Koumura T, et al. Mitochondrial phospholipid hydroperoxide glutathione peroxidase inhibits the release of cytochrome c from mitochondria by suppressing the peroxidation of cardiolipin in hypoglycaemia-induced apoptosis[J]. Biochemical Journal, 2000, 351(1): 183-193.

[109] Dixon S J, Lemberg K M, Lamprecht M R, et al. Ferroptosis: an iron-dependent form of nonapoptotic cell death[J]. Cell, 2012, 149(5): 1060-1072.

[110] Galluzzi L, Pedro J M B, Vitale I, et al. Essential versus accessory aspects of cell death: recommendations of the NCCD 2015[J]. Cell Death and Differentiation, 2015, 22(1): 58-73.

[111] 黄开勋，徐辉碧. 硒的化学、生物化学及其在生命科学中的应用[M]. 武汉：华中科技大学出版社，2009.

[112] Liu J, Luo G, Mu Y. Selenoproteins and Mimics [M]. Hangzhou: Zhejiang University Press, 2011.

[113] Zhong L, Arnér E S, Ljung J, et al. Rat and calf thioredoxin reductase are homologous to glutathione reductase with a carboxyl-terminal elongation containing a conserved catalytically active penultimate selenocysteine residue[J]. Journal of Biological Chemistry, 1998, 273(15): 8581-91.

[114] Sandalova T, Zhong L, Lindqvist Y, et al. Three-dimensional structure of a mammalian thioredoxin reductase: Implications for mechanism and evolution of a selenocysteine-dependent enzyme[J]. Proceedings of the National Academy of Sciences of the United States of America, 2001, 98(17): 9533-9538.

[115] Zhong L, Holmgren A. Essential Role of Selenium in the Catalytic Activities of Mammalian Thioredoxin Reductase Revealed by Characterization of Recombinant Enzymes with Selenocysteine Mutations[J]. Journal of Biological Chemistry, 2000, 275(24): 18121-18128.

[116] Holmgren A. The function of thioredoxin and glutathione in deoxyribonucleic acid synthesis[J]. Biochemical Society Transactions,

1977, 5(3): 611-612.

[117] Luthman M, Eriksson S, Holmgren A, et al. Glutathione-dependent hydrogen donor system for calf thymus ribonucleoside-diphosphate reductase[J]. Proceedings of the National Academy of Sciences of the United States of America, 1979, 76(5): 2158-2162.

[118] Kumar S, Björnstedt M, Holmgren A. Selenite is a substrate for calf thymus thioredoxin reductase and thioredoxin and elicits a large non-stoichiometric oxidation of NADPH in the presence of oxygen[J]. FEBS Journal, 1992, 207(2): 435-439.

[119] Björnstedt M, Odlander B, Kuprin S, et al. Selenite incubated with NADPH and mammalian thioredoxin reductase yields selenide, which inhibits lipoxygenase and changes the electron spin resonance spectrum of the active site iron[J]. Biochemistry, 1996, 35(26): 8511-8516.

[120] Björnstedt M, Kumar S, Holmgren A. Selenodiglutathione is a highly efficient oxidant of reduced thioredoxin and a substrate for mammalian thioredoxin reductase[J]. Journal of Biological Chemistry, 1992, 267(12): 8030-8037.

[121] Chae H Z, Chung S J, Rhee S G. Thioredoxin-dependent peroxide reductase from yeast[J]. Journal of Biological Chemistry, 1994, 269 (44): 27670-27678.

[122] Sg R, Hz C, K K. Peroxiredoxins: a historical overview and speculative preview of novel mechanisms and emerging concepts in cell signaling[J]. Free Radical Biology and Medicine, 2005, 38(12): 1543-1552.

[123] Lu J, Holmgren A. The thioredoxin antioxidant system[J]. Free Radical Biology and Medicine, 2014, 66(8): 75-87.

[124] Terry C M, Clikeman J A, Hoidal J R, et al. Effect of tumor necrosis factor-alpha and interleukin-1 alpha on heme oxygenase-1 expression in human endothelial cells[J]. American Journal of Physiology, 1998, 274(3 Pt 2): H883-H891.

[125] André M, Felley-Bosco E. Heme oxygenase-1 induction by endogenous nitric oxide: influence of intracellular glutathione[J]. FEBS Letters, 2003, 546(2): 223-227.

[126] Jeney V, Balla J, Yachie A, et al. Pro-oxidant and cytotoxic effects of circulating heme[J]. Blood, 2002, 100(3): 879-887.

[127] Mostert V, Hill K E, Ferris C D, et al. Selective induction of

liver parenchymal cell heme oxygenase-1 in selenium-deficient rats[J]. Biological Chemistry, 2003, 384(4): 681-687.

[128] Ejima K, Layne M D, Carvajal I M, et al. Modulation of the thioredoxin system during inflammatory responses and its effect on heme oxygenase-1 expression[J]. Antioxidants & Redox Signaling, 2002, 4(4): 569-575.

[129] Trigona W, Mullarky I, Y, Sordillo L. Thioredoxin reductase regulates the induction of haem oxygenase-1 expression in aortic endothelial cells[J]. Biochemical Journal, 2006, 394(1): 207-216.

[130] Lee B C, Dikiy A, Kim H, et al. Functions and Evolution of Selenoprotein Methionine Sulfoxide Reductases[J]. Biochimica et Biophysica Acta, 2009, 1790(11): 1471-1477.

[131] Sharov V S, Ferrington D A, Squier T C, et al. Diastereoselective reduction of protein-bound methionine sulfoxide by nethionine sulfoxide reductase[J]. FEBS Letters, 1999, 455(3): 247-250.

[132] Hy K, Vn G. Methionine sulfoxide reductases: selenoprotein forms and roles in antioxidant protein repair in mammals[J]. Biochemical Journal, 2007, 407(3): 321-329.

[133] Kim H Y, Gladyshev V N. Different catalytic mechanisms in mammalian selenocysteine-and cysteine-containing methionine-R-sulfoxide reductases[J]. PLoS Biology, 2005, 3(12): e375.

[134] Vinokur V, Grinberg L, Berenshtein E, et al. Methionine-centered redox cycle in organs of the aero-digestive tract of young and old rats[J]. Biogerontology, 2009, 10(1): 43-52.

[135] Moskovitz J, Bar-Noy S, Williams W M, et al. Methionine Sulfoxide Reductase (MsrA) is a Regulator of Antioxidant Defense and Lifespan in Mammals[J]. Proceedings of the National Academy of Sciences of the United States of America, 2001, 98(23): 12920-12925.

[136] Moskovitz J. Prolonged selenium-deficient diet in MsrA knockout mice causes enhanced oxidative modification to proteins and affects the levels of antioxidant enzymes in a tissue-specific manner[J]. Free Radical Research, 2007, 41(2): 162-171.

[137] Johansson C, Lillig C H, Holmgren A. Human mitochondrial glutaredoxin reduces S-glutathionylated proteins with high affinity accepting electrons from either glutathione or thioredoxin reductase[J]. Journal

of Biological Chemistry, 2004, 279(9): 7537-43.

[138] Lundströml jung J, Birnbach U, Rupp K, et al. Two resident ER-proteins, CaBP1 and CaBP2, with thioredoxin domains, are substrates for thioredoxin reductase: comparison with protein disulfide isomerase[J]. FEBS Letters, 1995, 357(3): 305.

[139] Jiménez A, Pelto-Huikko M, Gustafsson J, et al. Characterization of human thioredoxin-like-1: Potential involvement in the cellular response against glucose deprivation[J]. FEBS Letters, 2006, 580(3): 960-967.

[140] Felberbaumcorti M, Morel E, Cavalli V, et al. The Redox Sensor TXNL1 Plays a Regulatory Role in Fluid Phase Endocytosis[J]. PloS One, 2007, 2(11): e1144.

[141] May J M, Mendiratta S, Hill K E, et al. Reduction of Dehydroascorbate to Ascorbate by the Selenoenzyme Thioredoxin Reductase [J]. Journal of Biological Chemistry, 1997, 272(36): 22607-22610.

[142] Tamura T, Gladyshev V, Liu S Y, et al. The mutual sparing effects of selenium and vitamin E in animal nutrition may be further explained by the discovery that mammalian thioredoxin reductase is a selenoenzyme[J]. Biofactors, 1995, 5(2): 99-102.

[143] May J M, Qu Z C, Nelson D J. Uptake and reduction of α-lipoic acid by human erythrocytes[J]. Clinical Biochemistry, 2007, 40(15): 1135-1142.

[144] Brown S E, Ross M F, Sanjuanpla A, et al. Targeting lipoic acid to mitochondria: synthesis and characterization of a triphenylphosphonium-conjugated alpha-lipoyl derivative[J]. Free Radical Biology & Medicine, 2007, 42(12): 1766-1780.

[145] Xia L, Björnstedt M, Nordman T, et al. Reduction of ubiquinone by lipoamide dehydrogenase. An antioxidant regenerating pathway [J]. FEBS Journal, 2001, 268(5): 1486-1490.

[146] Lee S, Kim S M, Lee R T. Thioredoxin and thioredoxin target proteins: from molecular mechanisms to functional significance[J]. Antioxidants & Redox Signaling, 2013, 18(10): 1165-1207.

[147] A M, H I. Redox control of cell fate by MAP kinase: physiological roles of ASK1-MAP kinase pathway in stress signaling[J]. BBA-General Subjects, 2008, 1780(11): 1325-1336.

[148] Sachsenmaier C, Radlerpohl A, Zinck R, et al. Involvement of

growth factor receptors in the mammalian UVC response[J]. Cell, 1994, 78(6): 963-972.

[149] Knebel A, Rahmsdorf H J, Ullrich A, et al. Dephosphorylation of receptor tyrosine kinases as target of regulation by radiation, oxidants or alkylating agents [J]. The EMBO Journal, 1996, 15 (19): 5314-5325.

[150] Guyton K Z, Liu Y, Gorospe M, et al. Activation of mitogen-activated protein kinase by H_2O_2. Role in cell survival following oxidant injury[J]. Journal of Biological Chemistry, 1996, 271(8): 4138-4142.

[151] Ziv E, Rotem C, Miodovnik M, et al. Two modes of ERK activation by TNF in keratinocytes: Different cellular outcomes and bi-directional modulation by vitamin D[J]. Journal of Cellular Biochemistry, 2008, 104(2): 606-619.

[152] Wang X, Martindale J L, Holbrook N J. Requirement for ERK activation in cisplatin-induced apoptosis[J]. Journal of Biological Chemistry, 2000, 275(50): 39435-39443.

[153] Sekine Y, Takeda K, Ichijo H. The ASK1-MAP kinase signaling in ER stress and neurodegenerative diseases[J]. Current Molecular Medicine, 2006, 6(1): 87-97.

[154] Sayama K, Hanakawa Y, Shirakata Y, et al. Apoptosis Signal-regulating Kinase 1 (ASK1) Is an Intracellular Inducer of Keratinocyte Differentiation[J]. Journal of Biological Chemistry, 2001, 276(2): 999-1004.

[155] Takeda K, Hatai T, Hamazaki T S, et al. Apoptosis Signal-regulating Kinase 1 (ASK1) Induces Neuronal Differentiation and Survival of PC12 Cells[J]. Journal of Biological Chemistry, 2000, 275(13): 9805-9813.

[156] Noguchi T, Takeda K, Matsuzawa A, et al. Recruitment of Tumor Necrosis Factor Receptor-associated Factor Family Proteins to Apoptosis Signal-regulating Kinase 1 Signalosome Is Essential for Oxidative Stress-induced Cell Death[J]. Journal of Biological Chemistry, 2005, 280(44): 37033-37040.

[157] Song J J, Lee Y J. Differential role of glutaredoxin and thioredoxin in metabolic oxidative stress-induced activation of apoptosis signal-regulating kinase 1[J]. Biochemical Journal, 2003, 373(3): 845-853.

[158] Lee Y J, Galoforo S S, Berns C M, et al. Glucose deprivation-

induced cytotoxicity and alterations in mitogen-activated protein kinase activation are mediated by oxidative stress in multidrug-resistant human breast carcinoma cells[J]. Journal of Biological Chemistry, 1998, 273(9): 5294-5299.

[159] Matsuzawa A. Thioredoxin and redox signaling: Roles of the thioredoxin system in control of cell fate[J]. Archives of Biochemistry and Biophysics, 2017, 617(2017): 101-105.

[160] Engelman J A, Luo J L. The evolution of phosphatidylinositol 3-kinases as regulators of growth and metabolism[J]. Nature Reviews Genetics, 2006, 7(8): 606-619.

[161] Katso R, Okkenhaug K, Ahmadi K, et al. Cellular Function of Phosphoinositide 3-Kinases: Implications for Development, Immunity, Homeostasis, and Cancer[J]. Annual Review of Cell and Developmental Biology, 2001, 17(1): 615-675.

[162] Stambolic V, Suzuki A, Pompa J L D L, et al. Negative Regulation of PKB/Akt-Dependent Cell Survival by the Tumor Suppressor PTEN[J]. Cell, 1998, 95(1): 29-39.

[163] Li D M, Sun H. TEP1, Encoded by a Candidate Tumor Suppressor Locus, Is a Novel Protein Tyrosine Phosphatase Regulated by Transforming Growth Factor β[J]. Cancer Research, 1997, 57(11): 2124-2129.

[164] Chalhoub N, Baker S J. PTEN and the PI3-kinase pathway in cancer[J]. Annual Review of Pathology-mechanisms of Disease, 2009, 4(1): 127-150.

[165] Engelman J, Luo J, Cantley L C. The evolution of phosphatidylinositol 3-kinases as regulators of growth and metabolism[J]. Nature Reviews Genetics, 2006, 7(8): 606-619.

[166] Lee S, Yang K, Kwon J, et al. Reversible Inactivation of the Tumor Suppressor PTEN by H_2O_2[J]. Journal of Biological Chemistry, 2002, 277(23): 20336-20342.

[167] Torres J, Pulido R. The Tumor Suppressor PTEN Is Phosphorylated by the Protein Kinase CK2 at Its C Terminus IMPLICATIONS FOR PTEN STABILITY TO PROTEASOME-MEDIATED DEGRADATION[J]. Journal of Biological Chemistry, 2001, 276(2): 993-998.

[168] Song Z, Saghafi N, Gokhale V, et al. Regulation of the activity of the tumor suppressor PTEN by thioredoxin in Drosophila melano-

gaster[J]. Experimental Cell Research, 2007, 313(6): 1161-1171.

[169] Jamaluddin M, Wang S, Boldogh I, et al. TNF-alpha-induced NF-kappaB/RelA Ser(276) phosphorylation and enhanceosome formation is mediated by an ROS-dependent PKAc pathway[J]. Cellular Signalling, 2007, 19(7): 1419-1433.

[170] Meyer M, Schreck R, Baeuerle P A. H_2O_2 and antioxidants have opposite effects on activation of NF-kappa B and AP-1 in intact cells: AP-1 as secondary antioxidant-responsive factor [J]. EMBO Journal, 1993, 12(5): 2005-2015.

[171] Matthews J R, Wakasugi N, Virelizier J L, et al. Thioredoxin regulates the DNA binding activity of NF-kappa B by reduction of a disulphide bond involving cysteine 62[J]. Nucleic Acids Research, 1992, 20(15): 3821-3830.

[172] Hirota K, Murata M, Sachi Y, et al. Distinct roles of thioredoxin in the cytoplasm and in the nucleus. A two-step mechanism of redox regulation of transcription factor NF-kappaB[J]. Journal of Biological Chemistry, 1999, 274(39): 27891-27897.

[173] Mattmiller S A, Carlson B A, Sordillo L M. Regulation of inflammation by selenium and selenoproteins: impact on eicosanoid biosynthesis[J]. Journal of Nutritional Science, 2013, 2: 1-13.

[174] Osburn W O, Kensler T W. Nrf2 signaling: An adaptive response pathway for protection against environmental toxic insults[J]. Mutation Research, 2008, 659(1): 31-39.

[175] Copple I M, Goldring C E, Kitteringham N R, et al. The Nrf2-Keap1 defence pathway: role in protection against drug-induced toxicity[J]. Toxicology, 2008, 246(1): 24-33.

[176] Brigelius-Flohe R, Flohe L. Basic Principles and Emerging Concepts in the Redox Control of Transcription Factors[J]. Antioxidants & Redox Signaling, 2011, 15(8): 2335-2381.

[177] Ghosh S, Mukherjee S, Choudhury S, et al. Reactive oxygen species in the tumor niche triggers altered activation of macrophages and immunosuppression: Role of fluoxetine[J]. Cellular Signalling, 2015, 27(7): 1398-1412.

[178] Hulbert A J. Thyroid hormones and their effects: a new perspective[J]. Biological Reviews, 2000, 75(4): 519-631.

[179] Visser T J, Doestobé I, Docter R, et al. Subcellular localization of a rat liver enzyme converting thyroxine into tri-iodothyronine and possible involvement of essential thiol groups[J]. Biochemical Journal, 1976, 157(2): 479-482.

[180] Toyoda N, Berry M J, Harney J W, et al. Topological analysis of the integral membrane protein, type 1 iodothyronine deiodinase (D1)[J]. Journal of Biological Chemistry, 1995, 270(20): 12310-12318.

[181] Beckett G J, Beddows S E, Morrice P C, et al. Inhibition of hepatic deiodination of thyroxine is caused by selenium deficiency in rats[J]. Biochemical Journal, 1987, 248(2): 443-447.

[182] Behne D, Kyriakopoulos A, Meinhold H, et al. Identification of type I iodothyronine 5′-deiodinase as a selenoenzyme[J]. Biochemical and Biophysical Research Communications, 1990, 173(3): 1143-1149.

[183] Baqui M M, Gereben B, Harney J W, et al. Distinct subcellular localization of transiently expressed types 1 and 2 iodothyronine deiodinases as determined by immunofluorescence confocal microscopy[J]. Endocrinology, 2000, 141(11): 4309-4312.

[184] Dubois G M, Sebillot A, Kuiper G G J M, et al. Deiodinase Activity Is Present in Xenopus laevis during Early Embryogenesis[J]. Endocrinology, 2006, 147(10): 4941-4949.

[185] Bassett J H D, Boyde A, Howell P, et al. Optimal bone strength and mineralization requires the type 2 iodothyronine deiodinase in osteoblasts[J]. Proceedings of the National Academy of Sciences of the United States of America, 2010, 107(16): 7604-7609.

[186] Persson-Moschos M, Alfthan G, Akesson B. Plasma selenoprotein P levels of healthy males in different selenium status after oral supplementation with different forms of selenium[J]. European Journal of Clinical Nutrition, 1998, 52(5): 363-367.

[187] Burk R F, Hill K E. Selenoprotein P-Expression, functions, and roles in mammals[J]. Biochimica et Biophysica Acta, 2009, 1790(11): 1441-1447.

[188] Burk R F, Hill K E, Awad J A, et al. Liver and kidney necrosis in selenium-deficient rats depleted of glutathione[J]. Laboratory investigation; a journal of technical methods and pathology, 1995, 72(6): 723-730.

[189] Burk R F, Hill K E, Motley A K. Selenoprotein metabolism

and function: evidence for more than one function for selenoprotein P[J]. Journal of Nutrition, 2003, 133(5 Suppl 1): 1517S.

[190] Wilson D S, Tappel A L. Binding of plasma selenoprotein P to cell membranes[J]. Journal of Inorganic Biochemistry, 1993, 51(4): 707-714.

[191] Burk R F, Hill K E, Read R, et al. Response of rat selenoprotein P to selenium administration and fate of its selenium[J]. American Journal of Physiology, 1991, 261(1): 26-30.

[192] Yan J, Barrett J N. Purification from Bovine Serum of a Survival-Promoting Factor for Cultured Central Neurons and Its Identification as Selenoprotein-P[J]. The Journal of Neuroscience, 1998, 18(21): 8682-8691.

[193] Wilson D S, Tappel A L. Subcellular fate of selenium from 75 Se-labeled plasma selenoprotein P in selenium-deficient rats[J]. Journal of Nutritional Biochemistry, 1993, 4(4): 208-211.

[194] Olson G E, Winfrey V P, Nagdas S K, et al. Apolipoprotein E Receptor-2 (ApoER2) Mediates Selenium Uptake from Selenoprotein P by the Mouse Testis[J]. Journal of Biological Chemistry, 2007, 282(16): 12290-12297.

[195] Burk R F, Hill K E, Olson G E, et al. Deletion of Apolipoprotein E Receptor-2 in Mice Lowers Brain Selenium and Causes Severe Neurological Dysfunction and Death When a Low-Selenium Diet Is Fed[J]. The Journal of Neuroscience, 2007, 27(23): 6207-6211.

[196] Schweizer U, Streckfuss F, P, Carlson B, et al. Hepatically derived selenoprotein P is a key factor for kidney but not for brain selenium supply[J]. Biochemical Journal, 2005, 386(2): 221-226.

[197] Olson G E, Winfrey V P, Hill K E, et al. Megalin Mediates Selenoprotein P Uptake by Kidney Proximal Tubule Epithelial Cells[J]. Journal of Biological Chemistry, 2008, 283(11): 6854-6860.

[198] Christensen E I, Willnow T E. Essential Role of Megalin in Renal Proximal Tubule for Vitamin Homeostasis[J]. Journal of the American Society of Nephrology, 1999, 10(10): 2224-2236.

[199] Motchnik P A, Tappel A L. Rat plasma selenoprotein P properties and purification[J]. BBA-General Subjects, 1989, 993(1): 27-35.

[200] Y S, T H, A T, et al. Selenoprotein P in human plasma as an extracellular phospholipid hydroperoxide glutathione peroxidase-Isolation and enzymatic characterization of human selenoprotein P[J]. Journal of Bi-

ological Chemistry, 1999, 274(5): 2866-2871.

[201] Ursini F, Maiorino M, Gregolin C. The selenoenzyme phospholipid hydroperoxide glutathione peroxidase[J]. Biochimica et Biophysi ca Acta, 1985, 839(1): 62-70.

[202] Rock C, Moos P J. Selenoprotein P protects cells from lipid hydroperoxides generated by 15-LOX-1[J]. Prostaglandins Leukotrienes and Essential Fatty Acids, 2010, 83(4): 203-210.

[203] Hafeman D G, Hoekstra W G. Lipid Peroxidation in Vivo during Vitamin E and Selenium Deficiency in the Rat as Monitored by Ethane Evolution[J]. Journal of Nutrition, 1977, 107(4): 666-672.

[204] Burk R F, Hill K E, Awad J A, et al. Pathogenesis of diquat-induced liver necrosis in selenium-deficient rats: Assessment of the roles of lipid peroxidation and selenoprotein P☆[J]. Hepatology, 1995, 21(2): 561-569.

[205] Traulsen H, Steinbrenner H, Buchczyk D P, et al. Selenoprotein P Protects Low-density Lipoprotein Against Oxidation[J]. Free Radical Research, 2004, 38(2): 123-128.

[206] Arteel G E, Mostert V, Oubrahim H, et al. Protection by Selenoprotein P in Human Plasma against Peroxynitrite-Mediated Oxidation and Nitration[J]. Biological Chemistry, 1998, 379(9): 1201-1205.

[207] Steinbrenner H, Bilgic E, Alili L, et al. Selenoprotein P protects endothelial cells from oxidative damage by stimulation of glutathione peroxidase expression and activity[J]. Free Radical Research, 2006, 40(9): 936-943.

[208] Saito Y, Yoshida Y, Akazawa T, et al. Cell Death Caused by Selenium Deficiency and Protective Effect of Antioxidants[J]. Journal of Biological Chemistry, 2003, 278(41): 39428-39434.

[209] Halliwell B, Gutteridge J M. Role of free radicals and catalytic metal ions in human disease: an overview[J]. Molecular Aspects of Medicine, 1985, 8(2): 89-193.

[210] Beckman J S, Beckman T W, Chen J, et al. Apparent hydroxyl radical production by peroxynitrite: implications for endothelial injury from nitric oxide and superoxide[J]. Proceedings of the National Academy of Sciences of the United States of America, 1990, 87(4): 1620-1624.

[211] Gladyshev V N, Jeang K T, Wootton J C, et al. A New Hu-

man Selenium-containing Protein[J]. Journal of Biological Chemistry, 1998, 273(15): 8910-8915.

[212] Korotkov K V, Kumaraswamy E, Zhou Y, et al. Association between the 15-kDa Selenoprotein and UDP-glucose: Glycoprotein Glucosyltransferase in the Endoplasmic Reticulum of Mammalian Cells[J]. Journal of Biological Chemistry, 2001, 276(18): 15330-15336.

[213] Kasaikina M V, Fomenko D E, Labunskyy V M, et al. Roles of the 15-kDa selenoprotein (Sep15) in redox homeostasis and cataract development revealed by the analysis of Sep 15 knockout mice[J]. Journal of Biological Chemistry, 2011, 286(38): 33203-33212.

[214] Ferguson A D, Labunskyy V M, Fomenko D E, et al. NMR structures of the selenoproteins Sep15 and SelM reveal redox activity of a new thioredoxin-like family[J]. Journal of Biological Chemistry, 2006, 281(6): 3536-43.

[215] Labunskyy V M, Yoo M H, Hatfield D L, et al. Sep15, a Thioredoxin-like Selenoprotein, Is Involved in the Unfolded Protein Response and Differentially Regulated by Adaptive and Acute ER Stresses[J]. Biochemistry, 2009, 48(35): 8458-8465.

[216] Kasaikina M V, Fomenko D E, Labunskyy V M, et al. Roles of the 15-kDa selenoprotein (Sep15) in redox homeostasis and cataract development revealed by the analysis of Sep 15 knockout mice[J]. Journal of Biological Chemistry, 2011, 286(38): 33203-12.

[217] Jablonska E, Gromadzinska J, Sobala W, et al. Lung cancer risk associated with selenium status is modified in smoking individuals by Sep15 polymorphism[J]. European Journal of Nutrition, 2008, 47(1): 47-54.

[218] Hatfield Dolph l, Schweizer U, Tsuji Petra a, et al. Selenium: Its Molecular Biology and Role in Human Health [M]. 3 ed. New York: Springer Nature, 2016.

[219] Guariniello S, Bernardo G D, Colonna G, et al. Evaluation of the Selenotranscriptome Expression in Two Hepatocellular Carcinoma Cell Lines[J]. Analytical Cellular Pathology, 2015, 2015: 419561-419561.

[220] Tsuji P A, Carlson B A, Salvador N S, et al. Knockout of the 15 kDa selenoprotein protects against chemically-induced aberrant crypt formation in mice[J]. PloS One, 2012, 7(12): e50574.

[221] Korotkov K V, Novoselov S V, Hatfield D L, et al. Mammali-

an selenoprotein in which selenocysteine (Sec) incorporation is supported by a new form of Sec insertion sequence element[J]. Molecular and Cellular Biology, 2002, 22(5): 1402-1411.

[222] Ferguson A D, Labunskyy V M, Fomenko D E, et al. NMR structures of the selenoproteins Sep15 and SelM reveal redox activity of a new thioredoxin-like family[J]. Journal of Biological Chemistry, 2006, 281(6): 3536-43.

[223] Korotkov K V, Novoselov S V, Hatfield D L, et al. Mammalian Selenoprotein in Which Selenocysteine (Sec) Incorporation Is Supported by a New Form of Sec Insertion Sequence Element[J]. Molecular and Cellular Biology, 2002, 22(5): 1402-1411.

[224] Hwang D Y, Sin J S, Kim M S, et al. Overexpression of human selenoprotein M differentially regulates the concentrations of antioxidants and H_2O_2, the activity of antioxidant enzymes, and the composition of white blood cells in a transgenic rat[J]. International Journal of Molecular Medicine, 2008, 21(2): 169-179.

[225] Reeves M A, Bellinger F P, Berry M J. The Neuroprotective Functions of Selenoprotein M and its Role in Cytosolic Calcium Regulation [J]. Antioxidants & Redox Signaling, 2010, 12(7): 809-818.

[226] Du X, Li H, Wang Z, et al. Selenoprotein P and selenoprotein M block Zn^{2+}-mediated Aβ42 aggregation and toxicity[J]. Metallomics Integrated Biometal Science, 2013, 5(7): 861-870.

[227] Pitts M W, Reeves M A, Hashimoto A C, et al. Deletion of selenoprotein M leads to obesity without cognitive deficits[J]. Journal of Biological Chemistry, 2013, 288(36): 26121-26134.

[228] Ozcan L, Ergin A S, Lu A, et al. Endoplasmic reticulum stress plays a central role in development of leptin resistance[J]. Cell Metabolism, 2009, 9(1): 35-51.

[229] Zhang X, Zhang G, Zhang H, et al. Hypothalamic IKKβ/NF-κB and ER Stress Link Overnutrition to Energy Imbalance and Obesity [J]. Cell, 2008, 135(1): 61-73.

[230] Raber J, Chen S, Mucke L, et al. Corticotropin-releasing Factor and Adrenocorticotrophic Hormone as Potential Central Mediators of OB Effects[J]. Journal of Biological Chemistry, 1997, 272(24): 15057-15060.

[231] Kooijman S, Jk V D H, Rensen P C. Neuronal Control of

Brown Fat Activity[J]. Trends in Endocrinology & Metabolism Tem, 2015, 26(11): 657-668.

[232] Contreras C, Gonzã L-G a I, Martã-Nez-Sã N N, et al. Central ceramide-induced hypothalamic lipotoxicity and ER stress regulate energy balance[J]. Cell Reports, 2014, 9(1): 366-377.

[233] Rezai-Zadeh K, Yu S, Jiang Y, et al. Leptin receptor neurons in the dorsomedial hypothalamus are key regulators of energy expenditure and body weight, but not food intake[J]. Mol Metab, 2014, 3(7): 681-693.

[234] Oldfield B J, Giles M E, Watson A, et al. The neurochemical characterisation of hypothalamic pathways projecting polysynaptically to brown adipose tissue in the rat[J]. Neuroscience, 2002, 110(3): 515-526.

[235] Kryukov G V, Castellano S, Novoselov S V, et al. Characterization of mammalian selenoproteomes[J]. Science, 2003, 300(5624): 1439-1443.

[236] Shchedrina V A, Yan Z, Labunskyy V M, et al. Structure-function relations, physiological roles, and evolution of mammalian ER-resident selenoproteins[J]. Antioxidants and Redox Signaling, 2010, 12(7): 839-849.

[237] Carducci M, Perfetto L, Briganti L, et al. The protein interaction network mediated by human SH3 domains[J]. Biotechnology Advances, 2012, 30(1): 4-15.

[238] Lu C, Qiu F, Zhou H, et al. Identification and characterization of selenoprotein K: an antioxidant in cardiomyocytes[J]. FEBS Letters, 2006, 580(22): 5189-5197.

[239] Hoffmann P R, Höge S C, Ping-an L, et al. The selenoproteome exhibits widely varying, tissue-specific dependence on selenoprotein P for selenium supply[J]. Nucleic Acids Research, 2007, 35(12): 3963-3973.

[240] Verma S, Hoffmann F K W, Kumar M, et al. Selenoprotein K knockout mice exhibit deficient calcium flux in immune cells and impaired immune responses[J]. Journal of Immunology, 2011, 186(4): 2127-2137.

[241] Shchedrina V A, Everley R A, Zhang Y, et al. Selenoprotein K Binds Multiprotein Complexes and Is Involved in the Regulation of Endoplasmic Reticulum Homeostasis[J]. Journal of Biological Chemistry, 2011, 286(50): 42937-42948.

[242] Du S, Zhou J, Jia Y, et al. SelK is a novel ER stress-regulated

protein and protects HepG2 cells from ER stress agent-induced apoptosis [J]. Archives of Biochemistry and Biophysics, 2010, 502(2): 137-143.

[243] Hogan P G, Rao A. Store-operated calcium entry: mechanisms and modulation[J]. Biochemical and Biophysical Research Communications, 2015, 460(1): 40-49.

[244] Huang Z, Hoffmann F W, Fay J D, et al. Stimulation of unprimed macrophages with immune complexes triggers a low output of nitric oxide by calcium-dependent neuronal nitric-oxide synthase[J]. Journal of Biological Chemistry, 2012, 287(7): 4492-4502.

[245] Podrez E A, Febbraio M, Sheibani N, et al. Macrophage scavenger receptor CD36 is the major receptor for LDL modified by monocyte-generated reactive nitrogen species[J]. Journal of Clinical Investigation, 2000, 105(8): 1095-1098.

[246] Meiler S, Baumer Y, Huang Z, et al. Selenoprotein K is required for palmitoylation of CD36 in macrophages: implications in foam cell formation and atherogenesis[J]. Journal of Leukocyte Biology, 2013, 93(5): 771-780.

[247] Hoffmann P R. Selenoprotein K and Protein Palmitoylation in Regulating Immune Cell Functions [M]. New York: Springer International Publishing, 2016.

[248] Walder K, Kantham L, Mcmillan J, et al. Tanis: A Link Between Type 2 Diabetes and Inflammation? [J]. Diabetes, 2002, 51(6): 1859-1866.

[249] Gao Y, Feng H C, Walder K, et al. Regulation of the selenoprotein SelS by glucose deprivation and endoplasmic reticulum stress-SelS is a novel glucose-regulated protein[J]. FEBS Letters, 2004, 563(1): 185-190.

[250] Zhao Y, Li H, Men L L, et al. Effects of selenoprotein S on oxidative injury in human endothelial cells[J]. Journal of Translational Medicine, 2013, 11(1): 287-287.

[251] Ye Y, Fu F, Li X, et al. Selenoprotein S Is Highly Expressed in the Blood Vessels and Prevents Vascular Smooth Muscle Cells From Apoptosis[J]. Journal of Cellular Biochemistry, 2016, 117(1): 106-117.

[252] Kim K H, Gao Y, Walder K, et al. SEPS1 protects RAW264.7 cells from pharmacological ER stress agent-induced apoptosis[J]. Biochemical and Biophysical Research Communications, 2007, 354 (1):

127-132.

[253] Fradejas N, Serrano-Pérez, Carmen M D, et al. Selenoprotein S expression in reactive astrocytes following brain injury[J]. Glia, 2011, 59(6): 959-972.

[254] Du S, Liu H, Huang K. Influence of SelS gene silence on beta-Mercaptoethanol-mediated endoplasmic reticulum stress and cell apoptosis in HepG2 cells[J]. Biochimica et Biophysica Acta General Subjects, 2010, 1800(5): 511-517.

[255] 刘红梅, 金剑波, 周军, 等. 硒蛋白 S 的结构、功能及与疾病的关系[J]. 化学进展, 2018, 30(10): 25-33.

[256] Lee J H, Park K J, Jang J K, et al. Selenoprotein S-dependent Selenoprotein K Binding to p97(VCP) Protein Is Essential for Endoplasmic Reticulum-associated Degradation[J]. Journal of Biological Chemistry, 2015, 290(50): 29941-29952.

[257] Curran J E, Jowett J B M, Elliott K S, et al. Genetic variation in selenoprotein S influences inflammatory response[J]. Nature Genetics, 2005, 37(11): 1234-1241.

[258] Fradejas N, Serrano-Pérez M D C, Tranque P, et al. Selenoprotein S expression in reactive astrocytes following brain injury[J]. Glia, 2011, 59(6): 959-972.

[259] Cui S, Men L, Li Y, et al. Selenoprotein S Attenuates Tumor Necrosis Factor-α-Induced Dysfunction in Endothelial Cells[J]. Mediators of Inflammation, 2018, 2018: 1625414.

[260] Kryukov G V, Kryukov V M, Gladyshev V N. New mammalian selenocysteine-containing proteins identified with an algorithm that searches for selenocysteine insertion sequence elements[J]. Journal of Biological Chemistry, 1999, 274(48): 33888-33897.

[261] Grumolato L, Ghzili H, Monterohadjadje M, et al. Selenoprotein T is a PACAP-regulated gene involved in intracellular Ca^{2+} mobilization and neuroendocrine secretion[J]. FASEB Journal, 2008, 22(6): 1756-1768.

[262] Labunskyy V M, Hatfield D L, Gladyshev V N. Selenoproteins: Molecular Pathways and Physiological Roles[J]. Physiological Reviews, 2014, 94(3): 739-777.

[263] Dikiy A, Novoselov S V, Fomenko D E, et al. SelT, SelW, SelH, and Rdx12: Genomics and Molecular Insights into the Functions of

Selenoproteins of a Novel Thioredoxin-like Family[J]. Biochemistry, 2007, 46(23): 6871-6882.

[264] Boukhzar L, Hamieh A, Cartier D, et al. Selenoprotein T Exerts an Essential Oxidoreductase Activity That Protects Dopaminergic Neurons in Mouse Models of Parkinson's Disease[J]. Antioxidants and Redox Signaling, 2016, 24(11): 557-574.

[265] Ghzili H, Grumolato L, Thou? Nnon E, et al. Role of PACAP in the physiology and pathology of the sympathoadrenal system[J]. Frontiers in Neuroendocrinology, 2008, 29(1): 128-141.

[266] Anouar Y, Ghzili H, Grumolato L, et al. Selenoprotein T is a new PACAP-and cAMP-responsive gene involved in the regulation of calcium homeostasis during neuroendocrine cell differentiation[J]. Frontiers in Neuroendocrinology, 2006, 27(1): 82-83.

[267] Sengupta A, Carlson B A, Labunskyy V M, et al. Selenoprotein T deficiency alters cell adhesion and elevates selenoprotein W expression in murine fibroblast cells[J]. Biochemistry and Cell Biology, 2009, 87(6): 953-961.

[268] Tanguy Y, Falluelmorel A, Arthaud S, et al. The PACAP-Regulated Gene Selenoprotein T Is Highly Induced in Nervous, Endocrine, and Metabolic Tissues during Ontogenetic and Regenerative Processes[J]. Endocrinology, 2011, 152(11): 4322-4335.

[269] Prevost G, Arabo A, Jian L, et al. The PACAP-regulated gene selenoprotein T is abundantly expressed in mouse and human β-cells and its targeted inactivation impairs glucose tolerance[J]. Endocrinology, 2013, 154(10): 3796-3806.

[270] Malik R, Selden C, Hodgson H. The role of non-parenchymal cells in liver growth[J]. Seminars in Cell & Developmental Biology, 2002, 13(6): 425-431.

[271] Vendeland S C, Beilstein M A, Ream W, et al. Rat Skeletal Muscle Selenoprotein W: cDNA Clone and mRNA Modulation by Dietary Selenium[J]. Proceedings of the National Academy of Sciences of the United States of America, 1995, 92(19): 8749-8753.

[272] Yeh J, Beilstein M A, Andrews J S, et al. Tissue distribution and influence of selenium status on levels of selenoprotein W[J]. The FASEB Journal, 1995, 9(5): 392-396.

[273] Sun Y, Gu Q P, Whanger P D. Selenoprotein W in overexpressed and underexpressed rat glial cells in culture[J]. Journal of Inorganic Biochemistry, 2001, 84(1): 151-156.

[274] Jeong D W, Kim T S, Chung Y W, et al. Selenoprotein W is a glutathione-dependent antioxidant in vivo[J]. FEBS Letters, 2002, 517(1): 225-228.

[275] Chung Y W, Jeong D, Noh O J, et al. Antioxidative role of selenoprotein W in oxidant-induced mouse embryonic neuronal cell death[J]. Molecules and Cells, 2009, 27(5): 609-613.

第三章　硒在家禽生产中的应用

硒是家禽营养中的必需微量元素，尽管在现代家禽生产中与严重硒缺乏相关的各种功能紊乱和疾病已不再发生，但在一些特定的时期，如应激期间，由于日粮硒供应的不理想以及机体抗氧化防御能力的降低，往往会导致家禽的免疫功能、生产性能和繁殖性能降低。现代遗传学技术在家禽生产中的应用持续改善着蛋鸡的产蛋性能以及肉鸡的生长性能，这种高产蛋率和高生长率所带来的负面效应使家禽对各种应激高度敏感，因而对日粮营养以及管理提出了更高的要求。目前 NRC 推荐的家禽硒需要量主要是依据维持、生长和生产需要确定的。事实上，如果考虑动物的健康以及动物产品中硒的含量，则需要量可能要更高。一些证据表明，家禽发挥最优免疫功能所需的硒要远高于生长发育所需的硒。因此，如何精准地满足现代家禽生产对硒的需要是进一步优化家禽生产性能的重要措施之一。随着家禽体内更多硒蛋白的发现和表征，将有助于更好地揭示硒在家禽营养的作用。在家鸡中已发现有 26 种编码不同硒蛋白的基因，其中超过一半的已知硒蛋白直接或间接参与了机体的抗氧化防御和细胞的氧化还原平衡，对动物的肠道健康、生殖以及免疫功能具有重要的调节作用。

在包括小麦、大麦、玉米和大豆在内的饲料原料中均含有硒，且主要以有机形式存在。然而，世界上大多数地区土壤中的硒浓度很低，因而饲料原料中的硒浓度也很低，不能满足动物对硒的需要，因此，在 20 世纪 70 年代开始在动物饲料中添加硒。在动物日粮中添加的硒一般为有机和无机硒两种形式。无机硒主要为硒酸盐和亚硒酸盐，价格相对较低，在家禽生产中被广泛应用。有机硒主要为富硒酵母。无机硒在家禽营养中的应用存在一些不足，体现在以下五个方面：①由于饲料中营养物质的相互作用，亚硒酸盐可被饲料中具有还原活性的物质还原为不可吸收利用的元素硒。配合饲料中存在的抗坏血酸和亚铁盐在饲料加工、储存以及进入肠道消化过程中均可以导致亚硒酸盐还原为元素硒，而抗坏血酸的氧化导致其本身活性丧失，亚铁盐的氧化会降低其肠道消化率[1]。尽管维生素 C 不是常规添加到家禽日粮中的营养添加剂，但在应激期间，特别是热应激期间，维生素 C 常被用作抗应激剂在日粮中添加。预混料经过长时间贮存后出现的粉红色颗粒就是由亚硒酸盐还原为元素硒所致，当动物采食这种预混料后会导致机体对硒的吸收率降低，因而粪硒排出量增加。大鼠日粮中亚硒酸盐和维生素 C 同时使用时，硒对肿瘤发生的保护作用明显降低，在人体试验中也发现二者同时使用降低了硒的利用效率。②饲料中水分活度较高时，亚硒酸盐有可能溶解形成亚硒酸而以蒸汽形式散失[2]，从而降低机体对硒的摄取利用。③进入肠道后，亚硒酸盐的促氧化效应可能对家禽的肠道健康产生不利影响。研究发现，日粮中添加亚硒酸钠导致鸡的十二指肠隐窝上皮细胞层出现空泡变性，回肠中退化和坏死的肠腺间出现过度的单核细胞浸润和聚集[3]。④亚硒酸钠形式的硒向发育中鸡胚的转运能力较低，因而有可能会影响到孵化率甚至新生雏鸡的存活率。孵化时高温、高湿以及氧气可利用性的增加，使得孵化过程常伴有氧化应激，因此，提高胚胎的抗氧化防御能力可以提高孵化率。由于亚硒酸钠形式的硒向鸡蛋的转运能力较低，因而也会影响到鸡蛋的品质[4]。⑤亚硒酸钠在机体中只能以 Sec 形式通过翻译掺入到硒蛋白中，几乎不能在机体中建立硒储备，因而不能满足机体在应激条件下对硒需要量的提高。过去几十年积累的大量有关硒营养的研究表明，进化过程中家禽的消化系统能更好地适应有机形式的硒（以 SeMet 形式存在）[5]，有机硒在十二指肠和空肠中具有更高的吸收效率（SeMet＞硒酸盐＞亚硒酸盐）。⑥亚硒酸钠容易导致动物中毒。当蛋鸡日粮中亚硒酸钠的添加水平达到 5 mg/kg 时产生毒性效应[6]，而在高生长速度的肉鸡日粮中达到 1 mg/kg 即可检测到亚硒酸钠的毒性迹象[7]。

与无机硒相比，有机硒可以很好地避免上述问题，在提高家禽精子质量、受精卵孵化率、蛋品质以及肉品质等方面具有诸多优势。

第一节 硒与种公鸡精子质量

种公鸡的最佳硒供给是保证精子质量进而确保繁殖力的重要因素。硒对种公鸡生殖功能的营养重要性与精液中高比例的多不饱和脂肪酸（PUFA）使其易于发生脂质过氧化有关。研究表明，种公鸡精液在体外贮存过程中总脂含量、总磷脂的比例、磷脂酰胆碱、磷脂酰乙醇胺和鞘磷脂的含量均显著降低，同时形态正常的可运动的精子数量明显减少[8]。精液中高比例的花生四烯酸在储存过程中尤其容易发生脂质过氧化，将种公鸡的精液在 20 ℃条件下培养 12 h 后花生四烯酸的含量显著降低，同时丙二醛的含量显著提高[9]，这一试验有力地证明了精液储存过程中 PUFA 的过氧化。同样，火鸡的精液在 37 ℃、外源性 Fe^{2+} 存在条件下孵育导致精液中磷脂酰丝氨酸和磷脂酰乙醇胺含量显著降低[10]。火鸡精液稀释后在 4 ℃下保存 48 h，总磷脂含量降低，其中磷脂酰胆碱含量降低程度最大，鞘磷脂、磷脂酰丝氨酸和磷脂酰肌醇含量降低程度相对较小[11]。

鸡精液中硒的浓度约为 47 ng/g，精子中硒浓度大约是精浆中的 8 倍，日粮有机硒添加可明显提高精液中的硒含量[12]。硒蛋白酶 GPX 在禽类的精浆和精子中均有表达，但存在种属差异，精浆中 GPX 的活性火鸡最高，鸭和鹅最低，精子中 GPX 的活性，鸭和鹅最高，珍珠鸡、火鸡和鸡较低[13]。在硒不足日粮中添加硒可显著提高肝脏、睾丸、精子和精浆中的 GPX 活性[14]，因而保护了精子的过氧化损伤。事实上，当动物处于应激状态时，GPX 对精子氧化损伤的保护作用表现得尤为明显。研究表明，在热应激条件下（鸡舍温度 33～36 ℃），公鸡日粮中添加有机硒使精液中的 GPX 活性提高 1 倍，脂质过氧化显著降低，显著提高了精液的质量（数量和活力）并以剂量依赖的方式降低死精的百分比[15]。而且在热应激条件下当有机硒和维生素 E 配合使用时效果更高，明显提高了精子的数量和活力以及精液中的 GPX 活性，同时降低了精浆中的 TBARS[16]。研究表明，当给公鸡饲喂低硒的基础日粮时，正常精子的百分比仅为 57.9%，精子的异常主要表现为中段弯曲（18.7%）和头部呈螺旋状（15.4%）。当在日粮中添加亚硒酸钠（0.3 mg Se/kg）时，正常精子百分率增加到 89.4%，弯曲中段和螺旋状头部精子的比例分别降低到 6.2% 和 1.8%，而在日粮中添加相同量的有机硒时，精液质量进一步提高，正常精子百分率达到 98.7%[17]。分别给公鸡饲喂 3 种不同硒水平（0.2 mg/kg、0.3 mg/kg 和 0.4 mg/kg）的亚硒酸钠或酵母硒，分别于试验第 1、2 和 3 个月采集精液，尽管该研究没有观察到硒添加形式对精子质量的显著差别，但精子畸形率随着硒添加量的提高显著降低[18]。对饲养 30 周后火鸡精液样本的研究发现，有机硒添加组精液保存 6 h 后活力下降 3.95%，而相同硒剂量（0.3 mg Se/kg）的无机硒添加组精子活力下降达 8.7%，两组火鸡精子的受精能力分别为 88% 和 90.5%[19]。而另一个研究发现，尽管硒添加不影响火鸡精浆的抗氧化能力，但显著提高了精子的数量和浓度[20]。在珍珠鸡日粮中用有机硒等量取代亚硒酸钠（0.3 mg/kg），精液量从 0.084 mL 提高到 0.1 mL，精子密度从 47.8 亿/mL 提高到 57.4 亿/mL，受精率从 84.5% 提高到 89.9%[21]。对雄鹅的研究还发现，日粮硒（0.3 mg/kg）和维生素 E（100 mg/kg）添加可增加人工精液收集时射精反应的频率和缩短射精反应的时间间隔[22]。在生精过程中，硒可能会影响公鸡精原干细胞的数量。研究发现，0.5 mg/kg 的硒添加与对照组相比明显提高了精原干细胞标志物的 mRNA 表达[23]。

日粮硒缺乏对雄性生殖器官有明显的损害作用。内源性和外源性途径以及上游调节因子 p53、Bax 和 Bcl-2 均参与硒缺乏诱导的公鸡睾丸细胞凋亡[24]。鸡日粮缺硒显著降低了睾丸 GSH-Px 活性和 Bcl-2 mRNA 水平。

公鸡日粮中添加 0.3 mg/kg 硒与添加 0.3 mg/kg 硒相比，睾丸中 SelW 和 GPX4 的基因表达显著增加，睾丸生精小管细胞的数量和支持细胞的活力在体外培养时明显提高[25]。如在第二章中所述，GPX4 除了具有抗氧化功能外，还是精子的结构组分，因此，睾丸中 GPX4 基因表达水平的提高有利于保证精子的正常结构。

除了影响精液的质量外，硒还可以影响其受精能力。分别给公鸡饲喂 3 种不同硒水平（0.2 mg/kg、0.3 mg/kg 和 0.4 mg/kg DM）的亚硒酸钠或酵母硒，发现随着硒添加水平的提高，受精率逐步提高，而且有机硒更加有效[18]。有机硒不仅能够提高受精率，而且能够延长可受精时间。维生素 E 和有机硒单独添加以及两者混合添加都能很好地维持老龄鸡群的生育能力[26]。禽类繁殖的一个独特特征是将精子储存在输卵管的储精管内，有机硒对储精管抗氧化能力的维持可能使得母鸡能够在更长的“可育期”内产生可育卵。一些研究也证实了在家禽输卵管的子宫-阴道部分存在抗氧化系统，在子宫-阴道交界处 GPX 活性比肝脏高 12 倍[27]。

综合上述研究结果，硒对保证公鸡生殖器官结构和功能完整以及精子的质量具有重要作用，硒缺乏会导致精子颈部受损甚至断裂（图 3-1）。有机硒由于能够更有效地转运到睾丸和精子，因而在保证精子质量方面作用更强，尤其是在应激期间表现得更加明显。

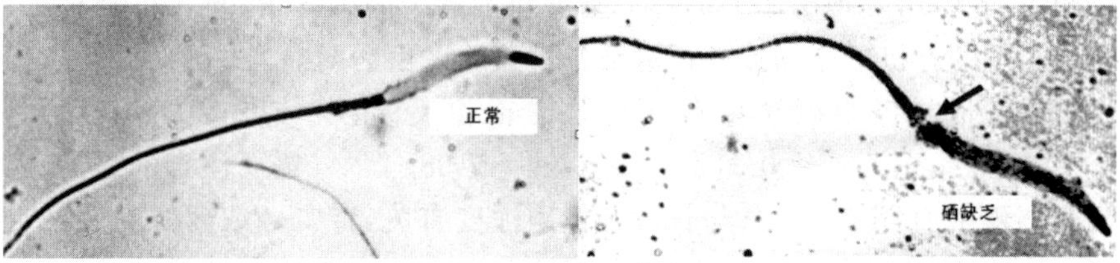

图 3-1　正常和硒缺乏时精子的结构[28]

第二节　硒与种母鸡

保证蛋黄和鸡胚组织中一定的硒含量对于维持胚胎发育过程中的抗氧化能力进而提高孵化率和新生雏鸡的存活率具有重要的意义。

一、硒对种蛋中硒含量的影响

种蛋中的硒浓度取决于母鸡日粮硒供给量和形态。一般而言，硒几乎是等量分布在蛋黄(58%)和蛋清(42%)中的[29]。而 SeMet 约占蛋清中总硒的 53%～71%,约占蛋黄中总硒的 12%～19%[30]。由于 SeMet 可非特异性地取代蛋氨酸而结合到蛋白质中,因而在蛋清中的含量较高。动物自身不能合成 SeMet,必须由植物性饲料提供,饲料中蛋氨酸含量提高会竞争性地降低 SeMet 在鸡蛋中的沉积。硒在蛋黄中的沉积率约为 13%～14%,在蛋清中的沉积率约为 8%～9%[29]。有机硒由于其能被更有效的吸收、转运和沉积,因而与无机硒相比可显著提高蛋黄和蛋清中的硒含量。将 48 只蛋鸡随机分为 3 组,一组饲喂包含 0.18 mg/kg 的基础日粮,另外 2 组分别添加 0.3 mg/kg 的亚硒酸钠和硒酵母,结果表明,硒酵母添加组蛋中的硒含量显著提高[31]。用作物施用硒肥生产的由小麦和野生大麦杂交的谷物作为日粮饲喂蛋鸡,与传统基础日粮添加亚硒酸钠饲喂的蛋鸡相比,其蛋黄中硒和维生素 E 的含量显著增加[32]。在蛋鸡日粮中分别添加硒代蛋氨酸、硒酵母和亚硒酸钠,并分别设计 0.1 mg/kg、0.3 mg/kg 和 0.5 mg/kg 三个硒添加水平,经 8 周的饲养试验发现,硒添加提高了鸡蛋中的硒含量,而且对硒添加表现出剂量依赖关系,最明显的是硒代蛋氨酸,其次是硒酵母,表现最不明显的是亚硒酸钠[33]。在另一项试验中也比较了亚硒酸钠、硒酵母以及 SeMet 的添加效果,亚硒酸钠和硒酵母设计 0.3 mg/kg 一个硒添加水平,SeMet 设计 0.1 mg/kg、0.3 mg/kg、0.5 mg/kg 和 0.7 mg/kg 四个硒添加水平,发现硒添加提高了蛋黄和蛋清中的硒含量,与无机硒相比,相同水平的有机硒添加提高蛋清硒含量的效果更为明显[34]。研究表明,有机硒添加量与蛋中硒含量存在显著的正相关[蛋黄 $y=1.01x+179.4$ ($r^2=0.96$);蛋清 $y=0.68x-24.8$ ($r^2=0.98$)][35]。需要注意的是,有机硒对鸡蛋中硒含量的提高与日粮背景硒浓度有关,背景硒浓度低则效果明显。一些研究表明,SeMet 向鸡蛋中转移硒的效率高于硒酵母,SeMet 类似物 2-羟基-4-甲基亚硒酸向鸡肉[36]或鸡蛋中转移硒的效率也高于硒酵母[37]。OH-SeMet 作为一种新型的有机硒源显示出比硒酵母更高的向鸡蛋中转运硒的效率。一些数据显示,其转运效率比硒酵母高 28.8%[37]。尽管有机硒比无机硒具有更高的小肠吸收率,但一些观点认为,化学结构的不同可能在转运效率差异方面起主导作用。在最近的一项试验中,蛋鸡日粮中 OH-SeMet 添加(0.2 mg/kg)提高了蛋黄中硒的含量[38]。

二、硒对种蛋孵化率以及新生雏鸡的影响

在对 21 周龄哈伯德种鸡(11 600 只母鸡和 1 530 只公鸡)的饲养研究表明,硒酵母添加较亚硒酸钠添加显著提高了种蛋受精率和卵孵化率[39]。

从生产和经济角度来看，硒酵母添加使整个生产期每只母鸡多孵化 5.63 只小鸡，潜在收益每只母鸡增加 1.17 美元。另一项对 22 周龄罗斯种鸡的研究也发现，硒酵母添加较亚硒酸钠添加可提高产蛋率和孵化率。在种蛋孵化过程中胚胎死亡是影响孵化率的重要因素。在对 23 周龄种母鸡所产蛋的孵化研究表明，胚胎死亡率在孵化的第一和第三周死亡率最高，而且随着鸡群年龄的增加胚胎死亡率降低。而对 27 周龄种母鸡所产蛋的孵化研究发现，豆油和鱼油添加组的胚胎死亡率与对照相比显著提高[29]，这与豆油和鱼油添加引发氧化应激有关。在添加鱼油的同时添加硒酵母可显著降低孵化第三周时的胚胎死亡率，提高孵化率和 1 日龄雏鸡的体重[12]。在对俄罗斯白鹅的研究也发现有机硒与无机硒相比可提高孵化率和新生雏鹅的体重。在珍珠鸡日粮中添加适量的硒酵母（0.3 mg/kg），产蛋量提高 22.5%，受精蛋的孵化率从亚硒酸组的 73.3%提高到硒酵母组的 83.1%[21]。一些观点还认为，硒进入卵细胞后可能对发育中的后代产生表观遗传效应。

鸡胚组织的脂质组分中含有高比例的多不饱和脂肪酸，对脂质过氧化非常敏感，因而需要抗氧化保护。母鸡的日粮结构对胚胎发育以及雏鸡出壳后早期抗氧化系统的发育起着决定性的作用。抗氧化剂胡萝卜素、维生素 E 和硒可通过母鸡日粮转移到蛋黄进而转移到胚胎组织。在孵化期间，脂溶性抗氧化剂以及蛋黄中的硒主要在孵化的最后一周转移到胚胎组织，而蛋清中的硒很可能在胚胎发育的前两周就被转移到胚胎中。硒依赖的 GPX 活性在胚胎发育的后半段持续增加，刚孵化出的雏鸡肝脏中的 GPX 活性是孵化前 10 天时的 3 倍[40]。胚胎中抗氧化剂的水平和抗氧化酶的活性与胚胎发育环境暴露于活性氧的程度是成比例的，总体表现为出生后为富氧环境做准备，抗氧化酶的基因表达也遵循这样的规律[41]。在早期神经发育过程中，将抗氧化系统未完善的大鼠胚胎暴露于高氧浓度（20%）导致胚胎中 GSH 浓度显著降低，神经管缺陷发生率显著增加[42]。人、兔、大鼠、小鼠和鸡胚细胞对氧诱导生长抑制的抗性研究表明，鸡胚细胞对氧诱导生长抑制的抗性最强，其抗氧化酶活性最低，GSH 含量最高[43]。随着孵化后鸡组织中主要天然抗氧化剂维生素 E 和类胡萝卜素含量的逐渐下降，硒依赖的抗氧化酶担负着最重要的抗氧化防御功能。从富含硒的种蛋中孵出的小鸡不仅在孵化期 GSH-Px 活性增加，而且也提高了在孵化后雏鸡肝脏的 GSH-Px 活性[35]。此外，由于母体有机硒的补充，孵化后几周雏鸡肌肉中 GSH-Px 活性也增加[44]。在对鸡进行的一项研究中，当硒停止补充时，前期通过有机硒添加的鸡在组织中积累的内源性硒可用于维持 GSH-Px 活性[45]。此外，肉用种鸡日粮添加相同水平（0.15 mg/kg）的亚硒酸钠、硒酵母和 SeMet 的研究发现，与亚硒酸钠相比，硒酵母或 SeMet 添加能显著提高肉用种鸡及其后代肝脏和肾脏的 TrxR1 活性[46]以及肝脏 GPX1 和 TrxR1 的 mRNA 水平，硒酵母或 SeMet 添加还显著提高了 1 日龄雏鸡肝脏 SelP 的浓度和 mRNA 水平[47]。因此，通过给母鸡补饲有机硒增强种蛋以及胚胎组织的 GSH-Px 活性对于雏鸡孵化后最初几周的生存能力具有积极的影响。雏鸡的抗氧化系统的增强也有利于其免疫系统功能的增强。此外，有机硒对维生素 E 的保护作用也为进一步提高雏鸡的抗氧化防御能力提供了可能。

母体硒营养还可能影响到出生后雏鸟的生长发育，同样，日粮有机硒的添加提高了鹌鹑蛋中硒的含量，而蛋中硒含量的提高与刚孵出鹌鹑的肝脏、大脑以及胸部和腿部肌肉中硒含量的增加有关，而且这种差异在孵化后 2 周表现最为明显[48,49]。肉用种鸡的一项研究设计在孵化前 14 天饲喂相同剂量的硒日粮，而在这之前一组饲喂高硒日粮，一组饲喂低硒日粮，发现前期高硒日粮组生产的种蛋孵化出的小鸡比前期低硒日粮组生产的种蛋孵化出的小鸡有更高的组织硒浓度[12]。高硒饲养(0.4 mg/kg)亲本孵化出的雏鸡肝脏、胸肌和全血中的硒含量分别是低硒饲养(0.03 mg/kg)亲本孵化出的雏鸡的 5.4、4.3 和 7.7 倍。当这两组后代保持低硒饲料饲喂时，高硒亲本雏鸡的组织硒浓度在孵化后 3～4 周内仍显著高于低硒亲本雏鸡，而且组织中 GPX 活性也比较高[44]。在一项肉用种鸡有机硒和无机硒(0.3 mg/kg)的对比研究中，有机硒添加提高了 1 日龄雏鸡的体重(41.1 g 比 40.4 g)和身长(18.6 cm 比 18.3 cm)[50]。因此，通过强化母体的硒营养状况是改善胚胎及雏鸡硒营养状况的有效策略。

上述研究提示，硒添加对种蛋孵化率以及新生雏鸡十分有益，这可能与种蛋中尤其是蛋黄中硒含量的提高有直接关系。但种蛋或蛋黄中硒浓度达到多少时可以最大限度地发挥这些优势仍然是一个需要回答的问题。研究表明，当标准的玉米-豆粕型日粮中天然硒含量为 0.1 mg/kg 和 0.17 mg/kg 时，蛋黄中的硒含量分别为 100～200 ng/g 和 300 ng/g(湿重)[35]，饲料原料产地的不同很大程度上影响着种蛋及蛋黄中的硒含量，而饲料原料中的硒含量又很大程度上取决于土壤条件、土壤中的硒含量以及存在形式。与商业生产的肉鸡和鸡蛋相比，野生动物的肉和在完全自然条件下生产的蛋的成分有很大的不同。在多数情况下，农业生产的后果是导致饲料和食品中的硒含量降低，原因包括土壤酸化、含硫和含磷化学合成肥料的施用等。因此，早在 100～200 年前采用自由放养或天然饲料饲养的动物生产出来的肉、蛋和奶要比现在集约化饲养方式下生产出来的动物产品中的硒含量高得多。由于缺乏与野生禽类的对比数据，人们往往倾向于认为对家鸡的研究结果一般也适用于其他禽类。这对于常量营养素如蛋白和脂质而言可能有一定的参考性，但对于鸡蛋中的许多微量营养素而言，则可能具有误导性。在对 14 种野生鸟类（黑嘴鸥、普通秧鸡、小黑背鸥、美国黑鸭、黄头黑鸟、布鲁尔黑鸟、家雀、谷仓燕、树燕、加拿大黑雁、美国乌鸦、画眉鸟、欧歌鸫和欧椋）鸟蛋黄中硒含量的研究发现，测定结果变化范围为 394～2 238 ng/g（湿重），平均值为 1 040 ng/g（湿重），比家鸡蛋黄中的硒含量高出近 10 倍，即使是蛋黄硒水平最低的物种，与家鸡相比也高出近 4 倍[51]。另一个对小白鹭、黑冠夜鹭和燕鸥的研究也发现蛋中硒含量非常高[52]。从这些有限的野生禽类的研究数据可以推测，提高家鸡鸡蛋以及蛋黄中硒含量的空间还很大，由此也可以推测通过提高家鸡蛋黄中硒含量进而提高种蛋孵化率也存在一定的空间。

第三节　硒在商品蛋鸡生产中的应用

一、硒对蛋鸡生产性能的影响

与未添加硒的对照日粮相比，有机硒添加（0.3 mg/kg）可使产蛋量增加 3.27%，饲料消耗减少 2.89%[53]。用有机硒部分或完全替代亚硒酸钠可增加产蛋量和蛋重[54]。将 48 只商品蛋鸡随机分为 3 组，一组饲喂包含 0.18 mg/kg 的基础日粮，另外 2 组分别添加 0.3 mg/kg 的亚硒酸钠和硒酵母，结果表明，硒添加不影响饲料采食量、蛋重比和生产性能，但与对照组相比，硒添加显著提高了蛋重、蛋壳重和蛋壳表面积，而且硒酵母添加组蛋壳破裂强度显著提高[31]。用作物施用硒肥生产的由小麦和野生大麦杂交的谷物作为日粮饲喂蛋鸡，与传统基础日粮添加亚硒酸钠饲喂的鸡相比，尽管其饲料消耗、蛋重、产蛋率无明显差异，但蛋黄中维生素 E 的含量以及肝脏和血浆的硒水平显著增加[32]。在蛋鸡日粮中分别添加硒代蛋氨酸、硒酵母和亚硒酸钠，并分别设计 0.1 mg/kg、0.3 mg/kg 和 0.5 mg/kg 三个硒添加水平，经 8 周的饲养试验发现，与对照相比硒添加提高了血清中的硒含量，但硒源和剂量对血清 GPX 活性无显著影响[33]。在另一项试验中也比较了亚硒酸钠、硒酵母以及 SeMet 的添加效果，亚硒酸钠和硒酵母设计 0.3 mg/kg 一个硒添加水平，SeMet 设计 0.1 mg/kg、0.3 mg/kg、0.5 mg/kg 和 0.7 mg/kg 四个硒添加水平，发现硒源（0.3 mg/kg）不影响蛋鸡的生产性能，但所有硒添加组均提高了蛋鸡的抗氧化防御能力（高的血浆 GPX 和 SOD 活性以及低的血浆 MDA 含量）以及肝脏、血浆、胸肌和腿肌中的硒含量。此外，与以 0.3 mg Se/kg 添加的 SS 相比，有机源（SY 和 SeMet）显示出更大的提高 GSH-Px 活性及腿和胸肌中 Se 含量的能力。此外，与无机硒添加相比，相同水平的有机硒添加提高腿和胸肌硒含量的效果更为明显[34]。在最近的一项试验中，蛋鸡日粮中 OH-SeMet 添加（0.2 mg/kg）不影响饲料采食量，蛋重和产蛋率，但改变了蛋黄中脂肪酸的组成，提高了蛋黄中维生素 E 的含量[38]。

二、硒对蛋壳质量的影响

蛋壳由大约 95％的矿物质和 5％的有机基质组成，有机基质由蛋白纤维构成，基质上沉积有钙质和结晶物。在蛋壳发育过程中，有机基质为钙盐的结晶化提供网格状结构并促进钙结晶化形成栅栏状构造，从而使蛋壳强度增加，不容易破裂。有机硒中的 SeMet 由于可以取代蛋氨酸结合到蛋白质中，因而可能会影响到蛋壳的质量。研究表明，蛋鸡日粮有机硒添加显著增加了蛋壳的破裂强度[55]，并增加了蛋壳的厚度、产蛋量和蛋黄重[56]。蛋鸡日粮中添加的亚硒酸钠 50％被有机硒取代后，蛋壳重量和厚度显著增加[57]。相同硒水平（0.2 mg/kg）的有机硒添加与亚硒酸钠添加相比，鸡蛋的重量和比重、蛋壳的厚度、哈氏单位以及蛋黄和蛋清的重量显著提高[58]。鹌鹑日粮中添加 60％的高硒小麦可使组织中的硒含量提高 3～13 倍，且蛋壳厚度显著增加[59]。在对鸡蛋不同部位硒沉积的研究发现，与对照（硒含量小于 0.1 mg/kg）相比，硒添加（0.5 mg/kg）显著提高了蛋壳、蛋清、卵黄膜、蛋黄和壳膜中的硒含量，分别从对照的 13 ng/g、30 ng/g、66 ng/g、85 ng/g 和 131 ng/g 提高到 77 ng/g、249 ng/g、262 ng/g、516 ng/g 和 854 ng/g[60]。这些数据表明，硒存在于包括壳和膜的蛋的所有部位，而且在壳膜中硒含量最高，与蛋黄中的硒含量相当。一些数据显示鹌鹑蛋蛋壳中的硒含量为 122.5～186.5 ng/g，约占鸡蛋总硒含量的 12％[61]。Golubkina 和 Papazyan（2006）[62]研究发现，在种蛋孵化和胚胎发育过程中，蛋壳（26％）和壳膜（39％）中硒的浓度显著下降。但有趣的是，孵化过程中富含有机硒的蛋壳中的硒浓度仅降低了 13％，而壳膜中硒含量几乎没有变化，这可能是由于蛋黄和蛋清中的硒含量足够用于胚胎发育所致[62]。此外，蛋鸡的骨髓可能在一定程度上沉积硒并用于蛋壳形成。研究发现，适宜的硒供应可使兔子的骨骼保持最大的弹性模量，而硒缺乏或过量均会降低骨骼的生物力学强度[63]。目前尚不完全了解硒对蛋壳和骨骼形成影响的分子机制。

三、硒对鸡蛋新鲜度的影响

鸡蛋新鲜度是决定消费者需求的重要参数之一，在储存过程中，鸡蛋新鲜度不可避免地要降低，蛋壳膜的组成和结构与鸡蛋的新鲜度密切相关，哈氏单位是评价鸡蛋新鲜度的常用指标。研究表明，蛋鸡日粮中亚硒酸钠和硒酵母添加(0.3 mg/kg)均提高了哈氏单位[64]，在包含 0.15 mg/kg 无机硒的基础日粮中添加有机硒改善了蛋黄的颜色和哈氏单位[65]。但也有一些研究并未观察到日粮硒对哈氏单位的影响，这可能与母鸡的年龄、日粮组成和鸡蛋存储条件不同有关。鸡蛋储存过程中脂质和蛋白质的氧化被认为是导致鸡蛋新鲜度降低的主要原因。如第二章所述，硒是一系列硒蛋白如 GPX、TrxR 和其他硒蛋白的组成成分，以 Scc 的形式存在于硒蛋白的活性位点，很多硒蛋白酶具有降低脂质过氧化的作用。因此，日粮有机硒添加引起的鸡蛋中硒含量的提高，可维持鸡蛋中的硒蛋白酶活性，而通过预防或减缓鸡蛋储存过程中脂质过氧化，可维持哈氏单位。另外，蛋白质的持水能力在鸡蛋储存过程中新鲜度的保持也起着重要作用。当蛋白质发生氧化时其持水能力降低[66]，蛋白质中蛋氨酸的反向氧化导致蛋白的结构和功能改变[67]。作为一种硒蛋白，蛋氨酸亚砜还原酶 B(MsrB)可将氧化蛋氨酸还原为活性形式，这可能是预防和修复氧化应激对蛋白质结构和功能改变的关键[68]。研究表明，有机硒可提高鸡蛋中的硒含量，防止鸡蛋储存过程中蛋白质的氧化，因而维持了蛋白质的保水能力[38]。

第四节　硒在肉鸡生产中的应用

硒是肉鸡的必需微量元素，也是各种组织和整个机体抗氧化系统的组成部分。因此，最佳的硒营养状态可为肉鸡提供足够的硒蛋白合成，有助于预防包括氧化应激在内的各种应激因素对肉鸡健康造成的不利影响。大量的研究表明，在肉鸡日粮中补充硒可提高其抗氧化防御能力，其中有机硒的这种积极效应更为明显。总体而言，硒对肉鸡的有益作用包括提高生长速度和饲料转化率，提高抗氧化能力和免疫功能，提高包括肌肉在内的组织中的硒含量，提高肌肉的保水能力和降低肌肉储藏过程中的滴水损失，增强抗病毒作用以及降低肉鸡死亡率。

一、硒对肉鸡抗氧化防御功能的影响

肉鸡的生长速度非常快，最快可达到 60 g/d，因而对各种应激的敏感性比较高，抗氧化防御是支持肉鸡生长和保证肉鸡健康的重要驱动力。研究表明，在孵化后 35 天日粮添加 0.3 mg/kg 面包酵母硒提高了红细胞，血浆和肝脏中的 GPX 活性[69]。Chen 等(2014)[70]研究表明，与亚硒酸钠相比，硒酵母添加提高了肉鸡血清 GPX 和 T-SOD 活性和总抗氧化能力(T-AOC)，降低了血清羟自由基的生成以及 MDA 含量。同样，0.35 mg/kg 的有机硒添加与无机硒相比提高了全血和组织中的 GPX 活性[71]。与未添加 SS 对照和肉仔鸡相比，添加 SeMet 可提高血浆中 GP，T-SOD 和 CAT 活性、GSH 浓度和 T-AOC，降低 MDA 产量[72]。添加 SeMet 也提高了胸肌中 GPX、T-SOD 和 CAT 的活性、金属硫蛋白和 GSH 的含量以及 T-AOC，降低了羰基蛋白的含量。与添加相同剂量的亚硒酸钠相比，添加 SeMet 提高了血浆中 GPX 和 CAT 的活性、GSH 含量以及 T-AOC[72]。同样，与无机硒处理组相比，L-SeMet 处理组的肉鸡血清、肝脏和胸肌中 GSH 浓度显著提高，肝脏中 SOD 活性显著提高，肾脏、胰腺和胸肌的 T-AOC 显著提高，肾脏和胸肌中 MDA 含量显著降低[73]。最近，一项实验研究了添加各种水平(0 μg/kg、100 μg/kg、200 μg/kg、300 μg/kg 或 400 μg/kg)的有机硒对商品肉鸡生长性能、胴体品质、氧化应激和免疫反应的影响，结果表明硒添加降低了血浆脂质过氧化水平，提高了血浆 GPX 和谷胱甘肽还原酶的活性，淋巴细胞增殖率随着日粮中硒水平的增加而线性增加[74]。在硒适宜的日粮中添加有机硒可进一步提高肉鸡的血浆、肝脏、肾脏、心脏和腿肌中的 GPX 活性[75]。在硒添加对其他微量元素的影响研究发现，有机硒添加不仅提高了组织中的硒含量，而且降低了组织中的 Cd 浓度，提高了组织中的 Zn、Cu 和 Fe 浓度[76]。在应激条件下，可能会导致机体的抗氧化功能降低，因而提高对硒的需要量。由于有机硒添加可使肌肉中的硒含量提高，因而可作为应激条件下必不可少的硒储备。Payne 和 Southern[45]的研究表明，日粮中补充有机硒而在鸡肌肉中积累的硒可用于日粮硒摄入降低或补充终止后 GPX 活性的维持。Zhao 等[77]的研究发现，有机硒源 OH-SeMet 具有更高的诱导肉鸡组织中 SelS 和 MsrB mRNA 的早期表达和 TrxR 活性以及增强 GPX4、SelP 和 SelU 的蛋白生成。

二、硒对肉鸡生长发育的影响

日粮中添加硒对肉鸡的生长、发育和健康的积极影响可能与多种硒蛋白的抗氧化作用、甲状腺激素的活化以及肠道健康和免疫力的改善有关。在病原挑战时期，许多营养物质需要从生长和发育中重新定向用于免疫系统的维持或激活。而硒的免疫优化作用可以节省因不必要的免疫系统激活而导致的营养物质浪费，从而使这些物质更多地被用于生长和发育。硒的抗病毒作用，包括防止呼肠孤病毒引起的肠道完整性受损，可能有助于改善饲料养分利用效率[78]。研究表明，肉鸡日粮中硒的添加改善了肉鸡增重[79]、饲料消耗，提高了饲料利用效率[71,80,81]及十二指肠相对质量[81]，降低了雏鸡的死亡率[82]。

大量研究证据表明，有机硒对上述关于肉鸡生长发育的有益作用明显优于无机硒。与添加无机硒相比，矿物质预混料中添加 SeMet 提高了鸡的体重[79]。在基础日粮硒水平为 0.15 mg/kg 的基础上添加 0.15 mg/kg 的亚硒酸钠和硒酵母，与无机硒添加相比，有机硒添加使增重提高 4.2%，饲料效率提高 9.8%，死亡率从亚硒酸钠添加组的 6.7% 降低到硒酵母添加组的 0.84%[83]。在孵化后前两周日粮添加有机硒或亚硒酸钠对雏鸡发育的影响研究表明，硒酵母添加改善了饲料消耗和饲料转化率，提高了十二指肠相对质量[81]。对 8 400 只雄性肉仔鸡的研究表明，日粮 0.3 mg/kg 的亚硒酸钠用硒酵母部分替代(0.1 mg/kg)提高了日增重和饲料转化率[84]。与日粮中添加无机硒相比，鸡日粮中添加有机硒可提高 T4 转化为 T3 的效率[85]。

在商业肉鸡生产中，应激条件是降低生产性能的常见因素，在应激条件下增加日粮抗氧化剂的添加有助于减缓有应激导致的不利影响。研究表明，在免疫应激条件下(大肠杆菌菌挑战)，硒添加组与对照组相比增加了鸡的体重(2 291 g 比 2 004 g)，改善了饲料转化率(1.84 比 1.98)，降低了死亡率(5.7%比 18.3%)，而且有机硒添加与无机硒添加相比，降低了死亡率，改善了增重[86]。一项研究设计三组日粮，一组为标准日粮组，二组用腐败油替代正常油脂(1%替代)，三组用霉变玉米替代正常玉米(20%替代)，三组日粮分别添加 0.3 mg/kg 的亚硒酸钠和硒酵母，研究发现，硒酵母添加分别使鸡的增重提高 39 g、58 g 和 68 g，而亚硒酸钠添加不具有这样的效应[87]。在热应激条件下，日粮中添加硒可提高平均日采食量、肝脏和胸肌中硒浓度以及肝脏 GPX 活性[88]，日粮中无机硒被硒酵母替代可提高血液 GPX 活性，降低热休克蛋白水平[89]。当肉鸡受到冷应激并饲喂受黄曲霉毒素污染的饲料时，有机硒与维生素 E 混合使用可降低腹水发生率[90]和肺动脉高压综合征发生率[91]以及由此引起的肉鸡死亡率。同样，当肉鸡受到热应激时，有机硒与维生素 E 混合使用提高了骨骼肌中 GPX 何 SOD 的活性，降低了骨骼肌的脂质过氧化水平[92]。

三、硒对肉鸡血液和组织硒浓度的影响

NRC 推荐的日粮硒水平为 0.15 mg/kg(DM 基础)。依据生化和分子生物标记物评估硒需要量已成必然。如果按照肝脏和血浆中的酶活性评价,目前肉鸡的最低硒需求量为 0.15 mg/kg;如果按照胰腺数据来评价,应将硒需求量提高到 0.2 mg/kg[93]。对硒源、硒补充时间、研究的年份、鸡的种类、数量、年龄和公母、分析方法、组织类型多种因素对组织硒积累的综合分析表明,鸡的类型、硒的来源、组织的类型以及所使用的分析方法等多种因素都会影响鸡组织中的硒含量,血浆中硒含量与肝脏($r=0.845$)和肾脏($r=0.953$)硒含量呈正相关,全血中硒含量与肝脏($r=0.929$)、肾脏($r=0.802$)和蛋黄($r=0.999$)硒含量呈正相关,肝脏硒含量与胸肌($r=0.790$)、腿肌($r=0.938$)和肾脏($r=0.988$)硒含量呈正相关,进一步对不同组织中硒的积累速率的分析表明,积累速率较高的是腿肌($b=0.727$),肝脏($b=0.570$)和肾脏($b=0.499$),而全血($b=0.326$)和血浆($b=0.106$)中硒的积累速率较低或几乎不发生积累[94]。对 810 只肉用公鸡进行了硒添加研究,对照不添加硒,另外两组分别添加硒酵母和富硒藻类(0.3 mg Se/kg),结果发现,42 日龄时胸肌硒浓度从对照组的 52.1 μg/g 增加到硒酵母组的 217.4 μg/g,腿肌中硒含量对照组的 71 μg/g 增加到硒酵母组的 249.7 μg/g[95]。在 21 天的肉鸡饲养试验中,添加不同水平的三种硒源(亚硒酸盐,硒酵母和 OH-SeMet)的研究发现,与对照组(0 添加)相比,不同硒源和添加水平均提高了肌肉硒含量,提高程度由低到高依次为亚硒酸盐、硒酵母和 OH-SeMet,添加亚硒酸盐、硒酵母和 OH-SeMet 的总硒消化率分别为 24%、46%和 49%,有机和无机硒源相比存在显著差异[96]。另一项关于三种硒源对肉鸡组织硒含量的影响研究发现,与对照(0 添加)和亚硒酸盐添加饲养 21 天后的胸肌硒含量显著降低,而两种有机硒源添加 21 天维持或提高了胸肌硒含量。与无机硒和硒酵母相比,OH-SeMet 表现出独一无二的在组织中沉积 SeMet 和总硒的能力[97]。

四、硒对鸡肉品质的影响

硒对鸡肉品质的影响和肌肉组织能否建立硒储备有着密切的关系。只有 SeMet 能够非特异性地取代蛋氨酸而结合到肌肉蛋白当中,结合的多少取决于肌肉蛋白中蛋氨酸的含量以及日粮提供的 SeMet 水平。显然无机形式的硒不具有这样的作用。尽管 SeMet 形式的硒是唯一能够在肌肉

蛋白中沉积的硒形式，但应该注意的是，只有一小部分蛋氨酸池可以被 SeMet 替代，蛋白质的周转阻止了 SeMet 在生物体内积累到毒性水平[98]。事实上，在肌肉中蛋氨酸的含量总是远远超过 SeMet 的含量，人的骨骼肌中的 SeMet/Met 比率约为 1∶7 000，鸡胸肌中二者的比例小于 1∶6 000[99]。

鸡肉在储存过程中会因脂质过氧化而产生异味，是影响其品质的一个因素。肌肉的脂质过氧化通常用 TBARS 指标评价。研究表明，日粮添加 0.1～0.3 mg/kg 的硒酵母可降低冷冻保存 16 周后胸肌和腿肌中的 TBARS，和添加 100 IU/kg 维生素 E 具有同等的抗氧化效应[100]。日粮添加 0.3 mg/kg SeMet 使冷冻鸡肉的脂质过氧化水平（TBARS）降低[101]。与添加相同水平的无机硒相比，有机硒添加降低了冷冻[73]和 3～5 ℃下储存 5 天[102]鸡胸肉中的 MDA 水平。

鸡肉在储存过程中会因蛋白质氧化而产生异味，同时蛋白质的氧化会导致其持水能力降低，是影响其品质的另一个因素。滴水损失是评价蛋白质氧化程度的常用指标。早在 1996 年，Edens 就发现日粮有机硒添加可降低鸡肉的滴水损失。后来大量的研究证实了有机硒的这种效应。Naylor 等（2000）[80]对肉用公鸡的研究表明，日粮有机硒添加降低了鸡肉的滴水损失。Periae 等（2007）[50]肉用种鸡的研究表明，日粮硒酵母添加不影响储存 24 h 和 48 h 后鸡肉的 pH，但降低了滴水损失。Wang 等（2011）[99]对肉鸡的研究表明，与无机硒相比，添加相同水平的有机硒降低了储存 24 h 和 48 h 后鸡肉的滴水损失（4.33 h 与 5.22 h）。硒蛋白 MsrB 可防止蛋白质发生氧化，有机硒提高了鸡肉中的硒含量，使肌肉中的 MsrB 活性增加，导致蛋白质氧化减少[77]，这可能是滴水损失降低的主要原因。

第五节 富硒鸡蛋和肉

在过去几年中，饮食与人类健康之间的关系受到了广泛关注，人们意识到饮食不均衡会导致严重的健康问题。并非所有人都食用相同的食物，人们以多种方式满足其营养需求。在我们饮食中常见的食品成分中，天然抗氧化剂被认为特别重要。众所周知，在正常生理条件下和应激条件下产生的自由基都可能对体内的生物大分子产生破坏作用。抗氧化保护对于预防或大幅减少自由基及其代谢产物所造成的损害至关重要。食物为人类提供了天然抗氧化剂的主要来源，包括维生素 E、类胡萝卜素、类黄酮和硒。临床研究表明，饮食中补充有机硒可将癌症死亡率降低 2 倍，硒摄入不足与人类的许多疾病的发生有关，包括遗传缺陷、生育力下降、心血管疾病、神经退行性疾病。硒摄入不足可导致机体对各种病毒和细菌的防御能力降低。有几种方法可以提高人的每日硒摄入量，主要包括饮食中直接补充硒制剂、土壤中施用硒肥、补充用富硒酵母生产的面包以及生产富硒动物食品（肉、蛋、奶）。过去几年进行的各种研究得出的结果表明，通过在动物日粮中补充硒以生产富含硒动物食品可以成为提高人们硒摄入量的有效途径。

一、富硒鸡蛋

生产富含硒鸡蛋需要考虑以下几个方面的内容：①日粮硒向鸡蛋中转运的效率；②日粮中硒的存在形式；③有机硒的可获取性；④是否对蛋鸡产生毒性作用；⑤与蛋中其他营养素吸收的相互作用；⑥鸡蛋烹饪过程中硒的稳定性；⑦对鸡蛋外观和风味的影响；⑧生产富硒鸡蛋额外增加的成本与效益之间的平衡；⑨人们对富硒鸡蛋的接受程度和消费能力；⑩专利申请以防止技术复制。

目前全球有超过 25 个国家和地区生产富硒蛋，俄罗斯在这方面技术相对比较先进和成熟，产品种类比较多，一般一颗鸡蛋可为人类提供 20～30 μg 硒，有的富硒蛋还同时富含维生素 E。鸡蛋是大多数国家的传统且价格合理的食品，各个年龄段的人需要或多或少定期且适度地食用。食用富硒鸡蛋如果要达到硒的中毒剂量需要每天消耗超过 25 个，这种情况不太可能发生，因此富硒鸡蛋安全性很高。

各个国家对富硒鸡蛋生产的观察表明，富硒鸡蛋生产涉及的成本通常不超过饲料总成本的 2%～5%，由于使用有机硒来提高鸡蛋中的硒含量，因而蛋重、蛋壳质量、内部蛋品质（Hough Unit）也得到改善，提高了市场销售潜力，在已经存在的改良鸡蛋（omega-3，富含维生素 E，富含碘）中进一步添加硒可以进一步提高其质量和销售潜力，而价格不会大幅上涨。

二、富硒肉

各种肉类是人体营养中硒的重要自然来源。2003 年世界猪肉产量达到 9 580 万 t，禽肉产量达到 7 520 万 t，牛肉产量为 6 190 万 t。尽管肉是硒的良好来源。但是，肉中硒浓度因国家/地区和所用硒补充剂的不同而有很大差异。英国、澳大利亚和美国生产的猪肉中硒含量分别为 14 μg/100 g、9.4～20.5 μg/100 g 和 14.4～45.0 μg/100 g[103]，瑞典生产的猪肉中硒含量为 11.3 μg/100 g[104]。如前所述，亚硒酸盐或硒酸盐不能有效提高肉中硒浓度，动物肌肉中硒主要以SeMet形式存在，而动物机体自身不能合成SeMet，必须由日粮提供。肉中的硒91%以硒代氨基酸形式存在，其中SeMet会占到总硒的60%，高剂量的有机硒添加可使鸡胸和鸡腿肉中95%以上的硒以SeMet形式存在[105]。

相对于富硒鸡蛋，富硒肉基本未实现商业化生产，但有一定的潜力。

第六节 硒与家禽免疫

在商业家禽生产过程中，免疫应激的发生不可避免，这会导致动物的生产性能和生殖能力下降。在这种情况下，一些常量和微量元素的免疫调节特性对家禽的健康生产非常重要，事实上，日粮中所有的营养素在维持家禽最佳免疫功能方面都起着非常重要的作用，摄入不足或过量都会对机体的免疫功能产生负面影响。一位演奏家手里拿着一件知名品牌的乐器可以演奏出优美的音乐，但是在调音之前发出来的却是噪音，这正是免疫系统反应过度时表现出的状况。例如过敏就是一种免疫反应过度现象，免疫反应过度不仅不能很好地发挥免疫功能，过度的免疫反应还会额外增加对营养物质的消耗。另一方面，免疫反应不足或抑制也是一个问题，因为它不能充分保护机体免受细菌、病毒等病原对家禽机体造成的侵害。因此，免疫系统必须被调整到最佳状态才能在不对家禽生产和生殖造成大的影响的基础上发挥其最大的保护作用。已有的大量证据表明，硒是主要的免疫调节剂之一，机体内的免疫器官（胸腺、淋巴结、脾脏）以及免疫细胞（异嗜性粒细胞、巨噬细胞、自然杀伤细胞、树突状细胞、肥大细胞、淋巴细胞）中均含有硒，当硒供给不足时，会导致免疫功能降低，病原感染的风险增加。

一、先天免疫系统

先天免疫系统主要包括由皮肤和胃肠道黏膜构成的物理屏障、特定的分子（凝集素、防御素、急性期蛋白、溶菌酶）和先天免疫细胞（异嗜性粒细胞、巨噬细胞、自然杀伤细胞、树突状细胞、肥大细胞），这些组织、分子和细胞构成了机体防御病原入侵的第一道防线。

在病原入侵的情况下，病原相关的分子模式（PAMPs）被先天免疫细胞的模式识别受体（PRRs）识别（图3-1），吞噬细胞（异嗜性粒细胞、巨噬细胞、树突状细胞）通过氧化爆发释放ROS和RNS以及通过脱颗粒作用释放杀菌肽(杀菌素、防御素)杀死入侵病原。吞噬细胞也会加工抗原并将抗原呈递给获得性免疫系统的细胞。此外,补体系统的激活也会参与病原体的破坏和清除,肥大细胞释放的组胺和肝素也有助于病原的清除。研究较多的 PRRs 是 Toll 样受体(TLR),鸡的基因组至少编码 TLR15 和 ILR21 两种特异性受体[106],这与哺乳动物有很大区别,和鸡体内没有淋巴结有关。

与哺乳动物的嗜中性粒细胞相似,家禽的异嗜性粒细胞具有很强的吞噬能力。事实上,异嗜性粒细胞是首先被招募到感染部位的免疫细胞,在先天免疫和随后的炎症反应中起重要作用。异嗜性粒细胞可以表达在鸡中发现的部分 TLR,主要通过氧化爆发和脱颗粒杀死入侵微生物。和哺乳动物的嗜中性粒细胞相似,鸡的异嗜性粒细胞也可通过形成胞外陷阱来杀死病原微生物[107]。但是鸡的异嗜性粒细胞不表达主要组织相容性复合物Ⅰ和Ⅱ[106]。

鸡的巨噬细胞在机体防御微生物感染中起着核心作用,巨噬细胞装配有一系列受体,包括捕获受体、补体受体、Fc 受体、C 型凝集素受体和甘露糖受体。吞噬过程依次经历 5 个阶段:①在趋化信号作用下巨噬细胞向微生物入侵部位迁移;②微生物与巨噬细胞发生物理接触并以非特异性或受体介导的结合方式附着于吞噬细胞表面;③微生物的内吞作用(吞噬);④吞噬体与溶酶体融合;⑤氧化剂(超氧化物和羟自由基,过氧化氢,一氧化氮,次氯酸等)杀死微生物。研究表明,鸡巨噬细胞在 15 分钟内即可杀死 80%的沙门氏菌[108]。

先天免疫反应在没有病原入侵的情况下也可以启动。机体自身细胞在应激、损伤和坏死情况下会释放多种具有免疫调节活性的细胞内分子,被称为报警因子或损伤相关的分子模式(DAMPs),先天免疫细胞通过识别 DAMPs 启动免疫反应。

先天免疫反应的目的是为清除入侵的病原微生物或清除坏死的细胞以及修复损伤,这一过程中除了生成杀菌物质氧化剂外,还会生成一些炎性介质(类花生酸如氧脂、炎性细胞因子如 IL-1、IL-6 和 TNF-(等))。这些包括氧化剂在内的物质的过量或不恰当的生成可造成一系列有害影响,包括杀死宿主细胞、通过氧化降解细胞的必要成分直接损伤细胞和组织以及通过改变组织间液中蛋白酶/抗蛋白酶的平衡间接损伤细胞,还可以引起细胞突变和姐妹染色单体交换。例如,在疾病挑战情况下,可能会发生 ROS 的过量生成,它们可以从溶酶体泄漏到周围的胞质和细胞内空间,当胞质抗氧化能力低下时(如缺硒),可能会损害细胞的杀菌能力和新陈代谢。研究表明,巨噬细胞抗氧化酶活性较低时,其功能会受到干扰,体外和体内加硒均可上调巨噬细胞的 TrxR 和 GPX 活性[109]。最近的研究结果显示，TrxR1 在巨噬细胞基因表达网络中既是一个调节因子，又是一个调节靶点，提示硒代谢和免疫信号之间存在联系[110]。

NK 细胞最初被认为是一类具有溶解肿瘤细胞和病毒感染细胞能力的颗粒淋巴细胞。与典型淋巴细胞不同，NK 细胞体积较大，胞浆较多，具有电子致密颗粒。NK 细胞的杀伤机制是通过释放颗粒内容物（穿孔素和颗粒酶）到感染细胞表面来介导的，还可以通过识别和诱导抗体包裹的靶细胞的裂解来发挥其细胞毒性，并且可以被一些细胞因子非特异性地激活。NK 细胞不仅能够识别和清除转化的或病毒感染的细胞，还能够识别和响应细菌感染的细胞，因为 NK 细胞也有包括 TLRs 在内的能够感受和响应 PAMPs 的受体，TLR 激活后，NK 细胞是 IFN-γ 和粒细胞巨噬细胞集落刺激因子的重要来源[111]。

一旦病原体进入宿主，最初的非特异性反应就是炎症反应。在炎症反应期间，先天免疫细胞释放信号，将远处的免疫细胞招募到感染部位，并改变机体的代谢活动。通过提高代谢率（发热）、减少饲料摄入量、启动骨骼肌优先分解来支持肝脏糖异生和急性期蛋白的合成，创造一种不适宜入侵病原体生存的环境[112]。因此，炎症反应导致一系列行为、免疫学、血管和代谢反应的改变，表现为嗜睡、食欲减慢、生长速度减慢和肌肉蛋白质降解增加，甚至可能会导致生产力降低，发病率和死亡率提高。炎症反应期间急性期蛋白质合成增加，参与宿主保护的各个方面并有助于稳态的恢复。在禽类中已经鉴定出许多急性期蛋白，包括金属硫蛋白、转铁蛋白、α_1-酸性糖蛋白、铜蓝蛋白、纤连蛋白、血清淀粉样蛋白 A、甘露聚糖结合蛋白和 C 反应蛋白等[112]。

二、获得性免疫系统

获得性或特异性免疫系统包括细胞免疫和体液免疫，T 淋巴细胞介导细胞免疫，B 淋巴细胞通过释放抗体到血液中介导体液免疫，负责特异性地识别和清除各种抗原（图 3-2）。在鸟类中，T 和 B 细胞前体都起源于骨髓，T 细胞的发育发生在胸腺，B 细胞的发育发生在法氏囊。T 细胞、B 细胞以及抗原提呈细胞之间的相互作用，负责特异性免疫的发展。与先天免疫相比，特异性免疫需要较长的时间，对抗原有很高的特异性，并且具有记忆性。

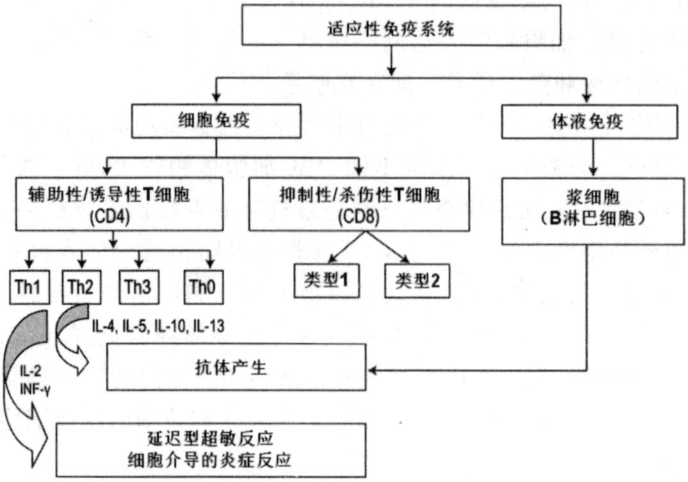

图 3-2　获得性免疫系统

先天和获得性免疫系统相当于人的两个手臂，作用是互补的，以相互协作的方式运作，通过细胞直接接触和涉及细胞因子和趋化因子等化学介质的相互作用共同发挥作用，为宿主提供有效的保护（图 3-3）。一方面，抗原被先天免疫细胞识别并呈递给获得性免疫系统的细胞，另一方面，获得性免疫系统的细胞（T 辅助细胞）调节吞噬细胞（巨噬细胞和异嗜性细胞）的杀菌活性。事实上，先天免疫系统对新的病原体做出快速反应，但这种保护反应在代谢和生理上对宿主来说可能代价比较高。获得性免疫系统可以在对宿主影响最小的情况下，以更具病原特异性的方式做出反应。

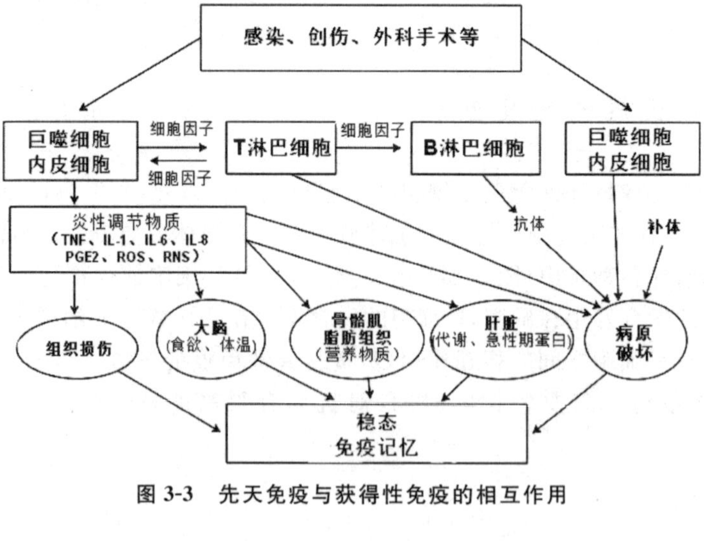

图 3-3　先天免疫与获得性免疫的相互作用

三、硒对家禽免疫功能的调节作用

动物免疫系统在现代畜牧生产系统中的地位和作用是难以估量的。饲料级抗生素的禁用使免疫功能成为决定动物生产效率的主要因素。通过营养调控免疫系统已成为现代畜牧生产系统中提高动物免疫力的重要手段。在调节动物免疫功能中发挥重要作用的营养素中，抗氧化剂硒一直以来备受关注。众所周知，动物硒缺乏会导致机体的抗氧化功能降低，因而细胞膜发生氧化损伤，而细胞膜的完整性对于细胞接收、响应和协调免疫反应所需的信息是非常重要的。因此，宿主的抗氧化状态是保证免疫系统最佳发挥的关键。

（一）硒对吞噬细胞功能的影响

如前所述，吞噬细胞在先天免疫中起着至关重要的作用。鸡日粮硒或维生素 E 缺乏导致腹腔巨噬细胞数量下降，对红细胞的吞噬功能降低[113]。鸡日粮硒缺乏降低了中性粒细胞和巨噬细胞对酵母和细菌的细胞内杀伤能力[114]。与硒适宜粒细胞相比，缺硒粒细胞中 GPX 活性降低，不能有效代谢 H_2O_2，而 H_2O_2 会对粒细胞中 NADPH 依赖的超氧阴离子生成系统造成损害，导致其活性丧失。抗氧化酶活性降低时，巨噬细胞的吞噬功能也会受到影响。硫氧还蛋白已被确认为一种淋巴细胞生长因子，在巨噬细胞与淋巴细胞之间的通讯方面可能起着关键的作用[109]。

一系列体外研究表明，硒添加增强了中性粒细胞的趋化性和随机迁移以及超氧化物的生成[115]，增强了巨噬细胞的吞噬作用，提高了 TNF、IL-1 和 IL-6 的生成[116]。但硒添加水平过高时，可导致细胞硒蛋白酶活性降低，杀菌活性降低，甚至导致细胞死亡。研究表明，每天 182 ng 的硒摄入降低了人中性粒细胞的吞噬功能，大于 5 μM 的亚硒酸钠添加导致 J774.2 巨噬细胞死亡[117]。硒可以影响中性粒细胞中 NO 水平和热休克蛋白（HSP）的基因表达水平，硒缺乏上调了鸡嗜异性粒细胞 HSP 和诱导型一氧化氮合酶（iNOS）的 mRNA 水平，并提高了 NO 水平[118]。硒在炎症过程中可促进抗炎和促炎氧脂生成的平衡，硒缺乏导致的硒蛋白酶活性的降低可提高环氧合酶（COX）和脂氧合酶（LOX）的基因表达，降低抗炎氧脂如脂氧素 A_4（LXA_4）和 9-氧代十八碳二烯酸（9-oxoODE）的生成[119]。硒处理增强了巨噬细胞中 PGE2 的水解，因而减缓了炎症[120]。硒还具有促进鸡树突状细胞的分化和成熟的作用[121]。

(二)硒对抗体生成的影响

早在 1972 年,人们就认识到硒可以增强体液免疫,刺激 IgM 和 IgG 的生成。缺硒会影响肠道黏膜免疫功能,减少十二指肠黏膜分泌型 IgA 的水平,增加促炎细胞因子如 IL-6 和 IFN-γ 的水平[122]。对小鼠的研究表明,硒可以增强对绵羊红细胞抗原刺激的初次免疫应答,而且在抗原刺激前以及刺激的同时使用硒时,初次免疫应答的增强作用最大[123]。对鸡的研究发现,当从孵化到孵化后 2 周给予硒缺乏或维生素 E 缺乏的日粮时,会引起抗 SRBC 的抗体滴度下降[124]。新城疫病毒挑战经常被用来评价各种抗氧化剂的免疫刺激效应,硒和维生素 E 能显著提高新城疫活疫苗免疫雏鸡的免疫应答,免疫前 14 天分别按 0.25 mg/kg 和 300 mg/kg 的比例添加到日粮,可显著提高血凝抑制效价和对新城疫强毒攻击的保护率[125]。白来航鸡日粮添加 0.1 mg/kg 和 0.8 mg/kg 硒,抗绵羊红细胞的抗体效价提高 77%,而且阻止了冷冻导致的抗体效价的降低[126]。日粮硒添加也提高了对沙门氏菌、霉菌毒素以及传染性法氏囊病的体液免疫应答。在抗体生成的效果上,有机硒的使用要优于无机硒。

(三)硒对淋巴细胞功能的影响

免疫器官的生长和发育状况直接影响着淋巴细胞的数量和功能。鸡在生长发育过程中缺乏硒会影响法氏囊的生长,如果硒和维生素 E 同时缺乏还会抑制胸腺生长。硒和维生素 E 缺乏时法氏囊和胸腺中淋巴细胞的数量减少,引起的法氏囊组织病理学变化在孵化后 10～14 天就可以检测到,表现为上皮退化和伴随的淋巴细胞枯竭。研究表明,饲喂低硒日粮的雏鸡表现出与免疫器官损伤相关的氧化应激,血清 IL-1β、IL-2 和 TNF 含量降低,胸腺、法氏囊和脾脏中的硒含量、总抗氧化能力、SOD 和 GPX 活性显著降低,黄嘌呤氧化酶活性和 MDA 含量显著提高,而且免疫组织出现病理损伤和 DNA 损伤[127],脾脏重量和相对重量显著减轻,脾细胞凋亡百分比显著增加,脾脏病变的特点是淋巴细胞耗尽和红髓充血[128]。低硒可能通过是阻滞细胞周期和降低 IL-2 的生成来抑制法氏囊和胸腺的发育[129]。低硒对免疫组织凋亡的诱导作用可能主要是由氧化应激引起的,低硒摄入雏鸡免疫器官(法氏囊、胸腺和脾脏)中自由基和 NO 含量及 iNOS 活性显著提高,同时 Fas 和 caspase-3 的 mRNA 表达显著增加[130]。SelW、Sel15 和 SelT 通过调节炎症相关基因,在保护家禽免疫器官免受炎症损伤中起着重要作用。SelW 在禽类的免疫器官中广泛表达,低硒日粮显著降低了鸡脾脏、胸腺和法氏囊等免疫器官中 SelW 的 mRNA 表达,并诱导了鸡免疫器官的促炎信号(COX-2、iNOS、NF-κB、前列腺素 E 合成酶和 TNF 的 mRNA 表达提高)[131]。硒缺乏导致免疫器官 Sel15 和 SelT 的水平降低，并且诱导了免疫器官的氧化应激[132,133]。

硒具有维持淋巴细胞增殖能力的重要功能。日粮或体外补充亚硒酸钠均能显著增强 C57B1/6J 脾淋巴细胞对有丝分裂原刺激的增殖反应，而缺硒则出现相反的反应[134]。鸡日粮补充硒可显著提高外周血淋巴细胞对植物凝集素的增殖反应[135]。硒对淋巴细胞增殖能力的影响与硒改变活化淋巴细胞表面高亲和力 IL-2 受体 IL-2R(IL-2R)的表达动力学有关。日粮或体外补硒可显著上调 ConA 刺激的小鼠淋巴细胞表面 p55 和 p70/75 IL-2 结合位点的表达[136]。硒或维生素 E 缺乏时淋巴细胞对抗原的增殖反应降低可能主要是由于脂质过氧化和膜结构(膜受体)损伤引起细胞间信号传导受阻导致的。研究表明，淋巴细胞培养液中加硒可降低 ConA 刺激的淋巴细胞的脂质过氧化反应，淋巴细胞中 LPO 水平与刺激的增殖反应呈负相关，SOD 具有促进淋巴细胞增殖的作用，而且表现出剂量依赖关系[137]。由于 T 淋巴细胞中多不饱和脂肪酸的含量较高，因而与 B 淋巴细胞相比更容易受到硒和维生素 E 缺乏的影响。

(四)应激条件下硒的免疫保护作用

已知的硒蛋白中有一半以上参与细胞信号转导、维持细胞氧化还原平衡和抗氧化防御，因此在各种应激条件下，硒对免疫的保护作用最为明显。

霉菌毒素具有免疫抑制作用，在各种霉菌毒素中，毒性和免疫抑制作用最强的是黄曲霉毒素。硒可降低黄曲霉毒素 B1(AFB1)诱导的鸡胸腺细胞的凋亡，减缓 AFB1 诱导的胸腺细胞 caspase-3 和 Bax 表达的提高和 Bcl-2 表达的降低[138]，阻止 AFB1 引起的外周血 T 细胞亚群百分比和血清 IL-2 和 IFN-γ 含量的下降[139]。日粮补硒减轻了 AFB1 引起的法氏囊组织病理学损伤，降低了法氏囊细胞的凋亡，减缓了 AFB1 诱导的法氏囊细胞 caspase-3 和 Bax 表达的提高和 Bcl-2 表达的降低[140]。AFB1 可降低回肠黏膜的体液免疫功能，导致 $CD3^+$、$CD4^+$、$CD3^+CD4^+$ 和 $CD3^+CD8^+$ 上皮内淋巴细胞和固有层淋巴细胞的百分率、$CD4^+/CD8^+$ 的比值以及 IL-2、IL-6 和 TNF-α 的 mRNA 水平明显降低[141]。硒添加可保护回肠黏膜的体液免疫功能不受 AFB1 诱导的损伤，与 AFB1 组相比，AFB1＋Se 组肉鸡 IgA(＋)细胞数、IgA、IgG 和 IgM 含量显著升高[142]。硒还可以通过增加脾脏相对重量和脾脏 T 细胞亚群百分比，减轻脾脏组织病理损伤，改善 AFB_1 所致的细胞免疫功能的降低[143]。线粒体肿胀分析显示，服用黄曲霉毒素 B1 14 天和 21 天的雏鸭线粒体通透性转换孔的开放程度增加，硒可显著减轻黄曲霉毒素 B1 的这些不利影响。此外，硒还具有减轻 AFB1 导致的雏鸭线粒体膜通透性转换孔开放程度的增加[144]。对小鼠的研究表明，硒可以对抗 T-2 毒素对 T 淋巴细胞的免疫毒性作用[145]。在仔猪脾淋巴细胞中，硒可通过提高 GPX 活性来减轻脱氧雪腐镰刀菌烯醇(DON)对抗氧化酶的损伤[146]。

越来越多的证据表明，热应激会导致禽类的生产性能降低，抗氧化功能降低，免疫功能受损。热应激条件下家禽的异嗜性粒细胞与淋巴细胞的比率改变，抗体产生减少，吞噬能力降低[147]。肉鸡在热应激条件下添加硒可提高腹腔渗出细胞(AEC)的数量、AEC 中巨噬细胞的百分比以及内化的调理和未调理的 SRBC，增加了抗 SRBC 抗体的产生[148]。热应激条件下肉鸡肝脏和淋巴器官的重量、IgM 和 IgG 含量以及抗 SRBC 抗体的效价显著降低，而硒添加表现出明显的保护作用[149]。在热带气候条件下，肉鸡日粮添加硒提高了血浆 GPX 和谷胱甘肽还原酶活性，降低了脂质过氧化水平，增强了细胞介导的免疫功能(淋巴细胞增殖率)[150]。日粮补硒可提高热应激雏鸡的平均日采食量、肝脏和胸肌硒浓度、肝脏 GPX 活性以及血清抗 H5N1 抗体效价[151]。HS 可诱导脾脏 TNF-α、IL-4、HSP27、HSP70、MDA 水平的升高和 TNF-γ、IL-2、GPX 和 SOD 水平的降低，而硒添加改善了上述这些变化[152]。

重金属污染是家禽生产中需要关注的一个重要因素。镉是一种有毒的重金属，可诱导氧化应激，并具有免疫抑制作用。镉诱导了 NO 介导的免疫器官凋亡，日粮添加硒可降低鸡免疫器官中 iNOS 的 mRNA 水平和活性、NO 的产生和免疫细胞的凋亡，减轻了免疫器官的超微结构损伤[153]。镉处理组鸡的脾淋巴细胞中 MDA 和 ROS 水平提高，CAT、GPX、SOD 活性降低，线粒体跨膜电位降低。与镉处理组相比，补硒显著提高了抗氧化酶的活性，降低了 MDA 和 ROS 的水平[154]。与对照组和单纯硒处理组相比，镉处理组鸡脾淋巴细胞 IL-2、IL-4、IL-10、IL-17 和干扰素-γ 的表达水平显著降低[155]。硒添加显著增加了鸡脾淋巴细胞硒蛋白 K、N、S 和 T 的基因表达，而镉处理则降低了这些硒蛋白的基因表达[156]。研究还表明，铅中毒可诱导鸡外周血淋巴细胞 HSP 和炎性细胞因子 mRNA 的表达，而硒可减轻铅诱导的 HSP 和炎性细胞因子的增加[157]。

第七节　硒与家禽健康

如前所述，硒是家禽的必需微量元素，硒的缺乏会导致一系列疾病，使生产性能降低，死亡率提高。

一、硒与渗出性素质病

渗出性素质病(ED)是所有与硒缺乏相关的疾病中研究最多的一种,维生素 E 和硒缺乏均会导致该病发生。维生素 E 和硒缺乏导致毛细血管通透性增加,使血浆成分漏出增多而引起胸腹部、腿部、颈部皮下发生浆液性渗出,剖开皮肤可见皮下有蓝绿色水肿液,蓝绿色的呈现主要是由于血红蛋白降解所致。腿部皮下组织更容易受到缺乏病变的伤害,并且在急性阶段还会表现出肌纤维退化,包括钙沉积,血管病变和出血。ED 可在任何年龄段发生,在胚胎期就可以发生,一般成年鸡和火鸡发生最为普遍。ED 的发生与肌肉硒含量降低以及肝脏 GPX 和维生素 E 降低有关,硒预防该病发生的功效要比维生素 E 高得多(200 倍)。

硒和维生素 E 缺乏可导致抗氧化系统破坏,浅表肌肉氧化还原平衡被打破,表现为 GSH 水平降低、GPX 活性降低、H_2O_2 水平提高[158]。早期研究表明,与 ED 相关的硒缺乏导致肌肉和肝脏中 7 种硒蛋白(GPX1、GPX4、SelW1、SelN1、SelP1、SelO 和 SelK)mRNA 水平的降低[159]。最近研究表明,硒缺乏引起的 ED 导致免疫器官(胸腺、脾脏和法氏囊)中 23 种硒蛋白 mRNA 水平的降低,其中,胸腺 DIO1、脾脏 TrxR 以及法氏囊 TrxR3 的 mRNA 水平显著降低[160]。另一项研究发现,与 ED 相关的硒缺乏导致免疫器官(胸腺、脾脏和法氏囊)中 SelT 的基因表达降低,CAT 活性降低,H_2O_2 和·OH 水平增加,IL-1R 和 IL-1β 的基因表达提高。因此,ED 发生的病理原因可能与硒和维生素 E 缺乏导致炎症发生有关[161]。此外,硒缺乏症可降低细胞活力,增加细胞内 ROS 水平并刺激血管平滑肌细胞凋亡[162]。

二、硒与营养性胰腺萎缩

营养性胰腺萎缩(NPA)似乎是唯一明确定义的硒缺乏综合征,因为该病的发生并不伴有其他抗氧化剂的缺乏。内质网破坏是缺硒雏鸡胰腺的主要超微结构病变。硒缺乏通过影响鸡胰腺 NO 和硒蛋白的生成而引起胰腺损伤。硒缺乏(日粮硒含量为 0.033 mg/kg)显著降低了胰腺中 25 种硒蛋白的基因表达,尤其是 TrxR2、GPX1、GPX3、SelI、DIO1、SelP、SelW1、SelO、SelT、SelM、SelX1 和 SPS2 的基因表达受硒缺乏的影响最大,硒缺乏显著提高了胰腺中 NO 水平及诱导型一氧化氮合酶(iNOS)的基因表达和活性[163]。硒缺乏引起的鸡胰腺萎缩与组织中硒蛋白编码基因的总体表达下调有关，包括胰腺中的18个基因，肌肉中的14个基因和肝脏中的9个基因，同样，胰岛素信号相关基因在胰腺，肝脏和肌肉中也被下调，导致鸡血脂异常，低胰岛素血症和高血糖症发生[164]。也有一些研究发现，硒缺乏时鸡肝脏TrxR1和SelS基因表达反而被上调，胰腺中的SepP1、GPX3和SelK的基因表达保持不变，肝脏、砂囊和胰腺中分别有33%、25%和50%的硒蛋白的基因表达下调[93]。关于出现这些现象的原因还不是很清楚，可能与硒缺乏的程度以及不同组织以硒蛋白表达的差异性有关，因为硒蛋白的表达具有明显的等级层次性，表现为在硒缺乏时一些硒蛋白会被优先合成，而且这种优先性还会表现在不同组织间。

三、硒与营养性脑软化病

人们普遍认为，营养性脑软化病(NE)通常是在日粮中亚油酸或花生四烯酸含量较高的基础上由维生素E缺乏引起。NE多发生于孵化后2～3周，主要症状包括共济失调、虚脱、双腿伸直、脚趾弯曲以及头部缩回并横向扭转。组织学损害主要在小脑，表现为小脑肿胀，可见红色或褐色坏死区域、皮质和白质缺血性坏死、毛细血管血栓、出血以及软化，这些损害也可以发生于大脑。脑软化病虽然多见于鸡，但火鸡和鹌鹑以及动物园内的一些鸟类品质也可发生。

对于NE的预防，增加日粮中维生素E的供给以及同时和其他抗氧化剂配合使用被认为是行之有效的方法。脑中多不饱和脂肪酸尤其n-6多不饱和脂肪酸的含量较高且维生素E含量较低，同时，脑由于生理活动所需会产生高水平的ROS，因而脑极易发生氧化损伤。通过使用维生素E以及其他抗氧化剂可以降低脑中ROS的生成，防止脑中多不饱和脂肪酸氧化，进而预防NE的发生。研究表明，硒缺乏会引起鸡脑中氧化损伤和钙稳态失衡，鸡脑中硒、GSH和GPX活性显著下降，脑细胞中线粒体结构损伤、核膜融合和核收缩[165]。最近的研究表明，硒缺乏激活了NF-κB，进而引起鸡脑中TNF-α、环氧合酶2(COX-2)、iNOS和前列腺素E合成酶的基因表达提高，这些基因的上调意味着发生了炎症反应，因此，炎症可能是NE发生的关键因素，该研究还发现，长期硒缺乏导致鸡脑中硒含量显著下降[166]。这和哺乳动物存在很大差异，哺乳动物在硒缺乏时会优先向大脑供给硒，通常不会发生明显的降低，因此鸡脑与哺乳动物大脑相比硒存留能力较弱。

四、硒与营养性肌肉营养不良

营养性肌肉营养不良（NMD）是维生素 E 和含硫氨基酸（半胱氨酸）同时缺乏的结果，表现为肌肉纤维退化。组织学检查显示 Zenker 变性，伴血管周浸润，有明显的浸润嗜酸性粒细胞、淋巴细胞和组细胞的积聚。光镜和电镜观察表明，纤维具有透明性和颗粒性变性。在透明质纤维中，最初的超微结构改变包括肌浆和肌原纤维的密度增加，肌浆网的扩张，肌膜下液泡的形成以及线粒体膜的破坏。在以后的阶段中，这些纤维的改变包括肌原纤维的破坏和溶解，核固缩和溶解，质膜的破坏以及基底层的持续存在和分散的粘附卫星细胞，以及最终被巨噬细胞侵袭。在具有颗粒变性的纤维中，超微结构观察包括肌浆密度降低，线粒体肿胀和变形以及肌原纤维溶解的多个病灶，最终合并产生广泛的溶解。

NMD 经常出现在小鸡、小鹅和小鸭中，通常从 3～5 周龄开始。肌肉纤维变性在胸部和腿部表现最为明显。在鸡和火鸡中，砂囊和嗉囊肌肉也会受到影响，导致饲料残留在嗉囊中，一般在发病后 3～10 天内死亡。在火鸡中，肌肉变性发生于 5～30 日龄，死亡率高达 70%。NMD 在乳房和大腿肌肉中最明显，会形成营养不良性变性纤维的淡条纹。研究表明，从孵化开始给小鸭饲喂氧化性日粮，第 10 天出现 NMD 的最初迹象，第 30 天超过一半的动物患有 NMD，患有 NMD 的动物第 14 天开始出现死亡，第 20 天时死亡率可达 22.5%，死亡动物的砂囊和心脏均可观察到了肌纤维退化病变。母体日粮中的过氧化物含量高或从维生素 E 含量低的鸡蛋中孵化出的鸡在孵化后的 7～15 天可能会出现 NMD。膜的过氧化损伤被认为是 NMD 发生的机制。

1～4 周龄患有遗传性肌营养不良的鸡肌肉和其他组织中 α 生育酚与 γ 生育酚的比值降低，生育酚氧化产物生育酚醌的含量提高，提示了肌膜的氧化应激[167]。一些研究还发现营养不良肌肉肌浆网膜上 Ca^{2+} 的主动转运障碍和膜对 Ca^{2+} 的通透性增加[168]。缺硒可导致骨骼肌钙离子外漏和钙化。最近的研究发现，硒缺乏导致典型的肌肉损伤并伴有 Ca^{2+} 渗漏，硒缺乏导致的氧化应激引起肌浆网和线粒体超微结构损伤，降低了 Ca^{2+} 通道 SERCA、SLC8A、CACNA1S、ORAI1、STIM1、TRPC1 和 TRPC3 的水平，提高了 Ca^{2+} 通道 PMCA 的水平，此外，SelW 基因敲除也导致肌肉中 Ca^{2+} 渗漏，说明 SelW 的氧化还原调节作用在硒缺乏诱导的肌肉 Ca^{2+} 外漏中至关重要[169]。日粮硒缺乏导致的肌肉细胞凋亡增加与这些肌肉中 GPX 活性的降低和脂质过氧化增强相关，缺硒反应以胸肌较强，大腿和翅肌次之，肌肉中4种内质网硒蛋白基因(SelN1、SelK、SelS和SelT)的分布及其对硒缺乏的反应与由此产生的氧化应激和细胞凋亡是相对应的,与其他骨骼肌相比,胸肌对氧化性细胞死亡的敏感性最强[170]。事实上,硒缺乏可导致鸡肌肉中19种硒蛋白的表达降低,其中11种为抗氧化硒蛋白,这些抗氧化硒蛋白中GPX3、GPX4和SelW的降低较为显著,似乎这些硒蛋白尤其是SelW起着关键的作用[171]。日粮硒缺乏降低了鸡翅肌、胸肌和大腿肌肉中SelW的mRNA表达,同时提高了这些组织中炎症相关基因的mRNA表达,而且炎症相关基因的mRNA表达水平与SelW呈显著的负相关[172]。日粮缺硒导致3日龄雏鸡胸肌肌纤维断裂和凝固性坏死,并伴有组织中MDA升高、总抗氧化能力降低、硒蛋白GPX1、GPX3、GPX4、SelP1、SelO、SelK、SelU、SelH、SelM和SelW1的mRNA水平降低以及炎症和凋亡指标p53、caspase-3、caspase-9、COX-2、PI3K、P-Akt、NF-(B、p-p38 MAPK、p-JNK和p-ERK的升高[173]。这些研究提示,硒蛋白参与的氧化还原调节以及对过氧化物的高效分解是预防NMD发生的机制。

NMD的发生还与鸡肉中蛋白质的氧化有关。在NMD中,蛋白质中二硫化物/巯基提高,低分子量蛋白质增多,说明蛋白质发生了氧化而降解。硒和维生素E缺乏导致肌肉GSH水平明显降低,而线粒体以及胞质中谷胱甘肽还原酶活性不降反增,说明GSH水平的降低并不是因为谷胱甘肽还原酶活性降低所致。因此,维生素E缺乏应该与GSH的降低有关。在NMD中,肌肉蛋白对半胱氨酸的特殊需求可以解释为维生素E缺乏导致蛋氨酸向半胱氨酸的转化降低。

五、硒与病毒感染

硒营养状况与家禽传染病的易感性和严重程度有关,病毒感染造成的氧化应激是发病率和严重程度提高的重要诱因。流感病毒感染过程中下调超氧化物的生成能显著降低病毒以及病毒复制造成的肺损伤[174]。2型猪圆环病毒感染引起GPX活性降低,生理浓度的SeMet可阻断氧化应激引起的病毒复制增加,当GPX基因敲除后硒的这种阻断作用丧失[175]。此外,硒营养状况低下引起的免疫功能降低,提高了机体对病毒感染的易感性,还有可能影响病毒毒株的致病性。在小鼠模型中研究发现,硒和维生素E缺乏导致一种良性的柯萨奇病毒B3株的毒力变强,并导致小鼠患心肌炎[176]。病毒致病性的改变是基因突变所致,从缺硒小鼠回收的病毒分离株的测序显示,柯萨奇病毒和流感病毒的基因组都发生了突变,而且这些变化与病毒致病性的增强有关[177]。研究表明,饮水中补硒提高了小鼠对柯萨奇病毒 B3 感染的抵抗力，在接种后第 14 天，补硒组小鼠的存活率为 58%，而对照组小鼠的存活率仅为 25%[178]。小鼠在感染柯萨奇病毒 B3 之前的 28 天分别给予低硒和硒充足日粮，感染后保持原有日粮不变，90 天后发现，低硒摄入小鼠血清 GPX 活性较低，心脏中病毒 RNA 水平较高，以心肌炎为特征的组织病理学改变明显，死亡率较高[179]。硒缺乏也可能导致人类免疫缺陷病毒（HIV）、甲型流感病毒、SARS 冠状病毒及埃博拉病毒的基因组发生突变，因而改变其致病性[180]。

参考文献

[1] Gosetti F, Frascarolo P, Polati S, et al. Speciation of selenium in diet supplements by HPLC-MS/MS methods[J]. Food Chemistry, 2007, 105(4): 1738-1747.

[2] Eisenberg S. Relative stability of selenites and selenates in feed premixes as a function of water activity[J]. Journal of AOAC International, 2007, 90(2): 349-353.

[3] Attia Y A, Abdalah A A, Zeweil H S, et al. Effect of inorganic or organic selenium supplementation on productive performance, egg quality and some physiological traits of dual-purpose breeding hens[J]. Czech Journal of Animal Science, 2010, 55: 505-519.

[4] Surai P F, Fisinin V I, Karadas F. Antioxidant systems in chick embryo development. Part 1. Vitamin E, carotenoids and selenium[J]. Animal Nutrition, 2016, 2(01): 5-15.

[5] Surai P F, Fisinin V I. Selenium in poultry breeder nutrition: An update[J]. Animal Feed Science and Technology, 2014, 191: 1-15.

[6] Ort J F, Latshaw J D. The Toxic Level of Sodium Selenite in the Diet of Laying Chickens[J]. Journal of Nutrition, 1978, 108(7): 1114-1120.

[7] Gowdy B, Mae K. Selenium supplementation and antioxidant protection in broiler chickens[J]. Formal Description Techniques IV, 2004: 99-108.

[8] Blesbois E, Grasseau I, Hermier D. Changes in lipid content of fowl spermatozoa after liquid storage at 2°C to 5°C[J]. Theriogenology, 1999, 52(2): 325-334.

[9] Surai P F, Speake B K. Distribution of carotenoids from the yolk

to the tissues of the chick embryo[J]. Journal of Nutritional Biochemistry, 1998, 9(11): 645-651.

[10] Maldjian A, Cerolini S, Surai P F, et al. The effect of vitamin E, green tea extracts and catechin on the in vitro storage of turkey spermatozoa at room temperature[J]. Poultry and Avian Biology Reviews, 1998, 9(4): 143-151.

[11] Douard V, Hermier D, Blesbois E. Changes in Turkey Semen Lipids During Liquid In Vitro Storage[J]. Biology of Reproduction, 2000, 63(5): 1450-1456.

[12] Pappas A C, Acamovic T, Sparks N H C, et al. Effects of Supplementing Broiler Breeder Diets with Organoselenium Compounds and Polyunsaturated Fatty Acids on Hatchability[J]. Poultry Science, 2006, 85(9): 1584-1593.

[13] Surai P F, Blesbois E, Grasseau I, et al. Fatty acid composition, glutathione peroxidase and superoxide dismutase activity and total antioxidant activity of avian semen[J]. Comparative Biochemistry and Physiology B, 1998, 120(3): 527-533.

[14] Surai P F, Kostjuk I, Wishart G J, et al. Effect of Vitamin E and Selenium Supplementation of Cockerel Diets on Glutathione Peroxidase Activity and Lipid Peroxidation Susceptibility in Sperm, Testes, and Liver[J]. Biological Trace Element Research, 1998, 64(1): 119-132.

[15] Ebeid T A. Organic selenium enhances the antioxidative status and quality of cockerel semen under high ambient temperature[J]. British Poultry Science, 2009, 50(5): 641-647.

[16] Ebeid T A. Vitamin E and organic selenium enhances the antioxidative status and quality of chicken semen under high ambient temperature[J]. British Poultry Science, 2012, 53(5): 708-714.

[17] Edens F W. Practical applications for selenomethionine: broiler breeder reproduction. in proceedings of the Nutritional Biotechnology in the Feed and Food Industries [C]. Nottingham, UK: Nottingham University Press, 2002.

[18] Davtyan D, Papazyan T, Nollet L. Dose response of Se added as sodiumselenite or Sel-Plex on male sperm quality and breeder productivity [C]. Verona, Italy, 2006.

[19] Dimitrov S, Atanasov V, Surai P F, et al. Effect of organic se-

lenium on Turkey semen quality during liquid storage[J]. Animal Reproduction Science, 2007, 100(3): 311-317.

[20] Slowinska M, Jankowski J, Dietrich G J, et al. Effect of organic and inorganic forms of selenium in diets on turkey semen quality[J]. Poultry Science, 2011, 90(1): 181-190.

[21] Papazyan T T, Royter Y S, Guseva N K. Effect of Sel-Plex organic selenium on guinea fowl reproduction. in proceedings of the Alltech's 22nd Annual Symposium [C]. Lexington, KY, USA, 2006.

[22] Jerysz A, Lukaszewicz E. Effect of Dietary Selenium and Vitamin E on Ganders' Response to Semen Collection and Ejaculate Characteristics[J]. Biological Trace Element Research, 2013, 153: 196-204.

[23] Shi L, Zhao H, Ren Y, et al. Effects of different levels of dietary selenium on the proliferation of spermatogonial stem cells and antioxidant status in testis of roosters[J]. Animal Reproduction Science, 2014, 149(3): 266-272.

[24] Huang Y, Li W, Xu D, et al. Effect of Dietary Selenium Deficiency on the Cell Apoptosis and the Level of Thyroid Hormones in Chicken [J]. Biological Trace Element Research, 2016, 171(2): 445-452.

[25] Khalid A, Khudhair N, He H, et al. Effects of Dietary Selenium Supplementation on Seminiferous Tubules and SelW, GPx4, LHCGR, and ACE Expression in Chicken Testis[J]. Biological Trace Element Research, 2016, 173(1): 202-209.

[26] Breque C, Surai P F, Brillard J P. Roles of Antioxidants on Prolonged Storage of Avian Spermatozoa In Vivo and In Vitro[J]. Molecular Reproduction and Development, 2003, 66(3): 314-323.

[27] Breque C, Brillard J P. Sperm storage in the avian oviduct: baselines for a complex antioxidant system in the sperm storage tubules [J]. Archiv Geflugelkunde 2002, 66: 83.

[28] Surai P F. Selenium in poultry nutrition 2. Reproduction, egg and meat quality and practical applications[J]. Worlds Poultry Science Journal, 2002, 58(4): 431-450.

[29] Pappas A C, Acamovic T, Sparks N H C, et al. Effects of supplementing broiler breeder diets with organic selenium and polyunsaturated fatty acids on egg quality during storage[J]. Poultry Science, 2005, 84(6): 865-874.

[30] Lipiec E, Siara G, Bierla K, et al. Determination of selenomethionine, selenocysteine, and inorganic selenium in eggs by HPLC-inductively coupled plasma mass spectrometry[J]. Analytical and Bioanalytical Chemistry, 2010, 397(2): 731-741.

[31] Invernizzi G, Agazzi A, Ferroni M, et al. Effects of Inclusion of Selenium-Enriched Yeast in the Diet of Laying Hens on Performance, Eggshell Quality, and Selenium Tissue Deposition[J]. Italian Journal of Animal Science, 2013, 12(1): 1-8.

[32] Tufarelli V, Cazzato E, Ceci E, et al. Selenium-Fertilized Tritordeum (×Tritordeum Ascherson et Graebner) as Dietary Selenium Supplement in Laying Hens: Effects on Egg Quality[J]. Biological Trace Element Research, 2016, 173(1): 219-224.

[33] Delezie E, Rovers M, Der Aa A V, et al. Comparing responses to different selenium sources and dosages in laying hens[J]. Poultry Science, 2014, 93(12): 3083-3090.

[34] Jing C L, Dong X F, Wang Z M, et al. Comparative study of DL-selenomethionine vs sodium selenite and seleno-yeast on antioxidant activity and selenium status in laying hens[J]. Poultry Science, 2015, 94(5): 965-975.

[35] Surai P F. Effect of selenium and vitamin E content of the maternal diet on the antioxidant system of the yolk and the developing chick [J]. British Poultry Science, 2000, 41(2): 235-243.

[36] Briens M, Mercier Y, Rouffineau F, et al. 2-Hydroxy-4-methylselenobutanoic acid induces additional tissue selenium enrichment in broiler chickens compared with other selenium sources[J]. Poultry Science, 2014, 93(1): 85-93.

[37] Jlali M, Briens M, Rouffineau F, et al. Effect of 2-hydroxy-4-methylselenobutanoic acid as a dietary selenium supplement to improve the selenium concentration of table eggs[J]. Journal of Animal Science, 2013, 91(4): 1745-1752.

[38] Tufarelli, Ceci E, Laudadio. 2-Hydroxy-4-Methylselenobutanoic Acid as New Organic Selenium Dietary Supplement to Produce Selenium-Enriched Eggs[J]. Biological Trace Element Research, 2016, 171(2): 453-458.

[39] Sefton A E, Edens F W. Sel-Plex improves broiler breeder performance. in proceedings of the Nutritional Biotechnology in the Feed and

Food Industry [C]. Lexington, Kentucky, USA, 2004.

[40] Surai P F. Tissue-specific changes in the activities of antioxidant enzymes during the development of the chicken embryo[J]. British Poultry Science, 1999, 40(3): 397-405.

[41] Ar A, Mover H. Oxygen tensions in developing embryos-system inefficiency or system requirement? [J]. Israel Journal of Zoology, 2013, 40(3-4):307-326.

[42] Ishibashi M, Akazawa S, Sakamaki H, et al. Oxygen-Induced Embryopathy and the Significance of Glutathione-Dependent Antioxidant System in the Rat Embryo During Early Organogenesis[J]. Free Radical Biology and Medicine, 1997, 22(3): 447-454.

[43] Yuan H, Kaneko T, Kaji K, et al. Species difference in the resistibility of embryonic fibroblasts against oxygen-induced growth inhibition [J]. Comparative Biochemistry and Physiology B, 1995, 110(1): 145-154.

[44] Pappas A C, Karadas F, Surai P F, et al. The selenium intake of the female chicken influences the selenium status of her progeny[J]. Comparative Biochemistry and Physiology B, 2005, 142(4): 465-474.

[45] Payne R L, Southern L L. Changes in glutathione peroxidase and tissue selenium concentrations of broilers after consuming a diet adequate in selenium[J]. Poultry Science, 2005, 84(8): 1268-1276.

[46] Yuan D, Zhan X, Wang Y. Effect of selenium sources on the expression of cellular glutathione peroxidase and cytoplasmic thioredoxin reductase in the liver and kidney of broiler breeders and their offspring [J]. Poultry Science, 2012, 91(4): 936-942.

[47] Yuan D, Zheng L, Guo X Y, et al. Regulation of selenoprotein P concentration and expression by different sources of selenium in broiler breeders and their offspring[J]. Poultry Science, 2013, 92(9): 2375-2380.

[48] Surai P F, Karadas F, Pappas A C, et al. Effect of organic selenium in quail diet on its accumulation in tissues and transfer to the progeny[J]. British Poultry Science, 2006, 47(1): 65-72.

[49] Surai P F. Selenium in Nutrition and Health [M]. Nottingham: Nottingham University Press, 2006.

[50] Periae L, Miloceviae N, Miliae D, et al. Effect of Sel-Plex in broiler breeder diets on quality of day-old chicks. in proceedings of the Alltech's 23rd Annual Symposium [C]. Lexington, Kentucky, USA, 2007.

[51] Papazyan T T, Lyons M P, Mezes M, et al. Selenium in poultry nutrition-Effects on fertility and hatchability [J]. Praxis Veterinaria, 2006, 54: 85-102.

[52] Lam J, Tanabe S, Lam M H W, et al. Risk to breeding success of waterbirds by contaminants in Hong Kong: evidence from trace elements in eggs[J]. Environmental Pollution, 2005, 135(3): 481-490.

[53] Sara A, Bentea M, Odagiu A, et al. Effect of Sel-Plex organoselenium on performance of laying hens. in proceedings of the Alltech's 23rd Annual Symposium [C]. Lexington, KY, USA, 2007.

[54] Rutz F, Pan E A, Xavier G B, et al. Meeting selenium demands of modern poultry: responses to Sel-Plex organic selenium in broiler and breeder diets. in proceedings of the 19th Alltech's Annual Symposium in Nutritional biotechnology in the feed and food industries [C]. Nottingham, UK, 2003.

[55] Paton N D, Cantor A J. Effect of dietary selenium source and storage on internal quality and shell strength of eggs[J]. Poultry Science, 2000, 70 (suppl.): 116.

[56] Pourreza J, Pishnamzi A. Effect of inorganic and organic selenium sources on egg quality and performance of laying hens [C]. Verona, Italy, 2006.

[57] Klecker D, Zantlokaul M, Zeaman L. Effect of organic selenium, zinc and manganese on reproductive traits of laying hens and cockerels on the quality parameters of eggs. in proceedings of the the 13th European Symposium on Poultry Nutrition [C]. Blankenberge, Belgium, 2001.

[58] Renema R A. Reproductive responses to Sel-Plex organic selenium in male and female broiler breeders: impact on production traits and hatchability. in proceedings of the 20th Alltech's Annual Symposium [C]. Nottingham, UK: Nottingham University Press, 2004.

[59] Stoewsand G S, Gutenmann W H, Lisk D J. Wheat grown on fly ash: high selenium uptake and response when fed to Japanese quail[J]. Journal of Agricultural and Food Chemistry, 1978, 26(3): 757-759.

[60] Karadas F, Pappas A C, Surai P F, et al. Increase of Se concentration in all parts of the egg as an effect of selenium supplementation in avian maternal nutrition. in proceedings of the Alltech's 21st Annual Symposium [C]. 2005.

[61] Surai P F, Karadas F, Pappas A C, et al. Selenium distribution in the eggs of ISA Brown commercial layers. in proceedings of the 20th Annual Symposium [C]. Lexington, KY, USA, 2004.

[62] Golubkina N A, Papazyan T T. Selenium distribution in eggs of avian species[J]. Comparative Biochemistry and Physiology B, 2006, 145(3): 384-388.

[63] Turan B, Balcik C, Akkas N. Effect of dietary selenium and vitamin E on the biomechanical properties of rabbit bones[J]. Clinical Rheumatology, 1997, 16(5): 441-449.

[64] Boruta A, Świerczewska E, Roszkowski T. Effect of organic and inorganic forms of selenium on the morphological composition of eggs and selenium content in egg mass[J]. Medycyna Weterynaryjna, 2007, 63(2): 238-241.

[65] Pan E A, Rutz F. Sel-Plex for layers: egg production and quality responses to increasing level of inclusion. in proceedings of the Alltech's 19th Annual Symposium on Nutritional Biotechnology in the Feed and Food Industries [C]. Lexington, KY, USA, 2003.

[66] Lund M N, Heinonen M, Baron C P, et al. Protein oxidation in muscle foods: a review[J]. Molecular Nutrition & Food Research, 2011, 55(1): 83-95.

[67] Kaya A, Lee B C, Gladyshev V N. Regulation of Protein Function by Reversible Methionine Oxidation and the Role of Selenoprotein MsrB1[J]. Antioxidants & Redox Signaling, 2015, 23(10): 814-822.

[68] Lee B C. Biochemistry and Function of Methionine Sulfoxide Reductase//Hatfield D L, Schweizer U, Tsuji P A. Selenium Its Molecular Biology and Role in Human Health[M]. New York: Springer International Publishing. 2016: 287-292.

[69] Arai T, Sugawara M, Sako N, et al. Glutathione peroxidase activity in tissues of chicken supplemented with dietary selenium[J]. Comparative Biochemistry and Physiology, 1994, 107A: 245-248.

[70] Chen G, Wu J, Li C. Effect of different selenium sources on production performance and biochemical parameters of broilers[J]. Journal of Animal Physiology and Animal Nutrition, 2014, 98(4): 747-754.

[71] Wang Y, Xu B. Effect of different selenium source (sodium selenite and selenium yeast) on broiler chickens[J]. Animal Feed Science and

Technology, 2008, 144(3): 306-314.

[72] Jiang Z, Lin Y, Zhou G, et al. Effects of Dietary Selenomethionine Supplementation on Growth Performance, Meat Quality and Antioxidant Property in Yellow Broilers[J]. Journal of Agricultural and Food Chemistry, 2009, 57(20): 9769-9772.

[73] Wang Y, Zhan X A, Zhang X, et al. Comparison of Different Forms of Dietary Selenium Supplementation on Growth Performance, Meat Quality, Selenium Deposition, and Antioxidant Property in Broilers [J]. Biological Trace Element Research, 2011, 143(1): 261-273.

[74] Rao S V, Prakash B, Raju M V, et al. Effect of Supplementing Organic Selenium on Performance, Carcass Traits, Oxidative Parameters and Immune Responses in Commercial Broiler Chickens[J]. Asian-Australasian Journal of Animal Sciences, 2013, 26(2): 247-252.

[75] Salman M, Muğlali O H, Selçuk Z. Investigations into effects on performance and glutathione peroxidase activity in broilers when increasing selenium contents of complete diets appropriate to animals′ selenium requirements by adding different selenium compounds (organic vs. inorganic)[J]. Dtw Deutsche Tier? rztliche Wochenschrift, 2009, 116(6): 233-237.

[76] Pappas A C, Zoidis E, Georgiou C A, et al. Influence of organic selenium supplementation on the accumulation of toxic and essential trace elements involved in the antioxidant system of chicken[J]. Food Additives and Contaminants Part A-chemistry Analysis Control Exposure & Risk Assessment, 2011, 28(4): 446-454.

[77] Zhao L, Sun L, Huang J, et al. A Novel Organic Selenium Compound Exerts Unique Regulation of Selenium Speciation, Selenogenome, and Selenoproteins in Broiler Chicks[J]. Journal of Nutrition, 2017, 147(5): 789-797.

[78] Edens F W, Burgos S, Read-Snyder J, et al. Sel-Plex maintains small intestine integrity in reovirus-infected broiler chickens. in proceedings of the 8th Asian Pacific Poultry Conference [C]. Bangkok, Thailand, 2007.

[79] Skrivan M, Marounek M, Dlouha G, et al. Dietary selenium increases vitamin E contents of egg yolk and chicken meat[J]. British Poultry Science, 2008, 49(4): 482-486.

[80] Naylor A J, Choct M, Jacques K A. Effects of selenium source and level on performance and meat quality in male broilers[J]. Poultry Science, 2000, 79: 117.

[81] Papazyan T, Surai P F. Effect of Se supplementation on chick growth and development. in proceedings of the Alltech's 23rd Annual Symposium [C]. Lexington, KY, USA, 2007.

[82] Stolic N, Radovanovic T, Stolic N, et al. Study of the improvement of the fattering chick feeding quality using organic selenium[J]. Biotechnology in Animal Husbandry, 2002, 18: 239-246.

[83] Vlahovic M, Pavlovski Z, Zivkovic B, et al. Influence of different selenium sources on broiler performance[J]. Yugoslav Poultry Science, 1998, 3: 3-4.

[84] Anciuti M A, Rutz F, Da Silva L A, et al. Effect of replacement of dietary inorganic by organic selenium (Sel-Plex) on performance of broilers. in proceedings of the 20th Annual Symposium: nutritional biotechnology in the feed and food industry [C]. Lexington, KY, USA, 2004.

[85] Valcić O, Jovanović I B, Milanović S. Selenium, thiobarbituric acid reactive substances, and thyroid hormone activation in broilers supplemented with selenium as selenized yeast or sodium selenite[J]. Japanese Journal of Veterinary Research, 2011, 59(2-3): 69-77.

[86] Edens F W. Involvement of Sel-Plex in physiological stability and performance of broiler chickens. in proceedings of the 17th Alltech's Annual Symposium on Biotechnology in the feed industry [C]. Nottingham, UK, 2001.

[87] Egorov I A, Papazyan T T, Ivachnick G V, et al. Sel-Plex organic selenium in commercial hen diets. in proceedings of the Alltech's 22nd Annual Symposium [C]. Lexington, KY, USA, 2006.

[88] Liao X, Lu L, Li S F, et al. Effects of selenium source and level on growth performance, tissue selenium concentrations, antioxidation, and immune functions of heat-stressed broilers[J]. Biological Trace Element Research, 2012, 150: 158-165.

[89] Mahmoud K Z, Edens F W. Influence of selenium sources on age-related and mild heat stress-related changes of blood and liver glutathione redox cycle in broiler chickens (Gallus domesticus)[J]. Comparative Biochemistry and Physiology B, 2003, 136(4): 921-934.

[90] Aftab U, Khan A A. Strategies to alleviate the incidence of ascites in broilers: a review[J]. Brazilian Journal of Poultry Science, 2005, 7(4): 199-204.

[91] Roch G, Boulianne M, De Roth L. Effect of dietary antioxidants on the incidence of pulmonary hypertension syndrome in broilers [J/OL]. https://en. engormix. com/poultry-industry/articles/effect-dietary-antioxidants-incidence-t33623. htm, 5/9/2007.

[92] Ghazi Harsini S, Habibiyan M, Moeini M M, et al. Effects of Dietary Selenium, Vitamin E, and Their Combination on Growth, Serum Metabolites, and Antioxidant Defense System in Skeletal Muscle of Broilers Under Heat Stress[J]. Biological Trace Element Research, 2012, 148 (3): 322-330.

[93] Li J, Sunde R A. Selenoprotein Transcript Level and Enzyme Activity as Biomarkers for Selenium Status and Selenium Requirements of Chickens (Gallus gallus)[J]. Plos One, 2016, 11(4):e0152392.

[94] Zoidis E, Demiris N, Kominakis A, et al. Meta-analysis of selenium accumulation and expression of antioxidant enzymes in chicken tissues[J]. Animal, 2014, 8(4): 542-554.

[95] Sevćikova S, Skivan M, Dlouhá G, et al. The effect of selenium source on the performance and meat quality of broiler chickens[J]. Czech Journal of Animal Science, 2018, 51(10): 449-457.

[96] Briens M, Mercier Y, Rouffineau F, et al. Comparative study of a new organic selenium source v. seleno-yeast and mineral selenium sources on muscle selenium enrichment and selenium digestibility in broiler chickens[J]. British Journal of Nutrition, 2013, 110(4): 617-624.

[97] Couloigner F, Jlali M, Briens M, et al. Selenium deposition kinetics of different selenium sources in muscle and feathers of broilers[J]. Poultry Science, 2015, 94(11): 2708-2714.

[98] Schrauzer G N. The nutritional significance, metabolism and toxicology of selenomethionine[J]. Advances in Food and Nutrition Research, 2003, 47: 73-112.

[99] Schrauzer G N, Surai P F. Selenium in human and animal nutrition: resolved and unresolved issues. A partly historical treatise in commemoration of the fiftieth anniversary of the discovery of the biological essentiality of selenium, dedicated to the memory of Klaus Schwarz (1914—

1978) on the occasion of the thirtieth anniversary of his death[J]. Critical Reviews in Biotechnology, 2009, 29: 2-9.

[100] Pesut O, Nollet L, Tucker L. Effect of organic selenium (Sel-Plex) in combination with alpha-tocopherol on fresh and frozen poultry meat. in proceedings of the Alltech's 21st Annual Symposium [C]. Lexington, KY, USA, 2005.

[101] Perez T I, Zuidhof M J, Renema R A, et al. Effects of Vitamin E and Organic Selenium on Oxidative Stability of ω-3 Enriched Dark Chicken Meat during Cooking[J]. Journal of Food Science, 2010, 75(2): 25-34.

[102] Dlouha G, Sevcikova S, Dokoupilova A, et al. Effect of dietary selenium sources on growth performance, breast muscle selenium, glutathione peroxidase activity and oxidative stability in broilers[J]. Czech Journal of Animal Science, 2018, 53(6): 265-269.

[103] Mcnaughton S A, Marks G C. Selenium Content of Australian Foods: A Review of Literature Values[J]. Journal of Food Composition and Analysis, 2002, 15(2): 169-182.

[104] Daun C, Johansson M, Onning G, et al. Glutathione peroxidase activity, tissue and soluble selenium content in beef and pork in relation to meat ageing and pig RN phenotype[J]. Food Chemistry, 2001, 73 (3): 313-319.

[105] Bierla K, Dernovics M, Vacchina V, et al. Determination of selenocysteine and selenomethionine in edible animal tissues by 2D size-exclusion reversed-phase HPLC-ICP MS following carbamidomethylation and proteolytic extraction[J]. Analytical and Bioanalytical Chemistry, 2008, 390(7): 1789-1798.

[106] Wu Z, Kaiser P. Antigen presenting cells in a non-mammalian model system, the chicken[J]. Immunobiology, 2011, 216(11): 0-1183.

[107] Chuammitri P, Ostojic J, Andreasen C B, et al. Chicken heterophil extracellular traps (HETs): Novel defense mechanism of chicken heterophils[J]. Veterinary Immunology and Immunopathology, 2009, 129 (1): 126-131.

[108] Qureshi M A, Hussain I, Heggen C L. Understanding immunology in disease development and control[J]. Poultry Science, 1998, 77 (8): 1126-1129.

[109] Ebert-Dumig R, Seufert J, Schneider D, et al. Expression of selenoproteins in monocytes and macrophages-implications for the immune system[J]. Medizinische Klinik, 1999, 9494: 29-34.

[110] Carlson B A, Yoo M, Conrad M, et al. Protein kinase-regulated expression and immune function of thioredoxin reductase 1 in mouse macrophages[J]. Molecular Immunology, 2011, 49(1): 311-316.

[111] Adib-Conquy M, Scottalgara D, Cavaillon J M, et al. TLR-mediated activation of NK cells and their role in bacterial/viral immune responses in mammals[J]. Immunology and Cell Biology, 2014, 92(3): 256-262.

[112] Korver D R. Implications of changing immune function through nutrition in poultry[J]. Animal Feed Science and Technology, 2012, 173(1-2): 54-64.

[113] Dietert R R, Combs G F, Lin H K, et al. Impact of Combined Vitamin E and Selenium Deficiency on Chicken Macrophage Function[J]. Annals of the New York Academy of Sciences, 1990, 587(1): 281-282.

[114] Larsen H J. Relations between selenium and immunity[J]. Norwegian Journal of Agricultural Sciences, 1993, 11: 105-119.

[115] Ndiweni N, Finch J M. Effects of in vitro supplementation with α-tocopherol and selenium on bovine neutrophil functions: implications for resistance to mastitis[J]. Veterinary Immunology and Immunopathology, 1996, 51(1): 67-78.

[116] Safir N, Wendel A, Saile R, et al. The Effect of Selenium on Immune Functions of J774.1 Cells[J]. Clinical Chemistry and Laboratory Medicine, 2003, 41(8): 1005-1011.

[117] Shilo S, Tirosh O. Selenite Activates Caspase-Independent Necrotic Cell Death in Jurkat T Cells and J774.2 Macrophages by Affecting Mitochondrial Oxidant Generation[J]. Antioxidants & Redox Signaling, 2003, 5(3): 273-279.

[118] Chen X, Yao H, Yao L, et al. Selenium Deficiency Influences the Gene Expressions of Heat Shock Proteins and Nitric Oxide Levels in Neutrophils of Broilers[J]. Biological Trace Element Research, 2014, 161(3): 334-340.

[119] Mattmiller S A, Carlson B A, Gandy J, et al. Reduced macrophage selenoprotein expression alters oxidized lipid metabolite biosynthesis from arachidonic and linoleic acid[J]. Journal of Nutritional Biochemis-

try, 2014, 25(6): 647-654.

[120] Kaushal N, Kudva A K, Patterson A D, et al. Crucial Role of Macrophage Selenoproteins in Experimental Colitis[J]. Journal of Immunology, 2014, 193(7): 3683-3692.

[121] Sun Z, Liu C, Pan T, et al. Selenium accelerates chicken dendritic cells differentiation and affects selenoproteins expression[J]. Developmental and Comparative Immunology, 2017, 77: 30-37.

[122] Liu Z, Qu Y, Wang J, et al. Selenium Deficiency Attenuates Chicken Duodenal Mucosal Immunity via Activation of the NF-κb Signaling Pathway[J]. Biological Trace Element Research, 2016, 172(2): 465-473.

[123] Spallholz J E, Martin J L, Gerlach M L, et al. Injectable Selenium: Effect on the Primary Immune Response of Mice[J]. Experimental Biology and Medicine, 1975, 148(1): 37-40.

[124] Marsh J A, Dietert R R, Combs G F. Influence of Dietary Selenium and Vitamin E on the Humoral Immune Response of the Chick[J]. Proceedings of the Society for Experimental Biology & Medicine Society for Experimental Biology & Medicine, 1981, 166(2): 228-236.

[125] Bassiouni A A, Zaki M M, Hady M M. Effect of vitamin E and selenium on the immune response of chickens against living Newcastle disease vaccine[J]. Veterinary Medical Journal Giza, 1990, 38: 145-155.

[126] Larsen C T, Pierson F W, Gross W B. Effect of dietary selenium on the response of stressed and unstressed chickens toEscherichia coli challenge and antigen[J]. Biological Trace Element Research, 1997, 58(3): 169-176.

[127] Zhang Z W, Wang Q H, Zhang J L, et al. Effects of Oxidative Stress on Immunosuppression Induced by Selenium Deficiency in Chickens[J]. Biological Trace Element Research, 2012, 149(3): 352-361.

[128] Peng X, Cui H, Fang J, et al. Low Selenium Diet Alters Cell Cycle Phase, Apoptotic Population and Modifies Oxidative Stress Markers of Spleens in Broilers[J]. Biological Trace Element Research, 2012, 148(2): 182-186.

[129] Peng X, Cui Y, Cui W, et al. The Cell Cycle Arrest and Apoptosis of Bursa of Fabricius Induced by Low Selenium in Chickens[J]. Biological Trace Element Research, 2011, 139(1): 32-40.

[130] Zhang Z W, Zhang J L, Gao Y H, et al. Effect of oxygen free

radicals and nitric oxide on apoptosis of immune organ induced by selenium deficiency in chickens[J]. BioMetals, 2013, 25: 355-365.

[131] Dong, Yu, Ziwei, et al. The role of selenoprotein W in inflammatory injury in chicken immune tissues and cultured splenic lymphocyte [J]. BioMetals, 2015, 58: 75-87.

[132] You L, Liu C, Yang Z J, et al. Prediction of Selenoprotein T Structure and Its Response to Selenium Deficiency in Chicken Immune Organs[J]. Biological Trace Element Research, 2014, 160(2): 222-231.

[133] Sun H, Deng T, Fu J. Chicken 15-kDa Selenoprotein Plays Important Antioxidative Function in Splenocytes[J]. Biological Trace Element Research, 2014, 161(3): 288-296.

[134] Kiremidjianschumacher L, Roy M, Wishe H I, et al. Regulation of cellular immune responses by selenium[J]. Biological Trace Element Research, 1992, 33(1): 23-35.

[135] Huang K, Chen W. Effect of selenium on T lymphocyte transformation rate and nature killer cell activities in chickens[J]. Journal of Nanjing Agricultural University, 1999, 22: 76-79.

[136] Roy M, Kiremidjian-Schumacher L, Wishe H I, et al. Selenium Supplementation Enhances the Expression of Interleukin 2 Receptor Subunits and Internalization of Interleukin 2[J]. Experimental Biology and Medicine, 1993, 202(3): 295-301.

[137] Sun E, Xu H, Liu Q, et al. The mechanism for the effect of selenium supplementation on immunity[J]. Biological Trace Element Research, 1995, 48(3): 231-238.

[138] Chen K, Shu G, Peng X, et al. Protective role of sodium selenite on histopathological lesions, decreased T-cell subsets and increased apoptosis of thymus in broilers intoxicated with aflatoxin B? [J]. Food and Chemical Toxicology, 2013, 59(3): 446-454.

[139] Chen K, Yuan S, Chen J, et al. Effects of sodium selenite on the decreased percentage of T cell subsets, contents of serum IL-2 and IFN-γ induced by aflatoxin B1 in broilers[J]. Research in Veterinary Science, 2013, 95(1): 143-145.

[140] Chen K, Fang J, Peng X, et al. Effect of selenium supplementation on aflatoxin B1-induced histopathological lesions and apoptosis in bursa of Fabricius in broilers[J]. Food and Chemical Toxicology, 2014,

74: 91-97.

[141] He Y, Fang J, Peng X, et al. Effects of Sodium Selenite on Aflatoxin B1-Induced Decrease of Ileac T cell and the mRNA Contents of IL-2, IL-6, and TNF-α in Broilers[J]. Biological Trace Element Research, 2014, 159(1): 167-173.

[142] He Y, Fang J, Peng X, et al. Effects of sodium selenite on aflatoxin B1-induced decrease of ileal IgA+ cell numbers and immunoglobulin contents in broilers[J]. Biological Trace Element Research, 2014, 160(1): 49-55.

[143] Chen K, Peng X, Fang J, et al. Effects of Dietary Selenium on Histopathological Changes and T Cells of Spleen in Broilers Exposed to Aflatoxin B1[J]. International Journal of Environmental Research and Public Health, 2014, 11(2): 1904-1913.

[144] Shi D, Liao S, Guo S, et al. Protective Effects of Selenium on Aflatoxin B1-induced Mitochondrial Permeability Transition, DNA Damage, and Histological Alterations in Duckling Liver[J]. Biological Trace Element Research, 2015, 163(1): 162-168.

[145] Salimian J, Arefpour M A, Riazipour M, et al. Immunomodulatory effects of selenium and vitamin E on alterations in T lymphocyte subsets induced by T-2 toxin[J]. Immunopharmacology and Immunotoxicology, 2014, 36(4): 275-281.

[146] Wang X, Zuo Z, Zhao C, et al. Protective role of selenium in the activities of antioxidant enzymes in piglet splenic lymphocytes exposed to deoxynivalenol[J]. Environmental Toxicology and Pharmacology, 2016, 47: 53-61.

[147] Habibian M, Sadeghi G, Ghazi S, et al. Selenium as a Feed Supplement for Heat-Stressed Poultry: a Review[J]. Biological Trace Element Research, 2015, 165(2): 183-193.

[148] Niu Z, Liu F Z, Yan Q, et al. Effects of different levels of selenium on growth performance and immunocompetence of broilers under heat stress[J]. Archives of Animal Nutrition, 2009, 63(1): 56-65.

[149] Habibian M, Ghazi S, Moeini M M, et al. Effects of dietary selenium and vitamin E on immune response and biological blood parameters of broilers reared under thermoneutral or heat stress conditions[J]. International Journal of Biometeorology, 2014, 58(5): 741-752.

[150] Rao S V R, Prakash B, Raju M V L N, et al. Effect of Supplementing Organic Selenium on Performance, Carcass Traits, Oxidative Parameters and Immune Responses in Commercial Broiler Chickens[J]. Asian-Australasian Journal of Animal Sciences, 2013, 26(2): 247-252.

[151] Liao X, Lu L, Li S F, et al. Effects of selenium source and level on growth performance, tissue selenium concentrations, antioxidation, and immune functions of heat-stressed broilers[J]. Biological Trace Element Research, 2012, 150(1): 158-165.

[152] Xu D, Li W, Huang Y, et al. The Effect of Selenium and Polysaccharide of Atractylodes macrocephala Koidz. (PAMK) on Immune Response in Chicken Spleen Under Heat Stress[J]. Biological Trace Element Research, 2014, 160(2): 232-237.

[153] Liu L, Zhang J, Zhang Z, et al. Protective roles of selenium on nitric oxide-mediated apoptosis of immune organs induced by cadmium in chickens[J]. Biological Trace Element Research, 2014, 159(1): 199-209.

[154] Liu S, Xu F, Yang Z, et al. Cadmium-Induced Injury and the Ameliorative Effects of Selenium on Chicken Splenic Lymphocytes: Mechanisms of Oxidative Stress and Apoptosis[J]. Biological Trace Element Research, 2014, 160(3): 340-351.

[155] Xu F, Liu S, Li S. Effects of Selenium and Cadmium on Changes in the Gene Expression of Immune Cytokines in Chicken Splenic Lymphocytes[J]. Biological Trace Element Research, 2015, 165(2): 214-221.

[156] Zhao W, Liu W, Chen X, et al. Four endoplasmic reticulum resident selenoproteins may be related to the protection of selenium against cadmium toxicity in chicken lymphocytes[J]. Biological Trace Element Research, 2014, 161(3): 328-333.

[157] Sun G X, Chen Y, Liu C, et al. Effect of Selenium Against Lead-Induced Damage on the Gene Expression of Heat Shock Proteins and Inflammatory Cytokines in Peripheral Blood Lymphocytes of Chickens[J]. Biological Trace Element Research, 2016, 172(2): 474-480.

[158] Avanzo J L, De Mendonca C X, Pugine S M P, et al. Effect of vitamin E and selenium on resistance to oxidative stress in chicken superficial pectoralis muscle[J]. Comparative Biochemistry and Physiology C-toxicology & Pharmacology, 2001, 129(2): 163-173.

[159] Huang J, Li D, Zhao H, et al. The Selenium Deficiency Dis-

ease Exudative Diathesis in Chicks Is Associated with Downregulation of Seven Common Selenoprotein Genes in Liver and Muscle[J]. Journal of Nutrition, 2011, 141(9): 1605-1610.

[160] Yang Z, Liu C, Liu C, et al. Selenium Deficiency Mainly Influences Antioxidant Selenoproteins Expression in Broiler Immune Organs [J]. Biological Trace Element Research, 2016, 172(1): 209-221.

[161] Bartholomew A, Latshaw D, Swayne D E. Changes in blood chemistry, hematology, and histology caused by a selenium/vitamin E deficiency and recovery in chicks[J]. Biological Trace Element Research, 1998, 62(1): 7-16.

[162] Wang Q, Huang J, Zhang H, et al. Selenium Deficiency-Induced Apoptosis of Chick Embryonic Vascular Smooth Muscle Cells and Correlations with 25 Selenoproteins[J]. Biological Trace Element Research, 2017, 176(2): 407-415.

[163] Zhao X, Yao H, Fan R, et al. Selenium Deficiency Influences Nitric Oxide and Selenoproteins in Pancreas of Chickens[J]. Biological Trace Element Research, 2014, 161(3): 341-349.

[164] Xu J, Wang L, Tang J, et al. Pancreatic atrophy caused by dietary selenium deficiency induces hypoinsulinemic hyperglycemia via global down-regulation of selenoprotein encoding genes in broilers[J]. PloS One, 2017, 12(8): e0182079.

[165] Xu S, Yao H, Zhang J, et al. The Oxidative Damage and Disbalance of Calcium Homeostasis in Brain of Chicken Induced by Selenium Deficiency[J]. Biological Trace Element Research, 2013, 151(2): 225-233.

[166] Sheng P, Jiang Y, Zhang Z, et al. The effect of Se-deficient diet on gene expression of inflammatory cytokines in chicken brain[J]. BioMetals, 2014, 27(1): 33-43.

[167] Murphy M E, Kehrer J P. Altered contents of tocopherols in chickens with inherited muscular dystrophy[J]. Biochemical Medicine and Metabolic Biology, 1989, 41(3): 234-245.

[168] Kurskiĭ, Grigoreva V A, Medovar E N, et al. [ATPase activity and processes of calcium transport in membranes of sarcoplasmic reticulum of skeletal muscles with E-avitaminotic dystrophy][J]. Ukrainskiiĭ biokhimicheskiiĭ zhurnal, 1978, 50(1): 85.

[169] Yao H, Fan R, Zhao X, et al. Selenoprotein W redox-regula-

ted Ca^{2+} channels correlate with selenium deficiency-induced muscles Ca^{2+} leak[J]. Oncotarget, 2016, 7(36): 57618.

[170] Yao H, Wu Q, Zhang Z, et al. Gene Expression of Endoplasmic Reticulum Resident Selenoproteins Correlates with Apoptosis in Various Muscles of Se-Deficient Chicks[J]. Journal of Nutrition, 2013, 143(5): 613-619.

[171] Yao H, Zhao W, Zhao X, et al. Selenium Deficiency Mainly Influences the Gene Expressions of Antioxidative Selenoproteins in Chicken Muscles[J]. Biological Trace Element Research, 2014, 161(3): 318-327.

[172] Wu Q, Yao H, Tan S, et al. Possible Correlation of Selenoprotein W with Inflammation Factors in Chicken Skeletal Muscles[J]. Biological Trace Element Research, 2014, 161(2): 167-172.

[173] Huang J, Ren F, Jiang Y, et al. Selenoproteins protect against avian nutritional muscular dystrophy by metabolizing peroxides and regulating redox/apoptotic signaling[J]. Free Radical Biology and Medicine, 2015, 83: 129-138.

[174] Suliman H B, Ryan L K, Bishop L R, et al. Prevention of influenza-induced lung injury in mice overexpressing extracellular superoxide dismutase[J]. American Journal of Physiology-lung Cellular and Molecular Physiology, 2001, 280(1): 169-178.

[175] Chen X, Ren F, Hesketh J E, et al. Selenium blocks porcine circovirus type 2 replication promotion induced by oxidative stress by improving GPx1 expression[J]. Free Radical Biology and Medicine, 2012, 53(3): 395-405.

[176] Beck M A, Levander O A. Host nutritional status and its effect on a viral pathogen[J]. The Journal of Infectious Diseases, 2000, 182: S93-S96.

[177] Beck M A. Antioxidants and Viral Infections: Host Immune Response and Viral Pathogenicity[J]. Journal of the American College of Nutrition, 2001, 20(5): 384.

[178] Ilback N, Fohlman J, Friman G. Effects of selenium supplementation on virus-induced inflammatory heart disease[J]. Biological Trace Element Research, 1998, 63(1): 51-66.

[179] Jun E J, Ye J S, Hwang I S, et al. Selenium deficiency contributes to the chronic myocarditis in coxsackievirus-infected mice[J]. Acta Virologica, 2011, 55(1): 23-29.

[180] Harthill M. Review: Micronutrient Selenium Deficiency Influences Evolution of Some Viral Infectious Diseases[J]. Biological Trace Element Research, 2011, 143(3): 1325-1336.

第四章　硒在猪生产中的应用

作为一种重要的蛋白质来源，猪肉是全球产量最大的肉类。2006 年的数据表明，全球猪肉产量约为 1 000 亿 t，而禽肉产量约为 8 000 万 t，牛肉产量约为 6 500 万 t，羊肉产量约为 1 300 万 t。在全世界范围内，中国是猪肉产量最高的国家，约占全球猪肉总产量的一半。2018 年的数据表明，中国猪肉产量约为 5 400 万 t，在各种肉类消费中，猪肉消费占比达到 62.9%，人均猪肉消费量达 40 kg。

对猪肉需求量的提高无疑对养猪业的发展起到了巨大的推动作用，不仅表现为生产规模的扩大，还表现为母猪生产力的提高（每头母猪每年可繁殖的断奶仔猪数）以及生猪生长速度、饲料利用率和瘦肉量的提高或改善。为了更准确地满足现代猪基因型不断变化的营养和代谢需求，人们不断开发各种日粮和饲养策略，不仅表现在能量、蛋白质和氨基酸方面，而且表现在包括矿物质和维生素在内的微量营养素方面。由于遗传潜力和生产性能的不断提高，现代养猪业在保障人类对肉类需求方面的意义无疑是非常重大的，但同时也使动物极易处于应激状态，对猪肉生产效率的影响也是不容忽视的。

如本书第三章所述，氧化应激同样也是导致猪免疫能力下降、繁殖受阻以及生长发育不良的重要诱因。因此，在猪日粮中更加科学合理地使用抗氧化剂对于维持动物机体内氧化还原平衡，进而保证动物生产和生殖功能的最优发挥是必不可缺少的，需要依据猪的繁殖潜力、生长阶段，以及所处环境的变化不断做出调整。在不同的日粮抗氧化剂中，硒具有特殊的地位，通过合成体内多种硒蛋白，除了调节抗氧化防御外，还参与调节包括激素代谢在内的多种代谢活动。利用其强大的抗氧化能力，硒可以保证种公猪的精液质量、精子活力以及受精能力。硒也可以通过胎盘从怀孕母猪转移到胎儿，通过乳腺转移到初乳和常乳中，这对于保证胚胎的正常发育以及新生仔猪的存活和健康生长具有重要意义。此外，日粮补硒对断奶后和生长肥育期的猪也具有重要意义。

第一节 硒需要量

如前所述，自然条件下配合日粮中的硒不能满足动物对硒的需要，必须在日粮中额外添加。早在1974年，美国FDA批准在所有猪日粮中添加硒，添加水平为0.1 mg/kg日粮干物质。由于该添加水平并不能防止断奶仔猪出现缺乏症，1982年，FDA将批准将硒添加水平提高到了0.3 mg/kg（20 kg以上的断奶仔猪），此后这一标准一直到现在。允许在所有猪的日粮中添加最多0.3 mg/kg的硒，这一添加水平是基于组织硒浓度、硒平衡和细胞GPX活性而确定的。其他国家也相继批准在猪日粮中添加硒，添加水平及批准时间因地区不同而有所不同。例如，1975年，丹麦批准在猪日粮中添加硒，添加水平为0.1 mg/kg，1991年作了上调，开食日粮上调为到0.4 mg/kg，生长和肥育猪日粮以及繁殖母猪和公猪日粮上调为0.22 mg/kg。目前，多数国家允许在猪日粮中添加最高0.5 mg/kg的硒。除了日粮中直接添加硒之外，一些国家，如芬兰和新西兰，也通过在土壤中施用硒肥的方法提高饲料作物中的硒含量，进而使动物从这些饲料作物中摄取更多的硒。

尽管目前普遍推荐的在猪日粮中的硒添加水平为0.3 mg/kg，但这是否是最佳硒摄入量还有待进一步确定。在生产实践中确定猪的最佳硒需要量有很多的困难：①低应激条件下获得的结果不一定适用于商业高应激条件。例如，在试验条件下测试猪的数量通常比商业实践中的要小得多，而且两种条件下猪的健康状况和猪舍内的微生物种群也有很大的不同，这些均会影响到试验结果的准确性；②由于硒是抗氧化防御的重要组成部分，其需要量可能会因日粮类型以及日粮中其他抗氧化剂和促氧化剂水平的不同而不同；③猪的品种和基因型不同，其生理状态和生产力不同，对应激的敏感性也不同，因而从生产力中等的动物身上获得的数据并不适用于生产力高、对应激敏感的动物；④评估标准不统一导致获得的结果差异很大，硒需要量多数是基于血浆、全血、组织中地硒水平和GPX活性确定的。但是，GPX只是在动物和人体中发现的25种硒蛋白中的一种，很可能并不是评估硒需要量的最佳指标，因此，有必要进一步寻找更为灵敏的确定硒需要量的生理指标；⑤在大多数研究中，硒需要量是基于日粮中添加亚硒酸钠形式的硒确定的，但现在人们知道，有机硒具有更高的生物利用效率，而且可以在组织中形成硒储备，在这种情况下，基于动物硒需要量而确定的日粮硒添加量应该与无机硒不同；⑥维持体内各种生理功能所需的硒有很大的差异，维持较高免疫能力所需的硒通常要高于生长发育所需的硒；⑦动物在特定生理阶段（围产期）以及处于应激状态时，对硒的需要量提高。

低硒日粮补硒对猪生长发育的各个方面都有影响，且可能表现受多种因素影响。例如，在缺硒和维生素 E 含量低的母猪日粮中添加 100 IU/kg 的 α-生育酚醋酸酯和不同水平的硒（0～0.1 mg/kg），发现在 4 周的试验期后，补硒对仔猪的平均日增重、采食量和饲料转化率没有显著影响，而血清 GPX 活性随补硒水平的增加呈线性上升，血清硒浓度与 GPX 呈显著的正相关（R＝0.81）[1]。根据肝脏 GPX1 和血浆 GPX3 活性的断点回归分析表明，断奶仔猪的硒需要量为 0.3 mg/kg 日粮[2]。在猪的不同部位，GPX1 和 GPX4 之间的相对活性分布不同，而且随年龄发生变化。对 1 日龄、28 日龄和 180 日龄硒充足公猪 8 种不同组织的 GPX1 和 GPX4 的分析表明，大多数组织中 GPX4 的 mRNA 表达和活性随年龄增加而增加，1 日龄至 28 日龄公猪 GPX 活性的提高高于 GPX4，在 0.2 mg/kg 基础上逐渐提高硒添加量到 0.5 mg/kg 并没有显著影响 GPX 的活性，因此，断奶仔猪饲粮中 GPX1 和 GPX4 在不同组织中的全活性表达对硒的需要量为 0.2 mg/kg[3]。目前对猪硒需要量的推荐范围是 0.15～0.3 mg/kg，但在商品猪生产系统中应根据应激水平的提高而增加。事实上，仅从生理指标来看，生长猪的硒需要量是相当低的，以血清 GPX 活性作为标准，生长阶段猪日粮中的硒添加水平为 0.1 mg/kg 时即可满足，育肥阶段猪日粮中的硒添加水平为 0.05 mg/kg 时即可满足，当饲喂亚硒酸钠时，生长猪血清 GPX 活性在日粮硒水平为 0.1 mg/kg 时达到平台期，而饲喂富硒酵母时，则需要硒水平为 0.3 mg/kg 时达到平台期，但在育肥期两种硒源使血清 GPX 活性达到平台期所需的硒水平均为 0.1 mg/kg。在长白母猪的大麦-燕麦型基础日粮中设置 0.1 mg/kg、0.5 mg/kg 和 0.9 mg/kg 三个硒水平，以血浆 GPX 活性作为评价指标，结果表明，妊娠和非妊娠母猪的硒需要量为 0.5 mg/kg，而且妊娠母猪的死胎率在该水平时最低。

第二节　公猪硒营养与精子质量

猪精子具有对脂质过氧化高度敏感的特点[4]。事实上，哺乳动物精子在结构和化学成分上是独一无二的，其膜磷脂部分中含有高比例的多不饱和脂肪酸 PUFA，这一特征决定了精子的生物物理特性，如精子膜的高流动性和灵活性，这些生物物理特性赋予了精子的特定功能，如运动和受精能力。普遍认为，DHA 是人、公牛、猴子、公羊和公猪等哺乳动物中最重要的精子 PUFA。人类精子中 DHA 含量与精子活力[5]和精子细胞的正常形态[6]呈正相关。公猪精子中 DHA 和 n-3 PUFA 含量与精子运动和活力及精子细胞的正常形态呈正相关[7]。哺乳动物精子中 C20-C22 PUFA 的比例占总脂肪酸含量 50%以上。然而，精子膜中高比例 PUFA 赋予的这种功能优势使精子对自由基的攻击非常敏感，导致膜发生脂质过氧化，引起精子活力下降，不可逆转地破坏了精子的果糖分解能力和呼吸活动，提高了精子细胞胞内酶的释放，加速了精子在体外储存过程中的衰老。在孵育前受损的精子细胞对过氧化的敏感性增加，而在洗涤的精子悬浮液中加入过氧化的 PUFA 可以迅速且永久地固定精子[8,9]。近年来对精液中的脂质过氧化的更详细的研究表明，哺乳动物精液中的脂质过氧化是导致人类不育和家畜精液保存过程中精子质量下降的重要原因之一[10,11]。Blininger 等[12]研究发现，TBARS 与猪精子活力呈负相关。储存的猪精液中 MDA 水平的增加与精子的运动性和细胞膜完整性的迅速丧失有关[13]。运动性降低可能是由于 ROS 引起的 ATP 利用障碍或鞭毛收缩器损伤所致。有必要指出的是，来自每头公猪的精子可以一种依赖于宿主的方式对不同的精液处理技术做出反应，包括脂质过氧化和 DNA 片段化[14]。

一、抗氧化系统

自由基的产生和抗氧化防御之间的平衡被认为是决定猪精液质量（尤其是受精能力）的重要因素。抗氧化剂和抗氧化酶两类物质在精液中构成一个完整的抗氧化系统，能够保护精子免受自由基及其代谢产生的有毒产物的伤害。脂溶性抗氧化剂包括维生素 E、类胡萝卜素、泛醌等，水溶性抗氧化剂包括抗坏血酸、尿酸、牛磺酸等，抗氧化酶包括 SOD、GPX 和 CAT。此外，还包括由谷胱甘肽系统（谷胱甘肽/谷胱甘肽还原酶/谷氧还蛋白/谷胱甘肽过氧化物酶）和硫氧还蛋白系统（硫氧还蛋白/硫氧还蛋白过氧化物酶/硫氧还蛋白还原酶）构成的硫醇氧化还原系统。这些抗氧化化合物主要从三个水平上保护细胞，第一级防御主要负责通过清除自由基生成的前体以及具有催化自由基生成的催化剂来防止自由基的生成。超氧阴离子是细胞在生理条件下产生的主要自由基，因此 SOD 被认为是第一级抗氧化防御的主要成分。SOD 将超氧阴离子歧化为 H_2O_2，再进一步在 GPX 或 CAT 的催化下还原为水。过渡金属离子能加速脂质氢过氧化物分解成醛类、烷氧自由基和过氧自由基等细胞毒性产物，因此，一些金属结合蛋白（转铁蛋白、乳铁蛋白、结合珠蛋白、血红蛋白、金属硫蛋白、铜蓝蛋白、铁蛋白、白蛋白、肌红蛋白等)也属于第一级抗氧化防御的成分。然而,细胞的第一级抗氧化防御并不能完全防止自由基的形成,一些自由基会逃过第一级抗氧化防御系统,引发脂质过氧化链式反应,并对 DNA 和蛋白质造成损伤。因此,第二级抗氧化防御的主要作用是中断脂质过氧化链式反应,参与的抗氧化剂包括维生素 E、泛醌、类胡萝卜素、维生素 A、抗坏血酸、尿酸等。谷胱甘肽和硫氧还蛋白系统在第二级抗氧化防御中也起着重要作用。第二级防御通过清除链式反应的过氧化中间产物来防止脂质过氧化的传播。例如,维生素 E 可以将脂质过氧化自由基还原为脂质氢过氧化物,脂质氢过氧化物进一步被 GPX 还原解毒,在这一反应过程中,维生素 E 被氧化为维生素 E 自由基,维生素 E 自由基再被抗坏血酸还原为活性形式。即使是两层抗氧化防御屏障也并不能完全阻止 ROS 和 RNS 对脂质、蛋白质和 DNA 的损伤,在这种情况下,第三级防御是基于对受损分子的清除或修复的系统,参与这一水平抗氧化防御的物质主要是一些与脂质、蛋白质和 DNA 水解和修复相关的酶类。

二、公猪硒营养与精子质量

硒对精子质量的重要作用在 20 世纪 80 年代初就被证明了。在轻度硒缺乏的情况下,硒会被优先保留在睾丸中,在哺乳动物的所有组织中,睾丸中的硒含量最高。静脉注射^{75}Se 后,^{75}Se 在肾脏中含量最高,其次是精囊和睾丸。进行性硒缺乏与精子细胞和精子形态改变有关,随后成熟生精细胞完全消失。由缺硒而导致的精子发生破坏在包括猪在内的几种动物中均有报道[15]。GPX4 在精子中还起着特殊的结构蛋白的作用,硒缺乏导致的 GPX4 降低会影响精子尾部的完整性和形态,从而影响精子的活力以及成熟精子细胞的数量和睾丸精子储备。维生素 E 可提高精子浓度,但对精子细胞结构异常没有影响。因此,硒与维生素 E 相比似乎发挥着更大作用,对精子发育和成熟是必不可少的。

公猪日粮中添加硒与添加维生素 E 相比,精子数量和浓度更高,异常精子数量减少,精子具有更高的活动能力,更高的受精能力,受精母猪的受精率也较高[16]。日粮添加硒的公猪与添加维生素 E 相比有更多的精子储备,而且日粮加硒与不加硒相比精子质膜与尾部的连接更加紧密,精子中 ATP 浓度更高,而且高于维生素 E 添加[15]。Jacyno 等[17]研究发现,与无机硒相比,有机硒添加对猪的射精量和活动精子百分率没有影响,但提高了精子的数量和浓度,顶体正常的精子比例较高,有轻微或严重形态异常的精子比例较低。Martins 等研究表明,与亚硒酸钠相比,日粮硒酵母添加显著提高了冷冻精液的 GPX4 活性，不影响 72 h 内冷冻精液的存活率，显著降低了精子头部畸形率[18]。日粮添加 0.6 mg/kg 的有机硒提高了猪精液的硒水平、GPX 活性、GSH/GSSG 的比值以及总抗氧化能力[19]。这些数据表明，有机硒提高精子质量的效果可能优于无机硒。需要注意的是，日粮硒添加对公猪繁殖是否有影响或影响程度与日粮中硒的基础水平直接相关。基础日粮硒水平相对较低(0.06～0.07 mg/kg)时，睾丸结构出现了特征性的有害变化。

毋庸置疑，硒对精子的抗氧化保护是通过提高硒依赖抗氧化酶的活性来实现的，尤其是 GPX4，精子中 GPX4 受日粮硒状况的影响，日粮有机硒添加可提高精子中 GPX4 的 mRNA 水平[20]。但也有研究表明，日粮缺硒时，附睾仍然可以有效地抵御不断增加的过氧化条件，缺硒动物的附睾头部显示了有限的脂质过氧化物的产生，总的 GPX 活性没有受到硒供应缺乏的显著影响，并且观察到非硒依赖性的 GPX5 mRNA 和蛋白水平的增加[21]。

三、精液抗氧化能力与精子质量

猪精液中的抗氧化剂-氧化剂平衡是精子许多生理过程的重要调节因子，包括精子与输卵管上皮细胞的粘附、获能和激活、与卵母细胞透明带的结合、顶体反应、穿透透明带以及与卵膜融合和穿透[22]。由于脂质和蛋白质氧化导致的精子膜功能受损，不可避免地影响上述过程，并降低体内受精成功的几率。因此，抗氧化保护是维持精子膜完整性、活力和受精能力的重要措施。

猪精浆中存在一种分泌型胞外形式的 Cu/Zn-SOD，与精子中 SOD 的抗原决定簇彼此相似，但浓度较低[23]。猪精浆中的 CAT 浓度相对较低，只有 H_2O_2 浓度大大高于生理水平时才发挥作用[24]，可能是含有 CAT 的过氧化物酶体在精子发生过程中从生殖细胞中消除的原因[25]。在猪精浆中 GPX 相对缺乏甚至检测不到[26]，而维生素 E 几乎检测不到[27]。需要提到的是，猪精浆中存在大量的 GSH[28]。

由于精浆中相对较低的 SOD、CAT 和 GPX 活性和维生素 E 的缺乏，猪精子对 ROS 诱导的毒性效应的抵抗力较差。在过冷却或冷冻过程中容易受到许多应激因素的影响，导致膜结构和功能完整性破坏、精子的运动性、线粒体功能、膜电位和生育力降低[29]。研究发现，在液氮中长期储存精液会导致大多数精子在解冻后活力、线粒体功能和质膜完整性降低，具体表现为顶体膜的渗透抗性显著降低，对脂质过氧化和 DNA 片段化的敏感性显著提高[30]。在猪精液稀释剂中添加海藻酸盐提高了 SOD 和 GPX 活性,降低了 MDA 水平,改善了解冻后精子的活力、功能完整性和抗氧化能力[31]。在猪精液稀释剂中添加维生素 E 明显抑制了精子存活率的降低,减少了精子的过氧化反应(TBARS 的产生),保存在处理稀释液中的精子迅速富含 α-生育酚,同时培养液中维生素 E 含量降低,在稀释液中加入 α-生育酚,可完全阻止精子在保存期内 DHA 的降低[32]。在解冻培养液中添加 GSH 减少了获能活精子的数量和膜蛋白中巯基改变的精子数量,降低了 ROS 生成和染色质凝聚,提高了体外卵母细胞穿透能力和解凝精子头的比例[33]。与对照组相比,亚硒酸钠、SeMet 和维生素 E 添加显著提高了猪精子的活力、存活率和顶体反应[34]。通过在稀释剂中添加抗氧化剂来减少猪精液在储存过程中的脂质和蛋白质氧化的各种尝试中,应非常小心,因为越来越多的证据表明,低水平的 ROS 参与了哺乳动物精子功能的生理调控。ROS 可通过各种信号通路介导猪精子的获能,当猪精子暴露在 ROS 生成系统并随后培养在获能介质中时,随着时间的推移,顶体反应显著增加[35,36]。一般来说,低浓度的 ROS 可以诱导精子激活、获能、精-卵细胞融合和顶体丢失,而高浓度的 ROS 会抑制精-卵细胞融合,降低精子活力和损伤精子 DNA。

第三节　母猪硒营养对胎儿和新生仔猪的影响

近年来,人们提出一种叫作生殖应激的假说,认为在包括发情、妊娠、分娩和哺乳等生殖过程中均会产生应激,包括胎儿激素分泌对母体造成的被动应激[37]。大量证据表明,氧化应激是哺乳动物生殖功能的重要影响因素,包括卵泡发育、排卵、受精、黄体生成、子宫内膜增厚和脱落、胚胎发育、着床以及早期胎盘的生长和发育[38]。事实上,由于胎盘线粒体活动增强,妊娠期被认为是一个持续的氧化应激期。

ROS 在雌性生殖道中具有双重作用,生理浓度的 ROS 起着关键的信号分子的作用,但过量 ROS 生成引起的氧化应激被认为是包括妊娠并发症在内的多种疾病发病的病理因素[39]。妊娠过程产生的 ROS 对胎盘功能(滋养层细胞增殖、分化和血管反应性)有显著影响[40]。怀孕母猪的特点是抗氧化防御能力降低,DNA 损伤增加。研究表明,从授精后 80 天一直到分娩,母猪血清维生素 E 水平下降,与妊娠第 30 天相比,妊娠末期(110 天)血浆维生素 E 浓度降低了 56%,从妊娠第 30 天开始一直到断奶期,淋巴细胞 DNA 损伤显著增加[41]。有研究表明,母猪血清中抗坏血酸的水平在妊娠的最初 60～80 天期间相对稳定，随后其浓度降低并持续到这个泌乳期，血清硒在妊娠 60d 左右开始下降，并在临近分娩时下降得更快，与此同时，GPX 活性也在下降[42]。环境应激因素可损害母猪的抗氧化防御系统，从而诱导产生氧化应激。例如，在日粮中添加鱼油会引起母猪的氧化应激并增加脂质过氧化，热应激会诱导母猪产生氧化应激。

在雌性动物（包括母猪）生殖过程中，抗氧化保护是极其重要的，日粮补硒是维持母猪抗氧化防御以及预防母猪妊娠后期氧化应激对胎儿的生长和健康以及仔猪的产后生长产生负面影响的重要手段[43]。当饲喂基础日粮时，母猪血清 GPX 活性从产前 6 天开始下降，并在整个哺乳期保持在较低水平，母猪日粮中添加硒可使产后 7 天和 14 天母猪血清硒浓度和血清 GPX 活性升高[44]。与饲喂补硒日粮的母猪相比，饲喂低硒日粮的母猪在妊娠期间全血和肝脏中的硒浓度降低，随着妊娠过程的发展，尽管两个处理的胎儿肝脏硒含量都降低了，但低硒日粮饲喂的母猪胎儿肝脏硒含量的下降幅度更大[45]。此外，饲喂低硒日粮的母猪胎儿尽管 GPX 活性没有降低，但肝脏匀浆中脂质过氧化指数较高，这可能说明 GPX 以外的硒蛋白受到影响，从而降低了胎儿对氧化应激的抵抗力。

母猪日粮硒缺乏时产仔数降低，仔猪死亡率增加[46,47]。硒和维生素 E 同时缺乏时，母猪生产的后代出生时生活力弱，哺乳能力低。低硒母猪日粮中添加 0.1 mg/kg 的硒提高了产仔数、产活仔数和断奶仔猪数，仔猪出生和断奶时的平均窝重也较高[48]。Pehrson 等[49]报道，即使母猪饲料中添加无机硒 0.35 mg/kg，新生仔猪血清中 GPX 的活性在出生后第一周也很低，作者认为新生仔猪的硒状况可能比维生素 E 状况对它们的健康和免疫力更重要。

在胎儿-新生儿过渡期间和出生时，与组织中 O_2 分压升高相关的循环和呼吸变化，加上刚出生时幼稚的抗氧化系统，会导致新生仔猪极易遭受严重的氧化应激。当新生仔猪暴露在高氧环境时，抗氧化防御能力降低，游离铁的水平提高，因而产生剧毒的羟基自由基[50]。由于胎盘转移有限，仔猪刚出生时组织维生素 E 水平较低（包括肝脏、脑、肾脏和心脏）。随着妊娠的发展，胎猪逐渐失去了合成维生素 C 的能力，仔猪在出生后第 1 周内几乎不合成维生素 C，完全依赖于胎盘转移的维生素 C 和初乳中提供的维生素 C。因此，氧化应激是妊娠母猪和新生仔猪的主要关注点，提高母猪和仔猪在这些关键期的抗氧化防御能力，对它们的健康、生产性能和繁殖性能具有重要意义。仔猪出生后第一周是抗氧化系统发育最重要的时期。仔猪血浆中维生素 E 水平和 GPX 活性在出生时很低，但到 3 日龄时可显著上升[51]。仔猪血浆中 SOD 活性在出生后第一周显著升高，血浆脂质和蛋白质的氧化显著降低，回肠 Nrf2 信号通路的表达显著增加[52]。从猪乳中获得重要的抗氧化剂维生素 E 和硒是提高新生仔猪抗氧化防御能力的重要途径。初乳中维生素 E 和硒水平在母猪间差异较大，但与常乳相比水平较高，产后 1 周，母猪乳中维生素 E 和硒的浓度分别下降了 86% 和 68%[53]。猪乳中硒和维生素 E 的含量随着胎次的增加而下降[54]，这表明年龄较大的母猪所生的小猪在出生时和出生后早期发育过程中可能面临更大的氧化应激风险，随着产次的增加，母猪可能需要更多的维生素 E 和硒来维持它们的体储备，以保证这些重要的抗氧化剂能够有效地转移给它们的后代。

新生仔猪的消化系统还不成熟，但在出生的头几周内活跃地发育。因此，初乳和常乳是促进这一发育过程极其重要的、必需营养素来源。如前所述，抗氧化剂维生素 E 和硒通过初乳从母猪传给子代来支持快速生长仔猪的抗氧化防御。

新生仔猪的免疫系统不成熟，自身主动免疫的发展可能需要几周的时间，因此对各种病原体极为敏感。在这一敏感时期，初乳和常乳中的抗氧化剂发挥着重要的免疫调节作用。

环境温度是影响仔猪生长发育的重要环境因子。由于仔猪出生时体内脂肪储备有限(仅占体重的 1%)，体温调节反应是维持体温的关键。仔猪的体温调节依赖于甲状腺激素功能，硒作为碘甲腺原氨酸脱碘酶的一部分，参与了仔猪甲状腺激素的激活和体温调节。此外，氨、一氧化碳和硫化氢等环境污染物会引起仔猪的氧化应激，因此增加了对额外抗氧化保护的需求。

第四节　断奶仔猪硒营养

追求更高的生长速度一直以来是养猪业发展的重要目标。随着遗传育种技术的发展，猪的遗传潜力在不断提高，利用营养的能力也在不断提高，因而生长速度也不断提高。然而，生长速度的提高伴随而来的对应激的敏感性提高，抗氧化防御在防止各种应激源的破坏性影响方面起着重要的作用。事实上，具有较高瘦肉组织生长潜力的基因型有时以较低的饲料摄入量为特征，因此需要比其他基因型更高的营养规格和更多的营养关注。满足现代快速生长仔猪日益增长的营养需求是营养学家面临的重大挑战。断奶时间、断奶后日粮选择和饲养策略等管理因素都会影响断奶仔猪的生长发育。

断奶对仔猪来说是一个巨大的挑战，断奶过程的成功与否决定了仔猪未来的生产性能和繁殖特性。断奶对猪来说会面临一系列应激，包括离开母猪和原有小猪产生的应激；从以液态奶为主的日粮转变为以饲料为主的日粮引起的消化生理和免疫状态的变化；转移到新的陌生的环境（猪舍温度的改变、不熟悉的小猪、可能过度拥挤）引起的应激。因此，成功的断奶是仔猪从断奶后到屠宰的整个生长过程能够高效利用饲料以及高效生长的关键。

仔猪开始断奶后，机体的硒储备会发生变化，血清和组织中的硒浓度下降。这些储备取决于猪乳的摄入量和乳中硒的浓度以及提供的代乳料中的硒状况。断奶后仔猪的食欲一般都很低，伴随着其他代谢挑战，这些储备也会下降。研究表明，在这一关键时期，硒缺乏在许多养殖场普遍存在，可能导致断奶后死亡率的增加。缺硒会影响断奶仔猪的生长发育，断奶仔猪体型越大、生长速度越快，越容易发生猝死。如果体内硒储备不足，断奶应激会加剧缺硒。因此，一些研究建议21日龄断奶仔猪在断奶后2周左右需要补充0.5 mg/kg的硒，此后硒添加量降低到0.35 mg/kg。

补硒对断奶仔猪影响的研究发现，硒添加不影响胰腺重量、蛋白质含量及胰蛋白酶、糜蛋白酶、α-淀粉酶和脂肪酶活性，胰腺中硒浓度随着硒补充水平的增加呈线性和二次曲线增加，胰腺硒水平与GPX活性呈正相关，干物质和氮的表观消化率随硒添加量的增加呈线性增加，但对乙醚浸出物的消化率没有显著影响[1]。在4周龄断奶仔猪日粮中添加0 mg/kg、0.3 mg/kg、0.5 mg/kg和1.0 mg/kg的亚硒酸钠，饲喂5周后发现，粪硒排出量和尿硒浓度随日粮硒水平的增加而增加，当日粮硒水平增加到1.0 mg/kg时，体内硒的留存量与进食量成正比，硒添加日粮中硒的表观消化率约为70%，而基础日粮中硒的消化率仅为30%～40%[55]。低剂量硒添加（0.1 mg/kg）增加了硒在体内的存留率，从45.5%提高到61.7%，但高剂量硒添加（0.5 mg/kg）降低了硒在体内的存留率，两种添加水平下硒在体内的存流量无明显区别[56]。

第五节　硒与猪的健康

临床缺硒在商品猪生产中很少见，但亚临床缺硒可能是导致猪生产性能和繁殖性能下降的原因。事实上，新生、出生后和断奶后的猪均可发生缺硒（Close，2003）。现代化猪生产模式与30年前相比截然不同，养殖模式的改变可能会导致生产过程中的硒缺乏或不足，主要有以下几方面的原因：①现代养猪对动物的限制消除了土壤作为猪的一种额外的硒来源；②不断改良的遗传学性状使猪的生长速度不断加快，瘦肉率和饲料转化效率不断提高，这无疑会提高猪对硒的需要量；③不断提高的饲料作物产量引起土壤硒的耗竭，最终导致植物中的硒含量降低，化学合成肥料的过度使用降低了植物对土壤硒的吸收，现代农业生产还会导致土壤 pH 降低，进一步降低了土壤硒的有效性以及植物对这些硒的吸收；④单一的玉米-豆粕型基础日粮的广泛使用排除了猪利用多样化植物来源硒的可能；⑤由于母猪硒摄入不足，初乳和常乳中的硒含量降低，因而降低了早期断奶仔猪体内的硒储备；⑥母猪生产力的提高、早期再繁殖以及母猪在猪群中保留时间更长，可能导致母体硒储备耗竭。

缺硒对猪的生产和繁殖具有诸多不利影响，缺硒的后果与维生素 E 缺乏相似。考虑到硒和维生素 E 在体内相互作用的复杂性，硒缺乏的临床症状会因同时缺乏维生素 E 而变得复杂。硒缺乏或不足首先会影响到猪的生产性能，饲料摄入量、饲料利用率、养分消化利用率鸡生长速度降低。严重硒缺乏可导致肝坏死，亚急性肝坏死常伴有腹水、黄疸、水肿和心肌病，急性肝坏死则会导致肝功能衰竭和动物猝死。肝坏死常发生在 3 月龄以下的小猪身上，在死亡前通常不表现临床症状，肝脏组织学检查表现出广泛的坏死，肝组织中硒浓度显著降低，可以与营养性肌营养不良和桑椹心病合并发生。硒缺乏可导致食道和胃黏膜溃疡，盲肠和结肠出血，引起肠道水肿，断奶后发生腹泻和便血。和鸡相似，硒缺乏也可引起猪发生 NMD，也称白肌病，以双侧骨骼肌苍白和营养不良为特征，全身肌肉无力，走路步态不稳，骨骼肌条纹消失、空泡化，有透明变性，最长肌、膈肌和内收肌最易受到影响，肝脏、骨骼肌、心肌和胰腺中的硒含量显著降低，常合并营养性肝病和桑椹心病。硒缺乏可导致猪发生桑椹心病，是一种营养性微血管病变，常发生在快速生长的猪身上，主要表现为血管和肌细胞病变，心肌内毛细血管以及毛细血管前小动脉纤维素样坏死，可与亚急性肝坏死一起发生，受影响的动物会突然死于没有先兆的急性心力衰竭。这种疾病的临床症状包括呼吸困难和严重的肌肉无力，肝脏和心脏中的硒浓度显著降低。硒缺乏或不足可导致母猪产后疾病（乳腺炎、子宫炎、母乳障碍）的发病率提高，主要和硒缺乏或不足引起的围产期母猪代谢功能紊乱和免疫功能障碍有关。

第六节　有机硒在猪营养中的应用优势

与在家禽日粮中使用无机硒一样，在猪的日粮中使用亚硒酸盐也具有很多局限性，如毒性相对较高、具有促氧化作用、无法在体内建立和维持硒储备、向动物产品转移的效率低以及与其他矿物质的相互作用等。将母猪日粮中的亚硒酸钠的水平从 0.1 mg/kg 提高到 0.9 mg/kg（以硒计）时，母猪临产时平均血硒浓度、初乳平均硒浓度以及新生仔猪平均血硒浓度只是适度增加[57]。有机硒可很好地避免上述问题。人体动力学试验研究表明，SeMet 的小肠吸收率高于亚硒酸钠（98%对 84%），肝脏对吸收后 SeMet 的摄取速度快于亚硒酸钠，SeMet 的粪便排泄量低于亚硒酸钠（4%对 18%），尿排泄量低于亚硒酸钠（11%对 17%），总排泄量低于亚硒酸钠（15%对 35%），SeMet 在体内的存留时间高于亚硒酸钠（363 天对 147 天）[58]，表明 SeMet 的生物利用效率及再利用效率要高于亚硒酸钠。因此，从营养学的角度讲，富含 SeMet 的有机硒更应该作为硒的一种营养形式，因为它可以更有效地维持机体最佳的硒营养状况，而亚硒酸钠更应该作为一种药物来治疗硒缺乏。与其他动物物种相似，有机硒在猪营养中的重要性，主要来自两个方面：首先，有机硒可以在组织中建立硒储备，这些硒储备在应激条件下可以用来维持和提高机体的抗氧化防御能力；其次，SeMet 形式的有机硒可以通过胎盘、初乳和常乳将硒有效地从母体转移到胎儿和新生仔猪。

母猪在分娩前 60 天开始饲喂 0.1 mg/kg 和 0.3 mg/kg 的亚硒酸钠或硒酵母直到 21 日龄断奶，结果表明，硒酵母中的硒可更有效地通过胎盘转运，使新生仔猪腰部和肝脏中的硒含量更高；硒酵母中的硒可更有效地转移到初乳和常乳中，使仔猪断奶时腰部硒浓度更高[59]。尽管将无机形式的硒从 0.1 mg/kg 提高到 0.3 mg/kg 也提高了新生仔猪腰部和断奶仔猪腰部的硒含量以及乳硒含量，但有机来源的硒的提高效果更为显著[60]，添加有机硒使泌乳第 7 天和第 14 天的乳硒含量提高了 2.5～3 倍（与无机硒添加相比）[44]。另一项 Mahan 和 Peters（2004）对母猪日粮添加有机硒和无机硒的比较研究表明，硒酵母中的硒可更有效地转移到初乳、常乳和母猪毛中；与亚硒酸钠相比，相同硒水平（0.3 mg/kg）的硒酵母添加显著提高了母猪肝脏、腰部和胰腺中的硒水平，硒酵母添加组的新生仔猪肝脏和腰部的硒水平是亚硒酸钠添加组的新生仔猪的 2 倍；进一步对新生仔猪全身硒含量的研究发现，母猪日粮未添加硒时新生仔猪全省硒含量为 0.075 mg/

头，添加 0.15 mg/kg 亚硒酸钠后为 0.079 mg/头，添加相同剂量的硒酵母后为 0.123 mg/头，提高了 64%；当母猪日粮硒添加水平提高到 0.3 mg/kg 时，硒酵母添加使新生仔猪全身硒含量增加 1 倍多，达到 0.2 mg/头，而同剂量亚硒酸钠添加仅使新生仔猪全身硒含量提高到 0.1 mg/头[47]。这些研究数据证实了有机硒可以更有效地通过胎盘转移给胎儿，从而维持发育中胎儿以及刚出生仔猪的硒营养状况，同时有机硒可以更有效地通过乳腺组织转移到初乳和常乳中，较高的乳硒浓度可以改善仔猪从出生到断奶时的硒状况，有助于降低饲养实践中经常遇到的断奶后死亡的发生和严重程度。由于有机硒中的 SeMet 可以通过非特定的方式取代乳蛋白中的蛋氨酸，因而可以提高猪乳中的硒含量，这和有机硒提高鸡蛋中以及牛奶中硒含量遵循相同的原理。对牛奶中硒存在形式的研究发现，硒含量的增加主要以 SeMet 的增加为主。奶牛日粮以 0.14 mg/kg 的硒酵母取代亚硒酸钠，可将牛奶中 SeMet 形式的硒从 36 ng/g 干样提高到 111 ng Se/g 干样，随着硒酵母的取代率进一步提高到 0.29 mg/kg 时，牛奶中 SeMet 形式的硒进一步增加到 157 ng Se/g 干样[61]。日粮添加有机硒可显著提高猪乳中的硒含量，大约是添加无机硒的 2.5～3 倍，而且显著提高了 7 日龄和 14 日龄仔猪血清中的硒含量[44]。

Fortier 等研究了日粮硒源对初产母猪胚胎发育的影响，试验从母猪发情开始到人工授精后 30 天，结果表明，与亚硒酸钠相比，硒酵母添加提高了子宫向胚胎的硒转移，并促进了胚胎的发育，整窝和单个胚胎的含硒量分别提高了 63%和 52%，硒酵母添加母猪胚胎的长度、重量和蛋白质含量均大于亚硒酸钠添加。这些研究结果表示，日粮有机硒对于有效地将硒转移到胚胎以及随后胚胎在妊娠早期的发育是至关重要的。Dalto 等[62]研究发现饲喂 0.3 mg/kg 硒酵母或亚硒酸钠的受孕母猪的扩张囊胚（5 日龄胚胎）中有 96 个差异表达基因，遗憾的是这种差异所代表的具体生物学过程没有被进一步探讨，但可以肯定的是这种差异必然有它的特殊生物学意义。Hu 等[63]研究了不同硒源对母猪血清和乳硒浓度以及抗氧化状态的影响，试验从妊娠的最后 32 天持续到哺乳期第 28 天，基础日粮硒含量为 0.042 mg/kg，在此基础上分别添加 0.3 mg/kg 的亚硒酸钠和 SeMet，结果表明，与亚硒酸钠相比，SeMet 添加显著提高了母猪血清硒浓度以及初乳和常乳中的硒含量，而且母猪血清 T-AOC 显著升高，MDA 水平显著降低，初乳和常乳中 T-AOC、GPX、SOD 和 GSH 水平显著升高，MDA 水平显著降低，说明有机硒添加可更有效地提高母猪的抗氧化防御能力，而这对于在分娩和断奶这两个母猪生命中最敏感的时期保证母猪的健康至关重要。在与 Hu 等（2011）相同试验设计的基础上，Zhan 等[64]研究了母猪日粮 Se-

Met 添加对仔猪硒营养状况和抗氧化功能的影响，SeMet 添加不仅显著提高了初乳和常乳中的硒含量以及 28 日龄断奶仔猪血清、肝、肾、胰腺、肌肉、胸腺和甲状腺中的硒含量，而且提高了仔猪肝脏、肾脏、胰腺、肌肉和血清中的 GPX 活性和 T-AOC、胰腺、肌肉和血清中的 SOD 活性以及肾脏、胰腺、肌肉和血清中的 GSH 水平，降低了仔猪肝脏、肾脏和胰腺中的 MDA 水平。Zhan 等[64]还研究了母猪日粮 SeMet 添加对仔猪甲状腺代谢和消化酶活性的影响，SeMet 添加提高了仔猪血清中的 T3 浓度，降低了 T4 浓度，提高了仔猪胰腺组织中蛋白酶、淀粉酶和脂肪酶的活性。血清 T3 水平的提高促进了蛋白质合成和能量产生，消化酶活性的提高改善了对营养物质的利用，这可能是仔猪从出生到断奶的日增重显著增加的原因。

上述大部分关于母猪繁殖中硒的母体效应的研究都没有显示出用有机硒取代无机硒对母猪及其后代生产性能和繁殖性能的优势，但这不不代表没有优势，原因可能有以下几个方面：①试验研究中通常使用的动物数量较少，因而无法显现母猪繁殖性能或子代生长发育可能存在的差异；②在动物数量较少的实验条件下，母猪或仔猪所受的应激会降低或消除，因而对抗氧化防御的需求不像实际生产过程中那么高。因此，要想真正了解有机硒对母猪繁殖性能及其后代生长性能的影响，需要在商业化生产过程中精心设计大型试验。事实上，用有机硒取代无机硒除了对猪营养有优势外，还对家禽营养（本书第三章所述）以及反刍动物营养（本书第五章所述）有诸多益处。因此，在不就的将来，有机硒的概念必然会被引入到畜牧业当中。

为数不多的研究比较探讨了无机硒和有机硒对断奶仔猪的影响。Huang 等[65]比较了有机硒和无机硒添加对 21 日龄至 42 日龄断奶仔猪的生长性能和腹泻发病率的影响，有机硒添加提高了平均日增重（291 g/d 比 274 g/d），降低了腹泻发生率（1.32％比 1.72％），每千克体重增长的饲料成本降低了 11％。与亚硒酸钠相比，在生长猪日粮中添加硒酵母显著增加了硒在体组织内的存留量[60]。Krska 等[66]研究了有机硒对猪肉颜色稳定性的影响，在屠宰前 60 天，猪日粮中添加高水平的维生素 E 和有机硒改善了屠宰后第 7 天猪肉的颜色稳定性（反射率），维生素 E 和有机硒对肉色的积极影响与降低猪肉的脂质过氧化水平有关。事实上，生长猪饲喂亚硒酸钠会增加肉的滴水损失，但有机硒代替亚硒酸钠可有效避免这一问题。

参考文献

[1] Adkins R S，Ewan R C. Effect of supplemental selenium on pan-

creatic function and nutrient digestibility in the pig[J]. Journal of Animal Science, 1984, 58(2): 351-355.

[2] Meyer W R, Mahan D C, Moxon A L. Value of Dietary Selenium and Vitamin E for Weanling Swine as Measured by Performance and Tissue Selenium and Glutathione Peroxidase Activities[J]. Journal of Animal Science, 1981, 52(2): 302-311.

[3] Lei X G, Dann H M, Ross D A, et al. Dietary Selenium Supplementation Is Required to Support Full Expression of Three Selenium-Dependent Glutathione Peroxidases in Various Tissues of Weanling Pigs[J]. Journal of Nutrition, 1998, 128(1): 130-135.

[4] Cerolini S, Maldjian A, Surai P F, et al. Viability, susceptibility to peroxidation and fatty acid composition of boar semen during liquid storage[J]. Animal Reproduction Science, 2000, 58(1): 99-111.

[5] Conquer J A, Martin J B, Tummon I, et al. Fatty acid analysis of blood serum, seminal plasma, and spermatozoa of normozoospermic vs. Asthernozoospermic males[J]. Lipids, 1999, 34(8): 793-799.

[6] Lenzi A, Gandini L, Maresca V, et al. Fatty acid composition of spermatozoa and immature germ cells[J]. Molecular Human Reproduction, 2000, 6(3): 226-231.

[7] Amin N, Kirkwood R N, Techakumphu M, et al. Lipid profiles of sperm and seminal plasma from boars having normal or low sperm motility[J]. Theriogenology, 2011, 75(5): 897-903.

[8] Castellano C A, Audet I, Bailey J L, et al. Dietary omega-3 fatty acids (fish oils) have limited effects on boar semen stored at 17 ℃ or cryopreserved[J]. Theriogenology, 2010, 74(8): 1482-1490.

[9] Satorre M M, Breininger E, Beconi M T. Cryopreservation with α-tocopherol and Sephadex filtration improved the quality of boar sperm [J]. Theriogenology, 2012, 78(7): 1548-1556.

[10] Naher Z, Ali M I, Biswas S K, et al. Effect of oxidative stress in male infertility[J]. Mymensingh Medical Journal, 2013, 22(1): 136.

[11] Chen S, Allam J, Duan Y, et al. Influence of reactive oxygen species on human sperm functions and fertilizing capacity including therapeutical approaches[J]. Archives of Gynecology and Obstetrics, 2013, 288 (1): 191-199.

[12] Breininger E, Beorlegui N B, Oflaherty C, et al. Alpha-to-

copherol improves biochemical and dynamic parameters in cryopreserved boar semen[J]. Theriogenology, 2005, 63(8): 2126-2135.

[13] Kumaresan A, Kadirvel G, Bujarbaruah K M, et al. Preservation of boar semen at 18 ℃ induces lipid peroxidation and apoptosis like changes in spermatozoa[J]. Animal Reproduction Science, 2009, 110(1): 162-171.

[14] Parrilla I, Olmo D D, Sijses L, et al. Differences in the ability of spermatozoa from individual boar ejaculates to withstand different semen-processing techniques[J]. Animal Reproduction Science, 2012, 132(1): 66-73.

[15] Marin-Guzman J, Mahan D C, Whitmoyer R. Effect of dietary selenium and vitamin E on the ultrastructure and ATP concentration of boar spermatozoa, and the efficacy of added sodium selenite in extended semen on sperm motility[J]. Journal of Animal Science, 2000, 78(6): 1544-1550.

[16] Marin-Guzman J, Mahan D C, Chung Y K, et al. Effects of dietary selenium and vitamin E on boar performance and tissue responses, semen quality, and subsequent fertilization rates in mature gilts[J]. Journal of Animal Science, 1997, 75(11): 2994-3003.

[17] Jacyno E, Kawecka M, Kamyczek M. Influence of inorganic Se^{+} vitamin E and organic Se^{+} vitamin E on reproductive performance of young boars[J]. Agricultural and Food Science, 2002, 11(3): 175-184.

[18] Martins S M M K, De Andrade A F C, Zaffalon F G, et al. Organic selenium supplementation increases PHGPx but does not improve viability in chilled boar semen[J]. Andrologia, 2015, 47(1): 85-90.

[19] Horky P, Jancikova P, Sochor J, et al. Effect of Organic and Inorganic Form of Selenium on Antioxidant Status of Breeding Boars Ejaculate Revealed by Electrochemistry[J]. International Journal of Electrochemical Science, 2012, 7: 9643-9657.

[20] Speight S M, Estienne M J, Harper A F, et al. Effects of dietary supplementation with an organic source of selenium on characteristics of semen quality and in vitro fertility in boars1[J]. Journal of Animal Science, 2012, 90(3): 761-770.

[21] Vernet P, Rock E, Mazur A, et al. Selenium-independent epididymis-restricted glutathione peroxidase 5 protein (GPX5) can back up

failing Se-dependent GPXs in mice subjected to selenium deficiency[J]. Molecular Reproduction and Development, 1999, 54(4): 362-370.

[22] Bailey J L, Bilodeau J, Cormier N. Semen cryopreservation in domestic animals: a damaging and capacitating phenomenon[J]. Journal of Andrology, 2000, 21(1): 1-7.

[23] Orzolek A, Wysocki P, Strzezek J, et al. Superoxide dismutase (SOD) in boar spermatozoa: purification, biochemical properties and changes in activity during semen storage (16 ℃) in different extenders [J]. Reproductive Biology, 2013, 13(1): 34-40.

[24] Vernet P, Aitken R J, Drevet J R. ANTIOXIDANT STRATEGIES IN THE EPIDIDYMIS[J]. Molecular and Cellular Endocrinology, 2004, 216(12): 31-39.

[25] Nenicu A, Luers G H, Kovacs W J, et al. Peroxisomes in Human and Mouse Testis: Differential Expression of Peroxisomal Proteins in Germ Cells and Distinct Somatic Cell Types of the Testis[J]. Biology of Reproduction, 2007, 77(6): 1060-1072.

[26] Cerolini S, Maldjian A, Pizzi F, et al. Changes in sperm quality and lipid composition during cryopreservation of boar semen[J]. Reproduction, 2001, 121(3): 395-401.

[27] Audet I, Laforest J P, Martineau G P, et al. Effect of vitamin supplements on some aspects of performance, vitamin status, and semen quality in boars[J]. Journal of Animal Science, 2004, 82(2): 626-633.

[28] Strzezek J. Secretory activity of boar seminal vesicle glands[J]. Reproductive Biology, 2002, 2(3): 243-266.

[29] Radomil L, Pettitt M J, Merkies K M, et al. Stress and Dietary Factors Modify Boar Sperm for Processing[J]. Reproduction in Domestic Animals, 2011, 46: 39-44.

[30] Fraser L, Strzezek J, Kordan W. Post-thaw sperm characteristics following long-term storage of boar semen in liquid nitrogen[J]. Animal Reproduction Science, 2014, 147(3): 119-127.

[31] Hu J, Geng G, Li Q, et al. Effects of alginate on frozen-thawed boar spermatozoa quality, lipid peroxidation and antioxidant enzymes activities[J]. Animal Reproduction Science, 2014, 147(3): 112-118.

[32] Chatterjee S, De Lamirande E, Gagnon C. Cryopreservation Alters Membrane Sulfhydryl Status of Bull Spermatozoa: Protection by Oxi-

dized Glutathione[J]. Molecular Reproduction and Development, 2001, 60 (4): 498-506.

[33] Gadea J, Gumbao D, Matas C, et al. Supplementation of the Thawing Media With Reduced Glutathione Improves Function and the In Vitro Fertilizing Ability of Boar Spermatozoa After Cryopreservation[J]. Journal of Andrology, 2005, 26(6): 749-756.

[34] Tareq K M A, Akter Q S, Khandoker M a M Y, et al. Selenium and Vitamin E Improve the In Vitro Maturation, Fertilization and Culture to Blastocyst of Porcine Oocytes[J]. Journal of Reproduction and Development, 2012, 58(6): 621-628.

[35] Awda B J, Mackenziebell M, Buhr M M. Reactive Oxygen Species and Boar Sperm Function[J]. Biology of Reproduction, 2009, 81(3): 553-561.

[36] Awda B J, Buhr M M. Extracellular Signal-Regulated Kinases (ERKs) Pathway and Reactive Oxygen Species Regulate Tyrosine Phosphorylation in Capacitating Boar Spermatozoa[J]. Biology of Reproduction, 2010, 83(5): 750-758.

[37] Lixin W, Hui Y, Lan X. Sow reproductive stress and its syndrome[J]. Animal Husbandry and Feed Science, 2009, 1(8): 20-24.

[38] Al-Gubory K H, Fowler P A, Garrel C. The roles of cellular reactive oxygen species, oxidative stress and antioxidants in pregnancy outcomes[J]. The International Journal of Biochemistry & Cell Biology, 2010, 42(10): 1634-1650.

[39] Rizzo A, Roscino M T, Binetti F, et al. Roles of reactive oxygen species in female reproduction[J]. Reproduction in Domestic Animals, 2012, 47(2): 344-352.

[40] Myatt L, Cui X. Oxidative stress in the placenta[J]. Histochemistry and Cell Biology, 2004, 122(4): 369-382.

[41] Berchieri-Ronchi C B, Kim S W, Zhao Y, et al. Oxidative stress status of highly prolific sows during gestation and lactation[J]. Animal An International Journal of Animal Bioscience, 2011, 5(11): 1774-1779.

[42] Mahan D C, Peters J C, Hill G M. Are antioxidants associated with pig and sow mortalities? in proceedings of the Swine Nutrition Conference [C]. Indianapolis, Indiana, 2007.

[43] Zhao Y, Flowers W L, Saraiva A, et al. Effect of social ranks

and gestation housing systems on oxidative stress status, reproductive performance, and immune status of sows[J]. Journal of Animal Science, 2013, 91(12): 5848-5858.

[44] Mahan D C. Effect of organic and inorganic selenium sources and levels on sow colostrum and milk selenium content[J]. Journal of Animal Science, 2000, 78(1): 100-105.

[45] Hostetler C E, Kincaid R L. Gestational changes in concentrations of selenium and zinc in the porcine fetus and the effects of maternal intake of selenium[J]. Biological Trace Element Research, 2004, 97(1): 57-70.

[46] Mahan D C, Penhale L H, Cline J H, et al. Efficacy of supplemental selenium in reproductive diets on sow and progeny performance [J]. Journal of Animal Science, 1974, 39(3): 536-543.

[47] Mahan D C, Peters J C. Long-term effects of dietary organic and inorganic selenium sources and levels on reproducing sows and their progeny[J]. Journal of Animal Science, 2004, 82(5): 1343-1358.

[48] Mihailovic M, Velickovski S, Ilic V. Reproductive performance in sows fed a diet low in selenium[J]. Acta Veterinaria, 1989, 39: 19-24.

[49] Pehrson B, Holmgren N, Trafikowska U. The Influence of Parenterally Administered α-Tocopheryl Acetate to Sows on the Vitamin E Status of the Sows and Suckling Piglets and Piglets After Weaning[J]. Journal of Veterinary Medicine A: Physiology, Pathology, Clinical Medicine, 2001, 48(9): 569-575.

[50] Saugstad O D. Oxidative stress in the newborn : A 30-year perspective[J]. Neonatology, 2005, 88(3): 228-236.

[51] Loudenslager M J, Ku P K, Whetter P A, et al. Importance of diet of dam and colostrum to the biological antioxidant status and parenteral iron tolerance of the pig[J]. Journal of Animal Science, 1986, 63 (6): 1905-1914.

[52] Yin J, Ren W, Liu G, et al. Birth oxidative stress and the development of an antioxidant system in newborn piglets[J]. Free Radical Research, 2013, 47(12): 1027-1035.

[53] Hakansson J, Hakkarainen J, Lundeheim N. Variation in Vitamin E, Glutathione Peroxidase and Retinol Concentrations in Blood Plasma of Primiparous Sows and their Piglets, and in Vitamin E, Selenium

and Retinol Contents in Sows' Milk[J]. Acta Agriculturae Scandinavica Section A-animal Science, 2001, 51(4): 224-234.

[54] Mahan D C. Effects of dietary vitamin E on sow reproductive performance over a five-parity period[J]. Journal of Animal Science, 1994, 72: 2870-2879.

[55] Mahan D C. Effect of inorganic selenium supplementation on selenium retention in postweaning swine[J]. Journal of Animal Science, 1985, 61(1): 173-178.

[56] Groce A W, Miller E R, Keahey K K, et al. Selenium supplementation of practical diets for growing-finishing swine[J]. Journal of Animal Science, 1971, 32(5): 905-911.

[57] Blodgett D J, Schurig G G, Kornegay E T, et al. Failure of an enhanced dietary selenium concentration to stimulate humoral immunity in gestating swine[J]. Nutrition reports international, 1989, 40(3): 543-550.

[58] Duntas L H, Benvenga S. Selenium: an element for life[J]. Endocrine, 2015, 48(3): 756-775.

[59] Mahan D C, Kim Y Y. Effect of inorganic or organic selenium at two dietary levels on reproductive performance and tissue selenium concentrations in first-parity gilts and their progeny[J]. Journal of Animal Science, 1996, 74(11): 2711-2718.

[60] Mahan D C, Parrett N A. Evaluating the efficacy of selenium-enriched yeast and sodium selenite on tissue selenium retention and serum glutathione peroxidase activity in grower and finisher swine[J]. Journal of Animal Science, 1996, 74: 2967-2974.

[61] Phipps R H, Grandison A S, Jones A K, et al. Selenium supplementation of lactating dairy cows: effects on milk production and total selenium content and speciation in blood, milk and cheese[J]. Animal, 2008, 2(11): 1610-1618.

[62] Dalto B D, Tsoi S, Audet I, et al. Gene expression of porcine blastocysts from gilts fed organic or inorganic selenium and pyridoxine [J]. Reproduction, 2014, 149(1): 31-42.

[63] Hu H, Wang M, Zhan X, et al. Effect of different selenium sources on productive performance, serum and milk Se concentrations, and antioxidant status of sows[J]. Biological Trace Element Research, 2011, 142(3): 471-480.

[64] Zhan X, Qie Y, Wang M, et al. Selenomethionine: an Effective Selenium Source for Sow to Improve Se Distribution, Antioxidant Status, and Growth Performance of Pig Offspring[J]. Biological Trace Element Research, 2011, 142(3): 481-491.

[65] Huang R, Zhu W, Hang S, et al. Effect of Bio-Mos and Sel-Plex on growth performance and incidence of diarrhoea of weaning piglets. in proceedings of the Alltech's 20th Annual Symposium on Nutritional Biotechnology in the Feed and Food Industries [C]. Lexington, KY, USA, 2004.

[66] Krska P, Lahucky R, Kuchenmeister U, et al. Effects of dietary organic selenium and vitamin E supplementation on post mortem oxidative deterioration in muscles of pigs[J]. Archives Animal Breeding, 2001, 44(2): 193-202.

第五章　硒与反刍动物营养

与单胃动物一样，硒也是反刍动物的必需微量元素。如前所述，硒作为一系列硒蛋白的组成部分，对反刍动物的生产和繁殖都具有重要意义，在提高生产性能、增强免疫力、提高抗氧化能力、改善繁殖性能以及缓解应激方面发挥着重要的作用。值得注意的是，反刍动物硒缺乏或不足在全球范围内仍然持续存在，导致动物生长速度减慢、繁殖效率降低、免疫功能紊乱、对各种疾病如胎盘滞留、子宫内膜炎、乳腺炎、营养性肌营养不良的易感性增强，因而提高了这些疾病的发病率和严重程度。由于瘤胃的强还原条件，反刍动物对亚硒酸钠中硒的吸收远低于单胃动物，口服放射性标记亚硒酸钠在猪小肠的吸收率约为 85%，在大鼠小肠的吸收率约为 86%，而在绵羊小肠的吸收率仅为 34%。与单胃动物一样，当日粮中的硒添加形式为亚硒酸钠时，反刍动物通过胎盘和乳腺组织转运硒的能力也非常低，因而也容易引起新生动物硒缺乏。本章主要探讨反刍动物牛的营养与硒的关系，重点阐述奶牛硒营养。

第一节　硒需要量

NRC 推荐的动物对硒的需要量是根据维持、生长、妊娠、泌乳的需要量确定的。但是，在生产实践中还应该考虑到其他所有可能影响到牛对硒需要量的因素。例如，目前 NRC 推荐的犊牛和肉牛硒需要量为 0.1 mg/kg 日粮干物质[1]，但是对于一些肌肉相对发达的牛种（比利时蓝牛）而言，该推荐量不能满足需要，0.3 mg/kg 的硒供给量可能是合适的[2]。再例如，目前 NRC 推荐的奶牛硒需要量为 0.3 mg/kg 日粮干物质[1]，但是在一些特殊的生理时期，如围产期或应激期间，奶牛对硒的需要量可能更高[3]。一些研究证据已表明，在围产期或热应激期间适当提高硒的供给量（高于 0.3 mg/kg）对母牛的硒营养状况、抗氧化能力、免疫功能、能量代谢以及新生犊牛的免疫功能都有积极的影响[4]。此外，日粮中其他营养素会影响硒的吸收利用，当这些营养素导致硒的吸收利用降低时，则应该提高日粮硒的添加量。维生素 E 是影响日粮硒供给量的因素之一，由于二者在抗氧化功能方面的相互依赖性，低维生素 E 摄入会增加对硒的需要量[5]。富含碳水化合物、硝酸盐、硫酸盐、钙或氰化氢(三叶草、亚麻籽)的日粮可降低牛对硒的吸收利用。当日粮中硫的浓度超过 2.4 g/kg 时,可通过空间竞争作用降低硒的吸收,Fe^{3+}将硒沉淀为肠道不能吸收的复杂形式,因而降低了硒的吸收速率[6]。饲料中一定水平的钙(0.8%)可以使怀孕后期奶牛对硒的吸收达到最佳水平[7],集约化生产条件下在牛日粮中补钙可导致肌肉中硒含量显著降低[8]。在犊牛饲料中铅浓度较高的情况下,血清硒水平及组织硒含量都会下降。在牛的甲状腺中,缺碘导致硒蛋白 DIO1 的显著诱导,并伴随着 GPx 活性的升高,因而加剧了硒缺乏。日粮富含粗蛋白和纤维素对硒的吸收利用有积极的影响。此外,瘤胃环境会影响进入后肠道的硒的存在形式,当日粮中硒添加形式为亚硒酸盐或硒酸盐时,经瘤胃的强还原环境,一部分日粮形式的硒会被还原为元素硒,导致肠道对硒的吸收利用降低,粪硒排出量增加。

第二节　硒缺乏

从地理环境来看,硒缺乏比硒毒性更成问题。世界上许多国家,特别是在拥有成熟的集约化农业系统的地区,通常由于酸性条件或随着时间的推移不断耗竭而导致土壤缺硒,因而也导致在这些土壤中生长的农作物和牧草的硒水平较低,最终无论是直接放牧还是通过配合饲料饲养,都减少了牛对硒的摄入量。

目前,成年荷斯坦奶牛群硒缺乏的阈值被认为是 0.068 mg/kg,即当硒摄入量低于这一数值时奶牛处于硒缺乏状态,尽管这一水平的硒摄入没有影响奶牛的繁殖性能,但奶牛的免疫状态明显受到抑制[9]。根据奶牛免疫和生殖参数,以血液硒含量作为硒状况的评价指标时,血浆硒水平低于 0.04 mg/L 为缺乏,0.04～0.07 mg/L 为边缘水平,超过 0.07 mg/L 为充足水平。在欧盟,缺硒是一个严重的问题。捷克共和国奶牛的血液分析显示,大约 50%的抽样动物处于边缘或缺硒状态[10],而且在青年动物、肉牛和繁殖公牛中表现更为明显。在波兰、爱沙尼亚、斯洛文尼亚、德国、爱尔兰、希腊和土耳其进行的研究也报告了类似的发现,大多数奶牛处于边缘或缺硒状态。即使生产需求下降、不再泌乳的干奶期奶牛似乎也表现出边缘或缺硒状态[11]。在南美的研究揭示了奶牛硒缺乏随季节和地区的变化而变化,奶牛群硒缺乏状况最低时所占比例为 13%,而最高时可达 93%[12]。在亚洲的一些国家,如印度、泰国和新西兰,也特别容易受到低硒状态的影响。

对于优质高产奶牛而言，硒摄入量不足会导致多种健康问题，而且多数表现为难以诊断的亚临床症状，包括繁殖问题、免疫状况差、发病率增加（乳腺炎）、胎衣不下以及生育能力降低。在生产实践当中，这些问题中的任何一个都可能导致奶牛被淘汰，因而对奶牛养殖业造成巨大的经济损失。

一、硒缺乏与奶牛受孕和妊娠

硒对奶牛的生殖具有重要的作用。一方面，硒通过合成抗氧化硒蛋白保护了卵子膜免受自由基的攻击，另一方面，硒通过合成抗氧化硒蛋白降低了阻止受精卵着床的子宫感染的机会。自由基可能对奶牛卵泡孕酮的合成产生不利影响，导致孕酮的合成减少。研究表明，硒添加可显著提高硒缺乏奶牛循环中孕酮的水平[13]。硒摄入量不足可导致受孕失败率提高，产犊间隔增大。研究表明，与日粮添加 0.1 mg/kg 硒（产后血清硒水平为 0.08 mg/L）相比，硒缺乏（产后血清硒水平为 0.03 mg/L）奶牛群明显增加了每次怀孕的人工授精次数（2.58 次比 1.96 次）、空怀天数（152 天比 107 天）和产犊间隔（437 天比 392 天），提高了患卵巢障碍、流产或死胎的几率[14]。另一项研究表明，硒充足奶牛更有可能在整个季节保持良好的身体状况和受孕率（86%对 78%）[15]。对于奶牛养殖业而言，邻近分娩时犊牛死亡是一个重要的经济损失因素，特别是当死亡的小牛是母牛时损失更大，因为这影响到了牛群的更替。对美国一些规模化奶牛养殖场的调查发现，平均大约有 7%的犊牛在出生前 48 h 内死亡[16]。对中国 7 个规模化奶牛场的调查表明，犊牛死胎率平均大约为 6.4%[17]。尤其是头胎牛，分娩死胎率可以超过 10%。为数不多的研究发现，肠外补硒或维生素 E 可显著降低奶牛分娩时的犊牛死亡率，一般在预产期前 21 天和第 5 天进行比较理想[18]，这种方法同时也可以提高初乳中的硒水平，补充形式为有机硒时效果更好。

对于犊牛来说，注射硒和维生素 E 可以治愈或预防这个问题。水肿和肌肉无力的问题似乎与低抗氧化状态下的膜损伤有关。钙盐沉积在肌肉中，由于线粒体断裂，造成组织的苍白外观，因此得名。

二、硒缺乏与胎盘滞留

胎盘滞留是奶牛繁殖过程中的常见问题，一个相对较大的奶牛场一般会有 9%的奶牛在分娩后会出现胎盘滞留[19]，管理不善时可以高达

39%[20]。胎盘滞留可导致产奶量降低[21]、生育能力差、难产和死产率显著提高，而且与子宫炎的发病率提高有关[22]。患有胎盘滞留的奶牛导致受孕率大大降低了。据估计，每头患有胎盘滞留的奶牛给牛场造成的经济损失大约在130～285美元。

早在1969年，人们就发现胎盘滞留的发生与奶牛的抗氧化状态有关，当硒和维生素E摄入水平不足时，奶牛的抗氧化能力降低，胎盘滞留的发病风险提高。患有胎盘滞留的奶牛以及死产和生育能力差的奶牛，通常血清硒水平较低，受影响的牛群比例有时可高达40%。无论是通过日粮补硒还是肠外补硒，都能有效降低胎盘滞留的发病率。虽然目前还不清楚为什么奶牛在产犊后并不总是能够完全排出胎盘，但人们认为这与免疫抑制有关。当胎儿分娩后，胎盘会被机体免疫细胞识别为“异物”，激活的免疫细胞迁移到子宫组织，通过破坏子宫和胎盘之间的联系，从而使胎盘脱落而排出。然而，围产期奶牛，尤其是在分娩前后，奶牛的免疫系统通常处于抑制状态，如果前期奶牛硒摄入不足或缺乏时，免疫抑制作用则更强，导致免疫细胞不能迁移到子宫组织而使胎盘排出。

三、硒缺乏与乳腺炎

乳腺炎被认为是影响奶牛场经济效益（牛奶商品价值降低、劳动力和治疗成本提高、奶牛淘汰损失增加）、牛奶品质、产奶量及奶牛健康和生殖的最重要的因素。牛奶中体细胞计数（SCC）是评价是否发生乳房感染的最常用的简单方法。牛奶中的体细胞主要由免疫细胞和死亡脱落的乳腺上皮细胞组成。正常情况下，乳房未感染奶牛乳中的SCC低于10万/mL[23]，大多数是中性粒细胞。

乳腺炎通常是由通过乳头进入乳腺的细菌引起的。亚临床或隐形乳腺炎是一种低程度细菌感染，当牛奶中SCC高于25万/mL时，即可确定为乳房内亚临床感染。由于细菌的存在，亚临床或隐性乳腺炎有可能发展成亚急性或急性乳腺炎。需要注意的是，对乳房的粗暴处理或长时间挤奶有可能引起无菌性乳腺炎，无菌性乳腺炎奶牛SCC也提高。根据严重程度可将乳腺炎分为4级：亚急性乳腺炎（1级：奶牛看起来正常，但挤出的牛奶不正常，SCC明显增加）、急性乳腺炎（2级：有发烧症状，乳房变化明显且对刺激非常敏感，牛奶中有肿块甚至血迹）、严重急性乳腺炎（3级：除有急性乳腺炎的症状外，乳房呈紫蓝色，可危及生命，通常由革兰氏阴性细菌如大肠杆菌、假单胞菌和克雷伯氏菌引起）、慢性乳腺炎（4级：反复发作的急性或亚急性乳腺炎，通常是由金黄色葡萄球菌深度感染引起）。

充足和均衡的营养，特别是维生素 E 和硒[24]，是从营养角度预防奶牛乳腺炎发病风险的重要措施。硒添加可显著降低奶牛乳中的 SCC[25]。在牧场监测的牛奶中 SCC 随着血液中硒浓度的增加而降低，血浆 GPX 水平与牛奶中的 SCC 呈显著的负相关关系（R2＝0.82）[26]。硒缺乏的奶牛可通过正确的补硒显著降低临床乳腺炎以及有临床诊断价值个体的发病率。早期泌乳阶段是奶牛乳腺炎的高发期，一项研究表明，通过肠外补硒（0.35 mg）可以使乳腺炎的发病率从 70％降低到 20％[27]。对已发表的 13 篇有关硒与奶牛乳腺炎的文章的整理发现，硒添加可以使乳腺炎发病率从 32％降低到 10％[28]。目前认为，硒可以通过抑制牛奶液体成分中病原菌的增殖以及提高 GPX 的活性[29]或其他硒蛋白的合成[30]来抑制或减缓乳腺炎病例的炎症反应。

四、硒缺乏与营养性肌营养不良

营养性肌营养不良（NMD），俗称白肌病（钙盐沉积于肌肉线粒体导致肌肉组织呈现苍白外观），是反刍动物硒缺乏的常见后果，饲喂低硒或低维生素 E 日粮均可引起。由于心脏和骨骼肌最容易受到影响，受影响的动物表现为肌肉僵硬，犊牛还表现为严重的虚弱（肌肉无力）。由于硒缺乏导免疫系统受损，白肌病还可引起许多继发性感染。快速生长的幼龄动物通常表现更为明显的临床症状，死亡率可达到 30％。新生动物白肌病的发生通常是由于母体日粮硒摄入不足引起的，血清肌酸激酶（CK）升高和维生素 E 及硒水平降低可作为白肌病的临床诊断指标，这些指标对肠外治疗有很好的反应。患有白肌病的小牛 2 天内血清 CK 水平可升高到 29 280 U/L，通过治疗，5 天后可降到 166 U/L[31]。当母牛日粮硒摄入不足时，初乳和常乳中的硒水平通常较低，此外，当母牛日粮硒添加形式为无机硒时，向牛奶中低效的硒转移也会引起牛奶中的硒水平降低，这将导致新生犊牛白肌病的发病风险提高，死亡率增加。对于成年奶牛而言，增加行走时间（走更远的距离去挤奶）或其他应激因素可导致白肌病，表现为全身虚弱（肌肉无力）和水肿，患白肌病的妊娠奶牛在产前容易发生突然死亡，通过检测血清 CK 和乳酸脱氢酶的水平可以证实是否发生白肌病。对于犊牛而言，当血清硒水平低于 30 ng/mL 时，NMD 的发病风险会大大提高[32]，注射硒和维生素 E 可以治愈或预防白肌病。

五、硒缺乏与硒反应性衰弱症

犊牛除了容易患白肌病外，还容易患上一种称为硒反应性衰弱症（Se-

RU)的疾病，表现为生长受阻、消瘦、不能站立、消化良性腹泻，常常空嚼(磨牙)，多发生于放牧条件下的犊牛，是由于采食或饲喂在硒缺乏土壤上生长的草(料)所引起。该病的发生被认为与硒和甲状腺激素的相互作用有关。其他哺乳动物有一种酶途径，即使在低硒的情况下也能产生 T3 甲状腺激素，但这一途径在牛身上不存在，因而容易在低硒摄入的情况下导致甲状腺功能障碍，生长速度减慢甚至生长受阻。在母猪日粮中补硒，子代出生后 40 天内血清生长激素和促甲状腺激素水平提高[33]。怀孕奶牛产前注射硒使犊牛出生后生长速度和增重显著提高[34]。犊牛饲料中添加硒，120 天的试验期内增重提高了 17%，饲料效率提高了 12%[35]。

第三节　硒添加

通过各种不同的补硒方法可以使牛获得充足的硒摄入，如饮水中加硒、预混料中加硒、直接肌肉注射(亚硒酸钠、亚硒酸钡)、体内植入剂以及通过土壤施硒肥提高饲料作物的硒含量或饲喂生长在硒充足土壤上的饲草料。补硒时应考虑硒的补充形式，对于反刍动物而言，由于瘤胃环境的影响，对无机形式硒的吸收利用远低于单胃动物，而且无机硒只在小肠中的回肠吸收。有机硒受瘤胃环境的影响较小，容易被小肠吸收利用，而且小肠全段都可吸收。补硒后的效果评估依据是血液或组织硒状况，如血清(血浆)、全血、肝脏硒水平、血液 GPX 活性、肌肉、牛奶硒水平、硒蛋白表达等，在利用这些指标评价硒状况时需要考虑硒的添加形式、吸收利用、对补硒的效应时间等。例如，用红细胞硒水平评价时需要考虑红细胞的寿命，用组织硒水平评价时，需要考虑组织硒储备的时间(奶牛需要 60～90 天)。补硒的过程中还应该考虑对动物造成的潜在的、直接的毒性作用，对人体造成的潜在的、间接的毒性作用，以及对环境造成的现在的、直接或间接的污染[36]。

对于奶牛而言，一些研究给出了支持某些功能所需的最低血液硒水平。如果是为了预防乳腺炎，奶牛血液硒水平应该在 0.2～1 g/L。在较高的生理应激期(如妊娠、产犊、早期泌乳阶段)，奶牛每天的硒需要量为 5～7 mg。一些研究表明，血清硒浓度需要高达 0.25 mg/L 以确保硒在子宫内充分转移到发育中的犊牛，从而避免 NMD 发生，而达到这一血清硒水平需要奶牛日粮每天提供 6 mg 硒[37]。

一、硒添加对繁殖性能的影响

如前所述，硒缺乏会导致奶牛子宫内膜炎和胎盘滞留的发病率和严重程度提高。此外，与家禽和单胃动物猪一样，硒缺乏也会导致牛睾酮和精子合成功能障碍，从而导致雄性不育。硒会影响睾丸的总体形态和组织形态[38]，一些硒蛋白如 SPS-2 和线粒体被膜硒蛋白（MCSeP）定位于睾丸中[39]。饲喂富硒谷物[40]、日粮添加无机硒（亚硒酸盐）[41]或有机硒（硒酵母）[42]可以显著提高牛睾丸中的硒含量。精子活力降低（精子中段断裂）是硒缺乏对精子质量影响的主要特征[43]。

补硒后生育力的提高可归因于减少了怀孕第一个月的胚胎死亡。Ceko 等[44]报道，在大的健康的卵泡颗粒细胞中硒蛋白 GPX1 的基因表达显著增加，作者认为，硒蛋白在卵泡发育的后期起着重要的抗氧化保护作用。对绒山羊的研究表明，添加纳米硒可促进次级卵泡的发育[45]。

子宫炎、卵巢囊肿和胎盘滞留都会影响到母牛的繁殖性能。研究表明，补硒可降低子宫炎、卵巢囊肿[46]和胎盘滞留[6]的发生率。Komisrud 等[47]报道，对缺硒奶牛进行补硒可提高首次发情人工授精的成功率。硒的来源（亚硒酸钠或硒酵母）对奶牛受胎率和产犊间隔无明显影响[48]。然而，一些报道并未观察到补硒对生殖功能的积极影响，这可能与机体硒不足的严重程度、补硒的条件以及在这些情况下机体酶合成能力的不同有关。需要注意的是，牛的慢性硒中毒可以通过支持卵巢囊肿的生长和延长发情时间降低生育力[49]。

二、硒添加对乳硒水平及乳产量的影响

牛奶中的硒含量是评估牛群硒状况的一种简单方法，具有采样方便的优点。根据 Wichtel 等[50]研究评估的牛奶为硒状态参考水平：硒含量低于 0.12 μmol/L 为缺硒状态，大于 0.28 μmol/L 为适硒状态，介于 0.12～0.28 μmol/L 为临界状态。牛奶中的硒水平通常为乳清中最高（56.6%），乳脂中最低（10.1%）[51]，55%～75%的硒存在于酪蛋白中，17%～38%的硒存在于乳清中，7%的硒存在于脂肪中[52]，而且这些发现与摄入硒的类型无关，主要取决于奶牛日粮中硒的含量[53]。血浆和牛奶中的硒水平存在着显著的正相关关系：$y=0.37x-4.2$ [x 表示血浆硒水平（μg/L），y 表示牛奶硒水平（μg/L）][54]。通常，血浆硒水平是牛奶硒水平的 3～5 倍。分娩前通过口服途径补充硒的奶牛产生的初乳硒含量（170 μg/L）是未补充奶牛的 2

倍(87 μg/L)[55]。初乳中的硒含量通常是牛奶中硒含量的2～3倍。

牛奶中的硒含量受饲料中硒含量的影响,而饲料中的硒含量又取决于季节和耕作地区。在放牧季节进行的一项研究表明,14%的奶牛群所产的牛奶中硒含量处于临界水平,牛奶硒水平低主要与饲草质量差导致硒摄入量不足有关[56]。在比利时[57]、韩国[58]、希腊[59]和澳大利亚[60]进行的研究报告,牛奶中的硒含量分别为30 μg/kg、60 μg/kg、15 μg/kg和22 μg/kg。在比利时,人口消耗的硒有4%来自牛奶及奶制品[57],在韩国,这一比例高达7%[58]。牛奶中的硒含量与奶牛饲料中有机硒的含量成正比,平均而言,饲喂有机硒的奶牛比补充无机形式硒的奶牛多摄入0.37 μmol/L的牛奶硒[61]。与对照相比,在奶牛饲料中添加硒酵母或亚硒酸钠显著提高了牛奶中的硒含量[62]、多不饱和脂肪酸的百分比和亚油酸的含量[63]。与添加亚硒酸钠相比,硒酵母提高乳硒水平的效果更为明显,在奶牛日粮中添加硒酵母,牛奶中的硒浓度比添加无机硒提高了190%[64]。另一项研究也报道了相似的结果,这项研究还探讨了硒对免疫细胞功能的影响,发现硒摄入量和硒源(硒酵母和亚硒酸钠)不影响淋巴细胞亚群和中性粒细胞的吞噬活性[65]。作者前期的研究表明,与无机硒亚硒酸钠相比,泌乳中期奶牛日粮中相同剂量(0.3 mg/kg)的硒酵母添加显著提高了牛奶中的硒含量,添加30天时提高幅度为59%,60天时提高幅度为25%[66]。奶牛在产犊前21天日粮添加300 μg/kg硒,同时联合注射50 mg硒和300 IU维生素E,可使血浆硒含量保持在足够水平,同时提高了牛奶中的硒含量[26]。

大多数研究证据表明,日粮补硒对奶牛产奶量及乳成分(乳脂、乳蛋白、乳糖)没有显著影响[62,63,65]。作者前期的研究表明,日粮添加无机形式的亚硒酸钠与添加有机形式的硒酵母相比对奶牛产奶量和乳成分也没有显著影响[66]。但也有一些报道表明,当以瘤胃丸剂形式补充碘、钴和硒时,会提高牛奶产量[67]。当硒和维生素E联合补充时,不仅提高了牛奶产量,而且提高了牛奶中粗蛋白、非脂固形物和乳糖的百分比,但该研究未发现硒和其他微量元素联合补充对产奶量的影响[68]。同样,Machado等[69]的研究也未发现注射硒、锌、锰、铜对产奶量有影响。

三、硒添加对新生犊牛的影响

在反刍动物中,硒从奶牛到新生犊牛的转移是通过胎盘和牛奶完成的,通过胎盘转移硒的效率通常要高于通过牛奶进行的硒转移[70],原因可能与血清硒浓度显著高于牛奶硒浓度有关。妊娠期的最后几个月是母体向胎儿转移硒的关键期,即使母体处于硒缺乏状态,硒也会高效地转移到胎儿，以确保新生犊牛有足够的硒摄入量[71]。妊娠期胎儿肾脏硒水平保持不变，而奶牛肾脏硒水平下降很好地说明了这一硒转移特性[72]。无机硒不能很好地转移到牛奶中，因而不能有效地维持犊牛充足的硒状态。与无机硒相比，在日粮中使用有机硒具有更好的向初乳或牛奶中转移硒的效率，因而可以保证犊牛快速生长和免疫力发育对硒的需求。McDowell 等[73]比较了注射无机硒或有机硒以及不注射对安格斯牛初乳中硒水平的影响，数据显示，对照组初乳硒水平为 28 μg/L，亚硒酸盐注射组为 46 μg/L，硒酵母注射组为 79 μg/L，有机硒注射与对照组相比初乳硒水平增加了 3 倍，与亚硒酸盐注射组相比增加了将近 1 倍，这很好地保证了犊牛从初乳中的硒摄入。

犊牛补硒一般不影响其生长性能。但一些研究发现补硒对硒缺乏犊牛的生长性能有积极的效果，分别在第 2 日、70 日、114 日和 149 日龄注射 0.05 mg/kg 硒显著提高了犊牛的平均日增重[74]。一般而言，日粮硒添加形式（无机和有机硒）对犊牛初生重、日增重和死亡率没有显著影响[48]。给新生犊牛每天补充 0.08 mg 硒对犊牛的体重增加、体高或体长发育没有影响，但补硒对犊牛的免疫系统有积极影响，增强了 30 日龄犊牛巨噬细胞的吞噬功能[75]。新生犊牛的免疫系统不成熟，产前母牛日粮中补硒可提高新生犊牛血清免疫球蛋白的浓度[76]，这有助于新生犊牛对抗出生后改变的环境条件。当环境温度低于 14.6 ℃时，新生牛犊表现出对寒冷的敏感性[77]，为了应对这种应激，棕色脂肪组织的代谢活动加强，生热作用提高，然而脂肪组织的代谢受 T3 激素调节，而 T3 激素又是一种硒依赖的激素[55]。因此，新生犊牛硒缺乏或不足不利于其适应寒冷环境。

综上所述，硒对犊牛的生长一般没有直接的促进作用。然而，它有助于消除所有可能延缓或抑制生长的制约因素。硒注射被认为是一种快速有效的补硒方法，可以迅速恢复犊牛硒状态，但是这并不是唯一的方法。生产实践中，为了避免犊牛缺硒，应将硒注射与自助获取日粮硒源相结合。

四、硒添加对血硒水平的影响

在血液中，SelP 中硒含量最高，一个 SelP 含有 10 个以 SeCys 形式存在的硒原子，SelP 由肝脏合成，负责外周组织中硒的供给。硒也可以与血液中的蛋白质，如 α 和 β 球蛋白、低密度脂蛋白（LDL）、极低密度脂蛋白（VLDL）和白蛋白结合。当然，几乎所有已知的硒蛋白都可以在血液中检测到。

牛全血中硒的适宜水平在 80～160 μg/L[78]，血浆（血清）中硒的适宜水平在51～85 μg/L。要达到这个适当的比率，日粮硒摄入量应该达到0.5 mg/kg。在奶牛[79]、犊牛[80]、青年母牛[81]和育肥公牛[40]日粮中补硒通常表现为血液中硒含量和GPX活性的增加。与无机形式相比，在日粮中添加有机形式的硒对血硒和GPX活性的提高更为明显。作者前期的研究表明，奶牛日粮中0.3 mg/kg的硒酵母添加与相同剂量的亚硒酸钠添加相比，全血硒含量和血清GPX活性显著提高[66]。据估计，这两种形式的硒在血液硒含量和GPX活性方面的效果差异分别为20%和16%[82]。围产期奶牛进入到妊娠后期及产后早期，由于胎儿的快速生长以及乳的合成和分泌，对硒的需要量大大提高，因此，在这一特殊时期，超营养的硒添加具有诸多有益的作用。妊娠最后8周添加超营养的硒酵母提高了奶牛产后48小时和14天时的全血和血清硒浓度，而且降低了产犊时和产后48小时的血清胆固醇浓度[83]。作者前期的研究也表明，产前4周奶牛日粮添加超营养的硒酵母提高了产后7天和21天的全血和血浆硒浓度以及产后7天的血浆和红细胞GPX活性[4]。日粮补硒对血硒水平的提高需要一个过程，尤其是红细胞硒或GPX活性，因为牛红细胞的平均寿命大约是120天。肌肉注射和日粮补硒相结合有助于小牛在育肥的头几周内保持正常的血硒水平和GPX活性[81]。

五、硒添加对生长性能的影响

硒参与甲状腺激素的新陈代谢。缺硒日粮通常会导致血液中T3激素减少，T4激素增加，T3/T4比值降低[71,72]。如本书第二章所述，T4全部在甲状腺合成，甲状腺只合成少量T3，大部分在甲状腺以外的组织中由T4通过脱碘代谢生成，T3是T4的活性形式，参与机体的生长调节。T4通过脱碘作用降解为T3需要硒蛋白DIO催化，当DIO活性降低时，T3生成不足，因而影响了动物的生长。任何导致T3生成不足的因素均可影响动物的生长，人类呆小症就是由于原料碘不足引起T3生成不足所致。

绝大多数关于硒添加对牛生长性能的研究没有显示出明显的促生长效应。皮下注射[74]或肌肉注射[84]硒对犊牛或肉牛的生长性能没有影响，在奶牛和犊牛日粮中添加亚硒酸钠或硒酵母不影响生长性能[48,85,86]，饲喂由富硒化肥生产的谷物组成的饲料对育肥公牛的生长没有显著影响[40]。奶牛瘤胃内推注硒对初生重和犊牛平均日增重没有明显影响[84]。饲喂含有超营养硒的日粮（矿物硒添加[87]、硒酵母添加[42]、富硒土壤生产的饲料作物[40]）不影响育肥公牛或青年公牛的屠宰重和胴体重。硒源（亚硒酸盐和硒酵母）对犊牛干物质摄入量和增重没有影响[81]。极少数的研究记录到补硒对动物的促生长效应。Wichtel 等[88]报道，在 5 月龄犊牛中瘤胃内推注硒使日增重增加了 20%。Castellan 等[74]报道，低硒犊牛补硒提高了日增重和生长率。Gleed 等[89]在患有低血压的牛的日粮中添加硒后，记录到了生长的增加。硒蛋白的合成具有不同的优先顺序，当硒的供应不足或缺乏时，机体会在牺牲一些硒蛋白合成的前提下优先保证另一些硒蛋白的合成，DIO 处于这一优先级的顶端，因而不容易受到硒缺乏或不足的影响，这可能是大多数研究并没有记录到补硒的促生长作用的原因。少数研究记录到的补硒的促生长作用可以解释为补硒的间接促生长作用，因为当硒不足或缺乏时，一些硒蛋白的合成不足(例如 GPX1 最容易受到硒缺乏的影响)会影响到动物的抗氧化和免疫等功能，而这些效应又影响了动物的生长。因此，动物生长对补硒的反应取决于动物当前的硒状况，当硒缺乏或不足时可能表现出间接的促生长作用，当硒充足时则不表现任何效应，而当动物处于严重硒缺乏状态时，直接的促生长作用可能才会显现。

六、硒添加对肌肉和器官中硒分布的影响

饲料作物和牧草中的硒含量直接取决于土壤中的硒含量。这造成了不同养殖地区动物肉中硒含量的不同。由于消化器官的形态和硒的代谢，肉中的硒含量也因物种而异。如本书第一章所述，亚硒酸盐在单胃动物和家禽消化道中的吸收率约为 80%，而在反刍动物消化道中的吸收率仅为 30%。吸收率的降低主要和瘤胃的强还原环境有关，亚硒酸盐和硒酸盐在瘤胃强还原作用下会还原为小肠不能吸收利用的元素硒，因而降低了其在消化道的吸收率。Galbraith 等[90]研究表明，与无机硒相比，有机硒在瘤胃微生物中的掺入程度较高，元素硒的形成较少，因而生物利用率较高，其中硒蛋氨酸的口服生物利用率更高。

在动物补硒过程中，富硒对肌肉和器官硒含量的影响是不同的。吸收的有机硒的即可直接结合到肌肉蛋白中(SeMet 取代 Met)，也可用于硒蛋白的从头合成，而无机硒只能用于合成硒蛋白[91,92]，补充无机硒对肌肉中的硒浓度没有显著影响，因为绝大部分的硒被用于合成到硒蛋白[41,42,87]。在补充有机硒的过程中，一部分遵循与亚硒酸盐相同的代谢途径，但一定量直接和非特异性地沉积在肌肉蛋白中。目前市售的有机硒主要为硒酵母，Kieliszeket 等[93]指出，硒与酵母细胞的结合在很大程度上取决于培养条件、培养基中硒的浓度和所使用的酵母，导致硒酵母中硒含量以及 SeMet 的含量存在较大差异，因而利用这些有机硒源补硒的效果不同。对补硒效果的综合分析显示，动物源性富硒食品在提高 GPx 活性方面比 SeMet 更

有效,会提高肝脏中的硒含量[94]。

牛的组织或器官中硒含量最高的是肾脏,肌肉中也含有高水平的硒。缺硒首先影响的器官是心脏、骨骼肌和肝脏[30]。补硒后,肾脏和肝脏含硒量明显增高[40,42],肾脏最高,其次是肝脏、睾丸和肺[42,95]。补充亚硒酸钠会使硒分布在整个生物体中,然而,大部分被吸收的硒存在于肝脏,肝脏被认为是硒的储存器官[96]。硒在肝脏中积累超过需要量时,一部分硒通过胆汁排出,另一个最重要的途径是通过肾脏排泄,当硒过量达到中毒剂量时,呼吸也是一条排泄途径[97]。

七、硒添加对肉品质、肉中脂肪酸模式和化学组成的影响

肉中的氧化反应是影响肉品质的最主要因素,脂肪氧化是肉品变质的主要原因。肉中的氧化反应会影响肉的颜色、风味、质地和营养价值。研究表明,硒具有保护肉类的感官特征及质地的效果[98],GPX4 是一种重要的抗氧化剂酶,可以防止脂质过氧化。此外,硒可能在脂质代谢的改变中发挥作用。据报道,在补充矿物质硒期间,牛肉中的胆固醇含量有所降低[87]。胆固醇是一种重要的生物化合物,存在于动物产品中,肉和肉制品中的胆固醇含量是不同的,一般每 100 g 肉中胆固醇含量低于 70 mg[99]。与肉类相比,可食用内脏中胆固醇的含量更高。事实上,胆固醇的氧化会产生一些化合物,这些化合物被发现具有细胞毒性、诱变性和致癌性,胆固醇氧化产物被认为是动脉粥样硬化的主要诱因[99]。因此,补硒后肉中胆固醇含量的降低使得这些肉品更具营养价值且更健康。然而,有关硒添加降脂的研究结果[40,87]在其他对牛[41,100,101]、兔[102]或猪[103,104]的研究中未发现显著效果。因此,利用不同来源硒和硒水平进行进一步比较研究,以及测定肉类中各种类型的脂质是必要的。另一项研究的发现比较有趣,使用富硒谷物对育肥公牛进行补硒后发现,该添加剂对肉色、pH 值、失水率、嫩度和氧化酸败无明显影响,这种添加剂降低了肉中脂肪含量[68]。在牛补硒过程中,肉中饱和脂肪酸、单不饱和脂肪酸和多不饱和脂肪酸的总量和组成不受硒源(亚硒酸钠或硒化酵母)的影响,有机硒提高了肉中的硒含量[105]。Netto 等[106]报道,肥牛的日粮中添加有机硒提高了肉中亚油酸和棕榈酸的含量。

就嫩度而言,文献报道结果不一致,原因可能与肉中不同的硒掺入率、硒来源、硒类型和给药途径有关。

八、硒添加对健康的影响

最佳的免疫功能和抗氧化能力是保证动物健康的前提，而营养状况与免疫和抗氧化功能密切相关，缺乏任何一种营养素都有可能导致相关的疾病发生。

硒缺乏可降低奶牛血液和乳中中性粒细胞对乳腺炎病原体的杀伤能力[107]，降低中性粒细胞中自由基的产量[108]。补硒有效提高了中性粒细胞的趋化性迁移和过氧化物的产量[109]，从饲喂较高硒水平日粮的奶牛血液中获得的中性粒细胞具有更高的产生过氧化物和杀死病原微生物的潜力[110]。Machado 等[69]报道，皮下注射微量矿物质（锌 300 mg、锰 50 mg、硒 25 mg、铜 75 mg）对奶牛乳腺健康有积极影响，乳中 SCC 降低，亚临床和临床乳腺炎的发病率降低。血液 GPX 活性与灌装乳中体细胞数呈明显的负相关[26]。日粮添加硒和/或维生素 E 显著降低了奶牛乳腺炎的发病率[111]，据估计，由此获得的经济效益为每头牛每天 0.21 美元[112]。妊娠后期饲喂超营养水平的硒酵母改善了产后奶牛的抗氧化状态和免疫反应[83]。产前 4 周添加超营养水平的硒酵母提高了产后奶牛的硒营养状况，有效减缓了产后早期氧化应激水平的提高[4]。日粮硒酵母和维生素 E 添加提高了公牛 NK 细胞的杀伤作用，增加了脾脏 NK 细胞的表达[113]。

硒添加可改善各种细胞和体液免疫应答。与日粮硒水平适宜相比，硒缺乏奶牛特定淋巴细胞亚群的成熟度降低[33]，抗原诱导的外周血淋巴细胞增殖率降低[114]。体外淋巴细胞培养试验表明，随着硒添加量的增加，淋巴细胞的增殖率提高[115]。与 B 淋巴细胞相比，T 淋巴细胞更易受到硒缺乏的影响，可能与 T 淋巴细胞膜内不饱和脂肪酸含量高和膜的流动性高有关。IL-2 主要由活化的 CD4 T 细胞和 CD8 T 细胞产生，是所有 T 细胞亚群的生长因子，并可促进活化 B 细胞的增殖。硒添加可使高亲和力 IL-2 受体提早表达，而硒缺乏延迟了该受体的表达[116]，表明硒对 T 细胞的增殖具有刺激作用。研究发现，产犊前给奶牛肌肉注射高剂量的硒提高了初乳中的抗体滴度[117]。与自由采食含 20 mg/kg 亚硒酸钠的矿物质预混料相比，肉牛和犊牛自由采食含 60 mg/kg 酵母硒或 120 mg/kg 亚硒酸钠的矿物质预混料明显提高了血浆和初乳中免疫球蛋白 G（IgG）和免疫球蛋白 M（IgM）含量[41]。奶牛日粮中添加酵母硒明显提高了血清中 IgG 含量[42]。体外细胞培养试验表明，加硒（100 mg/mL）明显提高了 B 淋巴细胞的增殖率和 IgM 产量[43]。然而，Leyan 等[44]的研究也发现，加硒对免疫球蛋白产量没有影响。补硒对母牛免疫被动转移有一定影响，奶牛补硒后血清和初乳中 IgG 含量升高，犊牛血清中 IgG 水平也较高[71]。初乳中添加亚硒酸盐提高了新生犊牛对 IgG 的吸收[118]。缺硒奶牛血液和乳中中性粒细胞对病原体的吞噬能力下降[119]，补硒可提高初乳的抗氧化性能[120]，初乳的氧化还原平衡对于免疫从初乳到犊牛的被动转移中起着重要作用。需要注意的是，尽管硒是各种免疫机制所必需的微量营养素，但过量的硒可能会对某些免疫功能产生有害影响。过量的硒可抑制 Nrf2 信号通路，导致抗氧化基因的表达下调，抗氧化酶的合成不足，最终导致 ROS 的生成增加，因而加强了对包括免疫细胞在内的细胞的氧化损伤。

第四节 硒与围产期奶牛

随着遗传育种技术的发展、营养调控措施的优化以及管理水平的提高，奶牛的产奶量在不断提高，已经远远超过了用于哺育后代所需的产奶量。奶牛为人类提供了丰富且营养价值高的牛奶，但为了产更多的奶，奶牛需要经历各种挑战（生理、营养、心理、管理）。在奶牛的整个泌乳周期中，围产期（产前 3 周到产后 3 周）是最为关键的时期。在这短短 6 周的特殊时期内，奶牛经历了从妊娠非泌乳状态转变为非妊娠泌乳状态。为了应对这一转变，奶牛机体发生了戏剧性的、剧烈的生理变化，包括氧化还原状态的改变（氧化应激）、能量代谢的改变（能量负平衡）、钙代谢的改变（低血钙）、免疫功能的改变（免疫力降低而炎症反应增强）。图 5-1 显示了围产期健康奶牛上述这些生理变化的理论模式。需要说明的是，奶牛在围产期所有这些变化都是不可避免的，在一定程度上都可视为从非泌乳状态到泌乳状态以及从怀孕状态到非怀孕状态的生理适应。然而，当这些生理变化的程度提高或持续时间延长，不能及时恢复到生理稳态时，则可视为生理适应失败或适应机制失调，最终将导致各种代谢性和感染性疾病（产后瘫痪、酮病、胎盘滞留、子宫炎、乳腺炎等）的发病率和严重程度提高。因此，在了解引起这些变化的原因以及在围产期出现的时间节点的基础上，通过营养调控和管理措施降低这些生理变化的程度和持续时间，进而使奶牛能够更好地适应从妊娠到泌乳的过渡具有至关重要的意义。

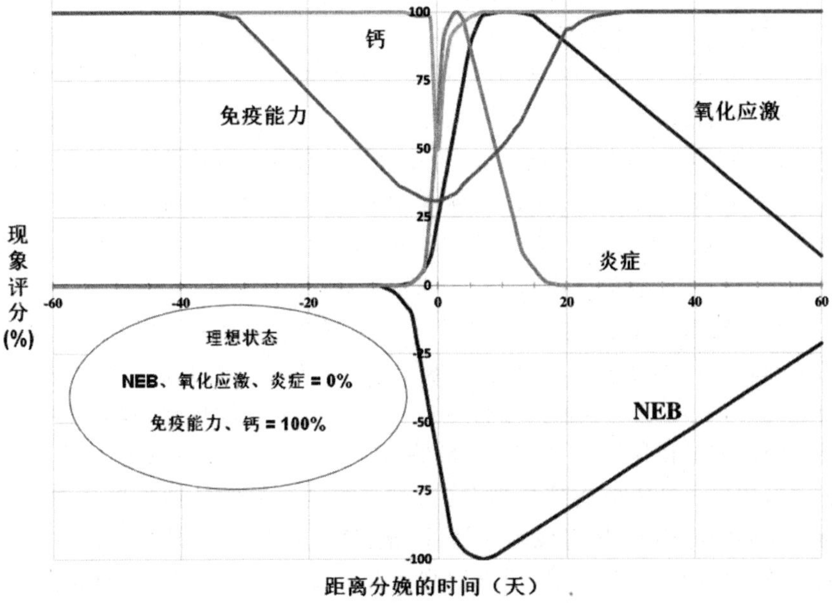

图 5-1 围产期健康奶牛生理变化的理论模式[121]

一、硒对围产期奶牛氧化应激的调节

(一)围产期奶牛氧化应激的发生

众所周知,氧化应激是由于 ROS 的生成超过机体的抗氧化清除能力时引起的,任何提高 ROS 生成以及降低抗氧化能力的因素均有可能最终导致氧化应激的发生。ROS 生成的主要来源是线粒体电子传递链,病原入侵后吞噬细胞的氧化爆发、嘌呤代谢过程中黄嘌呤氧化酶的激活、脂肪酸β-氧化过程中过氧化物酶体中氧化酶的激活以及脂解衍生的不饱和脂肪酸的氧化代谢也是 ROS 生成的来源。归结起来,引起商业生产条件下围产期奶牛氧化应激的因素包括环境应激(热应激)、营养应激(体况评分高、NEB)、代谢应激(包括乳腺和肝脏在内的组织代谢活动加强)和病原/疾病应激四类。研究表明,处于热应激条件下的奶牛产后氧化磷酸化水平提高、线粒体功能障碍提高、Nrf2 介导的氧化应激反应提高[122];体重和体况评分高(日粮能量浓度高)的奶牛 dROMs 和 TBARS 水平提高[123];产后早期严重 NEB 的奶牛生物抗氧化能力(BAP)降低[124];日粮中添加鱼油导致奶牛血浆 MDA、天门冬氨酸氨基转移酶(AST)和碱性磷酸酶(ALP)水平提高[125]；围产期奶牛产后 1 周 ROS、H_2O_2、MDA 水平提高，GSH、GPX、CAT、维生素 E 水平降低[4]；围产期奶牛与其他泌乳阶段相比血浆 ROS+RNS、氢过氧化物、15-F2t-异前列腺素水平提高，氧化应激指数(OSi)提高，抗氧化能力(AOA)降低[126,127]；肺蠕虫感染奶牛 TBARS、ROS、SOD 水平提高，CAT 水平降低[128]；新孢子虫血清反应阳性奶牛血清 ROS、TBARS、NO 水平提高，GST 和 T-AOA 水平降低[129]；酮病奶牛血浆 SOD、CAT、维生素 C、维生素 E 水平降低，H_2O_2 和 MDA 水平提高[130]；2 级子宫内膜炎奶牛晚期蛋白氧化产物(AOOP)水平提高[131]。

氧化应激与多种疾病的发生和发展有关，包括乳腺炎、酸中毒、酮病、肠炎，肺炎、呼吸道疾病、产犊后胎盘滞留、黄体活动受阻等[132,133]。雌性生殖过程中，ROS 通过各种信号转导途径在卵泡形成，卵母细胞成熟和胚胎发育中起调节作用[134]。然而，过量的 ROS 产生和氧化应激被证明与各种生殖障碍以及妊娠的病理生理有关，可影响到卵巢的各种生理功能，包括卵巢类固醇的形成、卵母细胞成熟、排卵、胚泡的形成、着床、黄体溶解和孕期黄体的维持。事实上，从卵母细胞成熟到受精和胚胎发育，氧化应激会在各个层面上影响雌性动物的生殖系统[135]。卵母细胞和其他卵泡细胞对氧化损伤敏感，可能导致原始卵泡的卵巢池耗竭并损害存活的卵母细胞[136]。越来越多的证据表明，氧化应激与奶牛的胚胎死亡率增加有关，并可能是其致死的原因[137]。与小母牛相比，泌乳母牛对氧磷酶-1(PON1)活性较低是由于怀孕期间与氧化应激相关的分娩过程中新陈代谢努力提高的结果[138]。与对照相比，妊娠晚期 OSI 较高的奶牛所生的小牛在出生时和整个研究过程中的体重都有所下降，犊牛血清 ROS、RNS 和 TNF-α 水平提高，而且犊牛对微生物激动剂免疫反应降低[139]。产后奶牛乳腺组织通过激活 Nrf2 信号通路被认为是维持细胞内适当的氧化还原平衡的关键[140]。围产期奶牛疾病发生率的增加，其中一部分原因是各种应激因素引起免疫功能不理想所致。尽管应激因素与免疫功能紊乱之间的复杂相互作用还远没有明确，但产犊前后 ROS 的产生增加和抗氧化防御能力不足在免疫系统损害中一定起着重要作用。

(二)硒对围产期奶牛氧化应激的减缓作用

如前所述，一系列的抗氧化剂可以添加到动物日粮中来提高抗氧化防御能力。硒作为至少 25 种硒蛋白的前体具有特殊的地位，在抗氧化防御网络中具有极其重要的作用。对于围产期奶牛，由于机体所面临的应激因素较多，而且较其他时间应激程度较高，因此对硒需要量很可能要高于其他阶段。作者前期的研究表明，在妊娠最后 4 周给硒充足的奶牛饲喂硒酵母改善了产犊前后的抗氧化状态，降低了氧化应激水平，表现为产前 7 天、产后 7 天和 21 天血浆 MDA 水平降低，产后 7 天和 21 天血浆 ROS 和 H_2O_2 浓度降低，血浆和红细胞 GSH-Px 活性升高，红细胞 GSH 浓度升高[4]。另一项研究表明，与添加无机硒的奶牛相比，添加 OH-SeMet 的奶牛血清 GSH-Px 和 SOD 活性以及总抗氧化能力提高[141]。在热应激期间，补充 OH-SeMet 的奶牛提高了血清和牛奶中的硒浓度，提高了总抗氧化能力，降低了血清 MDA、H_2O_2 和 NO 浓度[142]。

二、硒对围产期奶牛能量代谢的调节

（一）围产期奶牛能量代谢的改变

奶牛从妊娠期到哺乳期经历了能量代谢的动态变化，以支持胎儿的发育和泌乳的需求[143]。进入到妊娠晚期，奶牛对葡萄糖和氨基酸的需求明显增加，以支持胎儿的生长。随着牛奶生产的启动，加速了对这些营养素以及脂肪酸和矿物质的需求。然而在接近分娩时，奶牛由于干物质采食量降低而导致能量摄入不足。虽然在产后干物质采食量会提高，但滞后于乳产量的提高，因而在整个泌乳早期的能量摄入也不足。从接近产犊到整个泌乳早期，奶牛为了适应这种能量负平衡状态以确保满足维持、胚胎生长、乳房发育和泌乳的营养需求，启动了一系列协调配合的能量代谢调节机制。概括而言，围产期奶牛能量代谢的适应性调节包括以下几点[144]：①在脂肪组织中，脂肪的从头合成、脂肪酸吸收和酯化被钝化，脂肪分解加速，进入循环中的脂肪酸在肝脏和骨骼肌组织中被氧化，在乳腺组织中合成为乳脂；②在肌肉组织中，蛋白质合成减少，氨基酸动员增强，进入循环中的氨基酸被用来支持肝脏糖异生（丙氨酸）和牛奶蛋白质合成的增加；③在肝脏中，肝糖原分解增加，糖异生作用增强，脂肪酸氧化加强；④骨骼肌和脂肪组织对葡萄糖的利用降低，保证用于牛奶乳糖合成的葡萄糖供给。

1. 胰岛素抵抗

胰岛素在能量代谢调节中起着关键的作用。胰岛素通过糖原合成酶（即糖原磷酸化酶）的去磷酸化，增加糖原的产生，减少糖原的分解。胰岛素通过激活线粒体内的丙酮酸脱氢酶复合体来增加丙酮酸向乙酰辅酶 A 的转化。乙酰辅酶 A 可在三羧酸循环中进一步氧化或合成脂肪酸。胰岛素刺激脂肪组织中脂肪酸的合成，脂肪酸合成随乙酰辅酶 A 羧化酶磷酸化程度的增加而增加。胰岛素能诱导许多组织的蛋白质合成，负责核糖体

mRNA 的转录和转位。mRNA 转录增加包括葡萄糖激酶、磷酸激酶、肝脏中脂肪酸和白蛋白的合成以及脂肪组织中的丙酮酸羧化酶。胰岛素负责生长因子如类胰岛素生长因子-1(IGF-1)的合成。

胰岛素抵抗已成为了解围产期奶牛能量代谢的关键。胰岛素抵抗表现为胰岛素对葡萄糖的反应性降低(胰岛 β 细胞分泌胰岛素减少;胰岛素低反应性)和/或葡萄糖对胰岛素的敏感性降低(外周组织对葡萄糖的摄取降低;胰岛素低敏感性)。葡萄糖通过不同的葡萄糖转运蛋白(GLUT)进入细胞。脑和红细胞的 GLUT1,肾小管、小肠和肝脏上皮细胞的 GLUT2,神经元和胎盘的 GLUT3 可以在没有胰岛素存在的情况下转运葡糖糖,脂肪和肌肉细胞的 GLUT4 需要在胰岛素存在下转运葡糖糖。围产期奶牛表现为胰岛素耐受,这对于乳房这个不依赖胰岛素的器官来说是必要的,能够提供泌乳所需的足够的营养和能量。胰岛素抵抗现象在产犊后最为明显,与产前相比,产后胰岛素对葡萄糖的反应性明显降低,葡萄糖的清除明显增加[145,146]。在产后早期,响应促胰岛素剂(葡萄糖和丙酸酯)的胰岛素分泌受到抑制[147]。在哺乳期,乳腺表达不依赖胰岛素的 GLUT,比干乳期高 3 倍,这些受体在泌乳后期和干奶期奶牛的脂肪组织均有表达,但在泌乳高峰期奶牛的脂肪组织中未检测到。骨骼肌肉和脂肪组织中胰岛素敏感受体的激活在哺乳期和干奶期保持不变[148]。与干奶期相比,泌乳期奶牛外周组织的胰岛素敏感性没有变化,但泌乳中期的胰岛素清除率更高[149]。许多因素,如较大的体重、生长抑素、交感活动加强、炎症介质等,可抑制胰岛素的分泌。临近产犊时干物质采食量的降低以及产后能量负平衡是导致围产期奶牛胰岛素抵抗的主要原因。

2. 体脂动员

随着能量负平衡的发展,奶牛血清中棕榈酸、硬脂酸、油酸和亚油酸浓度增加[150,151],这意味着体脂动员增强。在脂肪分解增加的同时,奶牛产后肝脏积累了更多的棕榈酸、油酸和亚油酸[151]。肝脏甘油三酯和磷脂中观察到明显的棕榈酸、油酸和亚油酸的富集,而多不饱和脂肪酸则明显减少[150]。产后早期胰岛素分泌抑制消除了对脂解作用的抑制信号,使循环中脂肪酸含量提高,从而氧化和合成乳脂。棕榈酸通过刺激神经酰胺的合成使胰岛素基因表达降低,胰岛素的合成和分泌降低[152]。因此,泌乳早期胰岛素分泌降低可能与循环中棕榈酸浓度提高有关。单不饱和脂肪酸(棕榈油酸、油酸)可折中棕榈酸对胰岛 β 细胞的毒性作用。对泌乳中期奶牛的研究表明,与棕榈酸以及棕榈酸和硬脂酸混合添加相比,棕榈酸和油酸混合添加提高了循环中胰岛素的浓度[153]。奶牛日粮或脂解衍生的脂肪酸可改变外周组织胰岛素敏感性，其作用与脂肪酸的饱和程度、链长和浓度以及奶牛是否处于稳态控制有关。饱和脂肪酸可通过刺激脂质介质（神经酰胺、二酰甘油）的合成、炎症、氧化应激或内质网应激来降低胰岛素敏感性[154]。而 n-3 PUFA 通过刺激线粒体功能，减少 ROS 产生和阻止炎症来对抗胰岛素抵抗[155]，单不饱和脂肪酸也具有和 n-3 PUFA 类似的作用。奶牛静脉灌注饱和动物油脂引起胰岛素抵抗[156]。棕榈酸有利于营养物质分配给乳腺，而油酸有利于能量从牛奶生产转移到身体脂肪积累上[153]。限饲奶牛亚麻籽油（富含亚麻酸）皱胃灌注可增强胰岛素的抗脂解作用[157]。生长牛日粮 EPA 和 DHA 添加提高了胰岛素敏感性[158]。

3. 脂肪酸 β 氧化

脂肪酸 β 氧化的关键步骤是脂酰肉碱的形成，胞浆中的脂酰 CoA 在肉碱脂酰转移酶Ⅰ的催化下生成脂酰肉碱，脂酰肉碱在脂酰肉碱转位酶的作用下转移到线粒体，线粒体脂酰肉碱在脂酰肉碱转移酶Ⅱ的作用下再转变为脂酰 CoA，从而完成脂酰 CoA 由胞浆到线粒体的转移。这一过程中转移到线粒体的肉碱在肉碱转位酶的作用下由线粒体再转移到胞浆。在肥胖和Ⅱ型糖尿病的研究中，脂酰肉碱是脂质诱导的线粒体功能障碍的重要生物标志物[159]。脂酰肉碱的积累反映了脂肪酸氧化能力的降低和脂肪酸向合成甘油三酯和胰岛素抵抗脂质介质（神经酰胺和二酰甘油）的增强，关注的焦点集中在与胰岛素抵抗有关的长链脂酰肉碱上[159]。在非反刍动物来源的分化肌管中，C16∶0-肉碱处理抑制了胰岛素刺激的蛋白激酶 B（AKT）激活和葡萄糖摄取，脂酰肉碱转移酶抑制剂抑制了棕榈酸诱导的初级肌管胰岛素刺激葡萄糖利用的减少，而且还阻止了棕榈酸诱导的 ROS 积累[160]，这表明脂酰肉碱除了降低胰岛素敏感性外，还会促进氧化应激。围产期奶牛肌肉中长链酰基肉碱与中短链酰基肉碱水平均提高[161]。体况评分高的奶牛产后脂肪动员更强，血浆 C14∶0-、C16∶0-、C18∶0-和 C20∶0-肉碱水平更高，血浆总脂酰肉碱水平与循环中总脂肪酸浓度、总神经酰胺浓度和 C24∶0-神经酰胺浓度呈正相关[162]。

神经酰胺对胰岛素敏感性的影响是多方面的。过量的长链脂肪酸可促进胰腺 β 细胞凋亡（FA 过度积聚引起的凋亡，也称脂肪变性），从而导致胰岛素分泌不足，凋亡的部分原因是促进了神经酰胺的从头合成。神经酰胺可以使 AKT 失活，从而下调胰岛素刺激下肌管和脂肪细胞中 GLUT4 向质膜的转位。最近的证据还表明，胞外衍生的神经酰胺（LDL 中）也可能下调肌肉中的胰岛素信号[163]。奶牛血浆神经酰胺浓度随着从妊娠到哺乳期的转变而增加，产前体况评分高和产后循环中总脂肪酸浓度高的奶牛表现尤为明显[164]。产前体况评分高的奶牛产后肝脏总的和 C24∶0-神经酰胺浓度呈现进程性提高并伴有肝脂积累，所有奶牛骨骼肌和 LDL 组分中 C16∶0-神经酰胺浓度增加[165]。非妊娠和非泌乳奶牛静脉输注甘油三酯或限饲提高了循环中总脂肪酸浓度，同时观察到循环中和肝脏神经酰胺浓度的增加[166,167]，二者呈正相关，与胰岛素敏感性呈负相关[164,166]。组织饱和脂肪酸摄取量的增加很可能驱动了神经酰胺的从头合成。与不添加脂肪或硬脂酸相比，饲喂棕榈酸提高了泌乳中期奶牛血浆和肝脏神经酰胺水平[168]。神经酰胺从头合成抑制剂处理牛肝细胞防止棕榈酸孵育下细胞内神经酰胺的积累[169]。

二酰甘油（DAG）是骨骼肌和肝脏中蛋白激酶 C 的激活剂。这两个组织中 DAG 在骨骼肌和肝脏的积聚可降低胰岛素刺激的胰岛素受体底物酪氨酸磷酸化和三磷酸肌醇（IP3）激酶的活化，其结果是下调肌肉中 GLUT4 的转位和葡萄糖摄取，抑制肝糖原的合成和诱导糖异生。研究表明，与产前相比，产后肝脏 DAG 浓度增加并伴有甘油三酯积累，但产前过饲对肝脏 DAG 浓度没有影响[170]。由于没有证据支持产后奶牛肝脏出现胰岛素抵抗[171]，观察到的肝脏 DAG 浓度的升高可能仅代表围产期奶牛肝脏甘油三酯合成的中间产物，而不是胰岛素抵抗的原因。

脂肪因子瘦素由脂肪细胞产生，通过 Janus 激酶信号转导和转录激活途径介导其作用，具有调节能量代谢、胰岛素敏感性、甲状腺激素分泌、免疫力和食欲的功能。在妊娠晚期奶牛中，与瘦牛相比，肥胖奶牛的血浆瘦素水平更高[172]，而所有奶牛在泌乳开始时都经历了循环瘦素和白色脂肪组织瘦素 mRNA 表达的下降[173]。在妊娠晚期静脉输注甘油三酯乳剂会增加血液中脂肪酸和瘦素的浓度，但不会增加泌乳早期血清脂肪酸和瘦素浓度[174]。这些现象暗示，在反刍动物中，瘦素可能只在正能量平衡时有反应。尽管目前的证据表明瘦素在产后奶牛中的作用有限，但早期泌乳奶牛服用人的瘦素可以降低肝脏甘油三酯的水平[175]。

脂联素主要在脂肪组织中表达，随着肥胖程度的增加，脂联素的表达减少。脂联素具有增加胰岛素敏感性、降低循环葡萄糖、增强脂肪酸分解代谢和预防炎症的作用。脂联素的作用机制涉及过氧化物酶体增殖物激活剂受体-α 和腺苷酸活化蛋白激酶的激活。在肌肉中，下游的效应包括增强脂肪酸的转运和氧化，降低细胞内甘油三酯的含量以及增强胰岛素刺激的葡萄糖转运体到质膜的转位。围产期奶牛循环脂联素水平在分娩时最低[176]。经产过渡奶牛脂联素分泌受到抑制可能会限制胰岛素对脂肪酸分解代谢和葡萄糖的利用。此外，提高血浆脂联素水平与降低血浆长链脂酰肉碱浓度有关。值得强调的是，初产奶牛分娩后血清脂联素浓度似乎增加，更高的脂联素浓度可能会增强外周胰岛素敏感性和营养分配以促进生长[177]。与妊娠晚期相比，泌乳早期肌肉和肝脏中脂联素受体-1和脂联素受体-2的mRNA表达升高[176]。在脂肪组织中，观察到脂联素受体的表达没有变化[176]或降低[178]。

4. 糖异生

肝脏处于新陈代谢的“十字路口”，有能力“感知体内所有其他组织的能量需求，并相应地通过调整新陈代谢做出反应”。在妊娠最后21天，奶牛的葡萄糖的需要量估计为1 000～1 100 g/d，但在分娩后21天急剧增加到约2 500 g/d。大部分增加的葡萄糖需求必须通过肝脏糖异生作用来满足。产前11天时肝脏葡萄糖释放量为1 356 g/d，而产后11 d时高达2 760 g/d。在产前21和7天以及产后10和22天时，肝脏重量分别为9.0 kg、8.8 kg、8.8 kg和9.6 kg，干物质摄入量（DMI）分别为11.5 kg/d、12.3 kg/d、11.6 kg/d和16.0 kg/d，DMI增加38%只导致肝脏质量增加9%[179]。奶牛肝血流量从产前11 d的1 140 L/h增加到产后11 d的2 099 L/h，相对应的DMI从9.8 kg/d增加到14.1 kg/d，肝脏氧利用从4.4 mmol/g增加到8.6 mmol/g，DMI增加44%导致肝脏血流量增加84%，肝脏氧利用率增加95%[180]。显然，肝脏血流量的增加是对产后能量需求增加的主要反应。据估计，产后早期日粮可消化能来源的葡萄糖不能满足对葡糖糖的实际需要，缺口大约为500 g/d。因此必须通过提高肝脏糖异生活动来弥补。这里所说的糖异生活动提高包含两层涵义：提高对丙酸的利用能力；提高对其他糖异生底物（氨基酸、乳酸和甘油）的利用。

丙酸在瘤胃以及后肠道发酵产生，是糖异生的主要底物。牛日粮添加丙酸钠，丙酸盐对糖异生的碳贡献率由对照日粮的43.3%提高到67.1%[181]。高精料饲养与高粗料饲养相比，肝脏将丙酸盐转化为葡萄糖的效率更高[182]。饥饿或严重限饲降低了肝细胞转化丙酸盐为葡萄糖的能力[183,184]。奶牛在围产期，产后第1天和第21天肝脏将丙酸盐转化为葡糖糖的量分别比产前第21天提高19%和29%，而产后21天与产后56天相比无显著区别[185]。另一项研究表明，产后30天奶牛肝脏利用丙酸盐的能力是产后90天和180天的3倍[182]。对奶山羊的研究也发现，与非泌乳相比，泌乳阶段肝脏转化丙酸盐为葡萄糖的能力提高[186]。这些研究提示，在能量负平衡期间（产后早期），肝脏可更有效地利用丙酸盐进行糖异生，这和该阶段丙酸盐供应受限有关。

虽然在大多数情况下，氨基酸不是主要的糖异生底物，但氨基酸对反刍动物糖异生起着相当大的作用。除了完全生酮的亮氨酸和赖氨酸外，其他所有氨基酸对葡萄糖的合成都有净贡献，但丙氨酸和谷氨酰胺的贡献最大，大约占到所有生糖氨基酸产糖潜能的 40%～60%[187]。产后 1 天和 21 天丙氨酸转化为葡萄糖的能力分别是产后 21 天的 198%和 150%[185]。与丙酸相比，产前营养对丙氨酸转化为葡萄糖的影响不大，而且 NE_L 摄入量与丙氨酸转化为葡萄糖没有相关性。丙氨酸转化为葡萄糖的能力与丙酸转化为葡萄糖的能力相比有较大的相对变化，丙酮酸羧化酶(PC)活性增加。PC 活性的增加能有效地将转变为丙酮酸的碳转化为葡糖糖。产后 1 天肝脏中 PC 的 mRNA 水平是产前 28 天的 7.5 倍，产后 28 天时降至产前水平[188]。随着产后早期糖异生的增加，氨基酸可能扮演着特别重要的角色。骨骼肌是一个不稳定的氨基酸池，在围产期被动员以支持增加的糖异生，皮肤和内脏组织也可能发生适应性，为围产期的糖异生提供额外的氨基酸[189]。尿中 3-甲基组氨酸与肌酐的比值通常被用来评价骨骼肌蛋白的降解率，这一数值在奶牛产后 3 天增加了 3 倍多，在第 7～10 天虽然降低，但仍然是产前数值的 2 倍多[185]，表明肌肉蛋白库在泌乳的最初几天经历了大量的动员。

PC 活性的增加也会增加乳酸到葡萄糖的转化率。然而，利用乳酸进行糖异生主要代表的是碳的再循环，因为大多数循环乳酸要么是在外周组织分解葡萄糖的过程中形成的，要么是由内脏上皮组织部分分解丙酸形成的。虽然高精料饲喂时有 44%的葡萄糖碳是由乳酸盐贡献的[190]，但在过渡时期，奶牛典型日粮在瘤胃发酵过程中不会产生大量的乳酸盐。与泌乳早期相比，妊娠晚期乳酸盐对糖异生的贡献较大，因为在妊娠晚期，子宫和肌肉会释放乳酸[191]。

脂质动员过程中释放的甘油可能是奶牛适应泌乳过程的一个重要的糖异生前体。在脂质广泛动员(每天 3.2 kg 甘油三酯)的过程中，产后 4 天甘油最多可提供葡萄糖需求量的 15%～20%[191]。因此，甘油对糖异生的潜在贡献完全取决于围产期奶牛脂肪动员的程度。产后体脂动员速度随着产后时间的延长逐渐减慢，泌乳早期的典型值大约是 0.5～1.0 kg/d，因此，甘油最多可提供总葡萄糖需要量的 2%～5%。

影响糖异生作用的激素包括生长激素、胰岛素、胰高血糖素和皮质醇。生长激素、胰高血糖素增强糖异生和皮质醇增强糖异生，而胰岛素会降低糖异生作用。奶牛产犊时生长激素水平显著提高[192]，产犊时胰岛素水平达到峰值，但产后低于产前[193]，胰高血糖素水平基本保持不变[194]，皮质醇浓度在分娩前最后 3 天开始升高，分娩时达到峰值，通常产后 3～5 天下降到接近产前水平[195]。

(二)硒对围产期奶牛能量代谢的调节

H_2O_2 在细胞代谢中具有双重作用,低水平的 H_2O_2 是细胞内信号级联反应的第二信使,而其过量产生则可能导致细胞功能障碍、损伤和死亡。氧化剂和抗氧化剂之间偏向氧化剂的不平衡可能会导致氧化应激,而偏向还原剂的不平衡可能会导致还原应激。细胞氧化还原平衡对于维持和适应胰腺胰岛素分泌,提高靶组织对胰岛素的敏感性至关重要。胰腺 β 细胞的抗氧化防御能力较差,对氧化应激和硝化应激敏感[196]。生理上,主要抗氧化酶的低活性可能使 β 细胞对 H_2O_2 和其他 ROS 的信号效应敏感。将 β 细胞暴露于低微摩尔浓度的 H_2O_2 可以增加其细胞内钙浓度并刺激胰岛素分泌,而高浓度的 H_2O_2 会钝化胰岛素分泌[197]。GPX1、CAT、SOD1 或 Prx III 过表达保护了各种 β 细胞系免受氧化和硝化损伤[198,199],而 β 细胞特异性过表达 CAT 和金属硫蛋白诱导了糖尿病发生[200]。线粒体中超氧化物和 H_2O_2 的产生通过诱导质子泄漏和激活解耦联蛋白 2(UCP-2)降低 ATP 的生成,因而降低胰岛素分泌,耦联状态的 UCP-2,即谷胱甘肽化的 UCP-2,可抑制质子泄漏并增强胰岛素分泌[201]。

如前所述,能量代谢与胰岛素有着直接的关系,而胰岛素信号和胰岛素分泌与细胞的氧化还原状态有关。因此,硒及其合成的抗氧化硒蛋白可能通过调节细胞的氧化还原状态调节胰岛素信号和胰岛素分泌,进而调节碳水化合物和脂代谢。

目前关于硒对围产期奶牛胰岛素抵抗和能量代谢影响的研究甚少。作者前期的研究发现,产前超营养的硒酵母添加提高了产后早期奶牛血浆中的胰岛素和葡糖糖的水平,降低了 NEFA 水平(未发表),提示硒可以缓解围产期奶牛的胰岛素抵抗,进而在一定程度上缓解了过度的脂肪动员以及肝脏对碳水化合物和脂肪酸代谢的负担。虽然这在一定程度上不利于乳腺对葡萄糖的摄取利用,但可以避免因过度脂肪动员导致的肝脏负担加重,从而避免脂肪肝和酮病等代谢疾病的发生以及炎症反应增强而导致的感染性疾病发病风险的提高。如前所述,氧化应激会钝化胰岛素分泌,导致胰岛素抵抗。作者前期对围产期奶牛的研究表明,产前最后 4 周超营养的硒酵母添加提高了产后奶牛的硒营养状况,降低了产后早期奶牛的氧化应激水平[4],因而降低了胰岛素抵抗的程度,减缓了脂肪动员程度。虽然围产期奶牛产后早期表现的胰岛素抵抗是一种适应现象,有利于乳腺对葡萄糖的摄取利用,但同时也增加了各种代谢性和感染性疾病的发病风险。因此,从综合角度考虑,产前超营养硒添加一定程度上对胰岛素抵抗的缓解,有利于奶牛对围产期的成功适应。

需要注意的是，由于生理水平的 H_2O_2 作为第二信使介导胰岛素信号并刺激胰岛素分泌。因此过量的硒添加有可能因抗氧化酶的过量表达而解除 H_2O_2 作为第二信使介导胰岛素信号的作用，最终导致胰岛素抵抗。高水平的血浆硒和 SelP 已被发现与人类碳水化合物和脂质稳态受损有关。在动物模型中，膳食硒过量引起抗氧化硒蛋白的大量表达，导致胰岛素敏感性降低[202]。我们研究发现的产前超营养硒添加减缓胰岛素抵抗而不是增强，从另外一方面也说明了产前超营养的硒添加可能更好地满足了围产期奶牛对硒需要量的提高而非过量。

三、硒对围产期奶牛免疫功能的调节

（一）围产期奶牛免疫功能的变化

免疫系统是由免疫器官、免疫细胞及免疫分子组成的复杂网络，激活的免疫系统起着识别、对抗和清除抗原性异物，并与其他系统相互协调共同维持机体内环境稳定和生理平衡的重要功能。如本书第三章所述，依据对抗原性异物反应的速度和特异性，免疫系统可分为先天免疫系统和获得性免疫系统。通过免疫细胞（中性粒细胞、单核细胞、巨噬细胞）、体液因子（补体、溶菌酶）和细胞因子网络的协同作用，先天免疫系统可以解决 95％的感染性挑战[203]。先天免疫细胞通过 PRRs 识别 PAMPs 和 DAMPs，诱导两种反应，一种是炎症，另一种是吞噬。即使是机体从未遇到过的抗原性异物，这些反应也会发生。其发生是通过合成一些生化介质来驱动的，这些生化介质包括组胺和促炎细胞因子（PIC），可以吸引免疫细胞到达损伤部位并通过激活 NF-B 信号通路诱导局部炎症。在系统水平上，PIC 还可以诱导肝脏合成急性期蛋白（APPs），引起急性期反应。必须注意的是，免疫系统和内分泌系统之间存在双向通讯。在长期或慢性应激源存在时，下丘脑-垂体-肾上腺轴对免疫系统具有抑制作用。例如，糖皮质激素可通过下调中性粒细胞粘附分子（血管因子）L 选择蛋白和 CD18 的基因表达降低中性粒细胞的活化以及向感染位点的迁移速率[204]。代谢相关的激素（生长激素、促甲状腺激素、胰岛素）对免疫系统及其功能的发展具有积极作用[205]。例如，血糖提高增强了 中性粒细胞 CD66b（脱颗粒指示分子）和单核细胞 CD11b（参与黏附）的基因表达[206]。

在围产期，有很多大概率影响奶牛免疫系统的因素，既包括生物性因素（细菌、病毒、寄生虫），也包括非生物因素。众多的非生物因素，如营养水平、采食量、能量代谢、氧化应激、环境温度、隔离、分群、过度拥挤等，可能在影响免疫功能更为显著。事实上，奶牛在产前，免疫系统的一些功能会受到抑制，如中性粒细胞的吞噬能力[207]、淋巴细胞对有丝分裂原的反应以及产生抗体的能力[208]、外周血单个核细胞的DNA合成能力[207]、血浆中免疫球蛋白、IFN-γ[207]和补体[209]的浓度及溶菌酶的活性[210]。研究表明，妊娠末期能量负平衡可引起奶牛中性粒细胞功能降低[211]，分娩前采食量降低也可对奶牛的免疫功能产生不利影响[212]。Kehrli等的研究经常被用来支持围产期奶牛的免疫抑制现象，但他们发现免疫抑制（淋巴细胞生成、中性粒细胞趋化）出现在产后[213]，原因可能与试验动物是小母牛（而不是多胎奶牛）有关。在此基础上，Goff和Horst[214]用多胎奶牛进行了研究，并用从阉牛（当作没有怀孕的奶牛）获得的数据进行了矫正，结果发现，中性粒细胞吞噬和淋巴细胞增殖能力的明显降低始于产前3周左右。后来进一步的研究表明，在产犊前几天，中性粒细胞黏附分子L选择蛋白和CD18的表达降低[133]，对无菌刺激的外周免疫反应明显降低[215]。或多或少在同一时间，奶牛呈现出明显的系统性炎症反应[133,210,216-218]，这可以通过产犊后几小时急性期蛋白的提高达到峰值来证实，几乎所有奶牛都会发生这种炎症反应，与子宫是否出现全面感染无关。这种反应出现的时机类似于在成年动物中观察到的缺血性损伤。Trevisi等[219]的研究发现产后子宫创伤引起的血浆PIC的升高比较温和，而作者前期的研究发现产后血浆PIC显著提高（未发表），这可能与产后子宫创伤的程度不同有关。但无论温和还是显著，均表现为产后炎症反应增强。此外，Jahan等[215]研究发现，产犊前几天外周血白细胞在LPS刺激后释放PIC的能力显著增加。上述这些证据表明，奶牛免疫功能的改变始于产前，其中一些功能（细胞免疫反应）减弱，而炎症反应增强。因此，传统认为的围产期奶牛免疫功能抑制的观点可能并不全面，描述为免疫功能障碍或失调可能更为恰当。Minuti等[220]对健康奶牛循环中的白细胞进行了转录组学研究，观察到产犊后差异表达基因的数量增加，产后1周内数量最大，这些基因主要表现为上调，在产犊前，所有差异表达的基因都上调了，这些激活的路径包括与黏附、迁移、分化、吞噬等细胞活动相关的路径。这一研究从基因层面上也提示了围产期奶牛的免疫功能在产犊前后表现为失调，而不是简单的抑制。

如前所述，围产期奶牛要经受许多外部因素的影响，这些影响因素的数量和严重程度可能会逐渐降低免疫能力，同时增加对炎症的易感性（图5-2），导致产犊前后炎症反应发生[121]。如前所述，围产期奶牛临近产犊时采食量的明显降低以及产后早期采食量提高与乳生成的不同步导致体脂动员。严重的体脂动员可降低免疫力[207]。如前所述，围产期奶牛由于乳腺组织活动增强引起的ROS生成提高可通过激活NF-B信号通路诱导

PIC的释放(炎症)。炎症可降低采食量因而加重脂解[221],释放入血的血浆非酯化脂肪酸(NEFA)在肝脏进行β-氧化的过程中可增强ROS的生成[222],因而形成一种炎症与氧化应激之间的恶性循环。泌乳早期氧化应激水平较高时,乳腺促炎细胞因子的表达显著提高[223]。与泌乳中期和后期相比,围产期和泌乳早期奶牛外周血单核细胞TNF-α的基因表达显著提高,而TrxR1的活性降低[224]。血浆Ca^{2+}水平是包括免疫细胞在内的很多细胞代谢活动的关键因子,奶牛产后头两周血浆Ca^{2+}水平的降低与中性粒细胞功能的下降有关[225]。随着LPS的作用,血浆Ca^{2+}水平呈剂量依赖性降低[226],严重的低血钙发生在败血症状态或有严重组织损伤的情况下,如瘤胃酸中毒[227]。钙参与内毒素的解毒过程,血浆中钙的减少可能是内毒素血症期间的一种防御机制[228]。

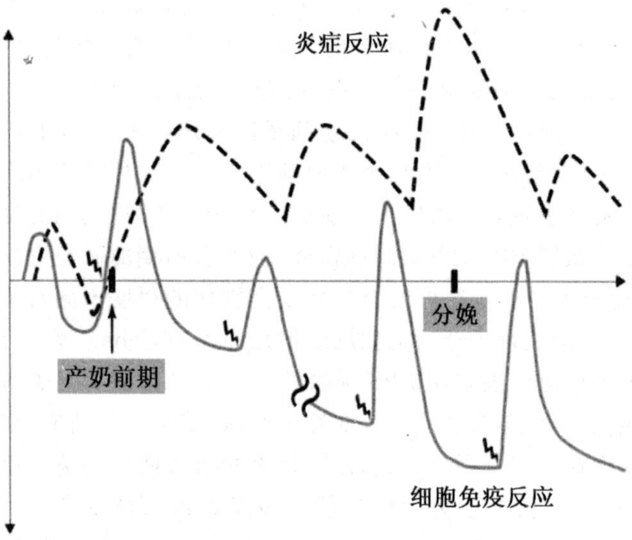

图5-2 围产期奶牛细胞免疫反应与炎症反应的可能关系[121]

肝脏在碳水化合物、蛋白质和脂质的新陈代谢、清除毒素和病原体以及调节免疫反应方面发挥多种功能。对于免疫反应,肝脏可以通过急性期反应对其他细胞产生的PIC做出反应,也可以对局部新陈代谢并可能导致局部组织损伤的抗原产生反应。感染、组织损伤、创伤或手术、肿瘤生长或免疫紊乱都可触发先天免疫反应,引起ARP。ARP期间,肝脏阳性急性期蛋白(posAPP:C反应蛋白、血清淀粉样蛋白A、结合珠蛋白、铜蓝蛋白等)的合成增加。posAPP具有抵抗病原体的作用(微生物及其产物的调理和捕获、补体激活、结合细胞核片段、中和酶、清除游离血红蛋白和自由基、调节宿主免疫反应),一些posAPP能够从血液中捕获矿物质(如Fe、Zn),以降低病原体的存活率。需要注意的是,肝脏在激活合成 posAPP 的同时,肝脏的其他功能由于一些蛋白质合成的降低而受到损伤,这些蛋白质被称为阴性急性期蛋白(negAPP:白蛋白、对氧磷酶、阿朴脂蛋白、亲脂分子的载体等)。negAPP 对于维持新陈代谢的正常活动是必不可少的,对于奶牛,产后 negAPP 的降低(白蛋白、对氧磷酶)或缓慢提高(胆固醇、视黄醇结合蛋白)会导致采食量和生产性能下降,身体储备动员增加和养分利用受损。

有效的即可控的炎症反应表现为适时发起、及时消退,在清除感染源和修复损伤的过程中起着关键的作用,通常是在不表现任何可见组织变化的情况下实现对感染源的及时清除[229]。而不可控或功能异常的炎症反应要么表现为持续的、低程度的慢性炎症,要么表现为高强度的急性炎症,不仅不能有效清除感染源或其他组织损伤源,持续存在的炎症还会加重组织损伤。许多医学证据表明,不可控的炎症反应是动脉粥样硬化、肥胖、糖尿病、败血症、乳腺炎等疾病发生的病理原因[230−232]。基于此,一些研究者针对围产期奶牛,肝活性指数(LAI)、肝功能指数(LFI)和产后炎症反应指数(PIRI)用于在没有临床症状的情况下判断动物的生理状况、肝功能和产后炎症反应的程度以及对需要进一步诊断的器官(子宫、乳腺、蹄)提供预警作用。PIRI 的低评分意味着泌乳早期存在亚临床问题。LAI 和 LFI 指数相关性很好,但 LFI 更容易分析,LFI 或 LAI 的低评分意味着在整个过渡期内较低的饲料摄入量、较高的体脂动员、较低的产奶量[233]和较短的反刍时间[234],反过来意味着更强的能量负平衡、更低的能量利用效率、更高的氧化应激以及更低的循环氨基酸利用。关于这些指数的详细说明可参考 Bertoni 和 Trevisi(2013)[216]和 Trevisi 等(2016)[235]的相关研究。

(二)硒对围产期奶牛炎症反应的调节

作为一种重要的日粮来源的抗氧化剂,微量元素硒对炎症具有重要的调节作用。硒对炎症反应的调节作用在癌症[236]、心血管疾病[237]、乳腺炎[238]、骨质疏松[239]等许多疾病模型中得到很好的证实。尽管硒对炎症反应具有重要的调节作用,但潜在的机制还不完全清楚。在免疫反应中,免疫细胞在受影响的区域产生炎症环境,以吸引更多的细胞来协调扩大免疫反应。炎症环境是由分泌的介质造成的,细胞因子和氧化脂质是两类介导炎症启动、发展和消退的重要物质。硒通过调节这些物质的释放或合成实现对炎症反应的调节。

1. 硒对细胞因子释放的调节

细胞因子是一类低分子量的蛋白质,介导不同细胞之间的通讯或以自分泌的方式发挥作用。细胞因子环境对巨噬细胞分化为 M1 或 M2 表型以及 Th 细胞分化为 Th1 或 Th2 细胞都是非常关键的。TNF-α 和 IL-1β 对 M1 巨噬细胞有很强的刺激作用,而 IL-10 和 IL-4 则是 M2 巨噬细胞活化所必需的。干扰素-γ 是 Th1 型分化的主要刺激因子,而 IL-10 则激活 Th2 型免疫应答。细胞因子的表达主要受信号转导与转录激活因子(STATs)的调节,但也受 NF-κB 的调节,两者都具氧化还原敏感性[240]。

硒对不同模型中的细胞因子模式有广泛的影响。在感染金黄色葡萄球菌的 RAW264.7 巨噬细胞中,补硒下调了 TNF-α、IL-1β 和 IL-6 mRNA 的表达,这归因于 NF-κB 和 MAPK(ERK、JNK、p38)信号通路的抑制[241]。在 DSS 诱导的小鼠结肠炎模型中,补硒可减轻炎症反应、炎性细胞浸润和病理损伤,同时,硒可降低 NF-κB 的活化和促炎细胞因子 TNF-α、IL-6 和单核细胞趋化蛋白-1 的释放,显著增加 IL-10 的分泌[242]。在雄性小鼠的红细胞中,硒降低了注射 LPS 后 TNF-α、单核细胞趋化蛋白-1 和 IL-6 的释放[243]。作者前期的研究表明,奶牛产前 4 周硒添加降低了产后 TNF-α、IL-1β 和 IL-6 的释放以及由此引起的 posAPP(结合珠蛋白、血清淀粉样蛋白 A)生成提高和白蛋白合成的降低,奶牛外周血单核细胞源性巨噬细胞硒耗竭后添加硒可降低 TNF-α、IL-1β 和 IL-6 的 mRNA 表达(未出版结果)。与日粮硒缺乏相比,从日粮硒适宜鸡分离的法氏囊具有较高的 IL-1β、IL-2、IL-6、IL-8、IL-10、IL-17、IFN-α、IFN-β 和 IFN-γ 的 mRNA 表达水平,而 TNF-α 的表达水平显著降低[244]。对缺硒和适硒小鼠的研究表明,适宜的硒水平提高了 7 种细胞因子(IFN-γ、IL-6、IL-1β、IL-10、IL-12、Cxcl1、TNF-α)中的两种(IFN-γ 和 IL-6)[245]的表达水平。在另一项研究中,亚硒酸钠可阻止汞诱导的 IL-6、IL-10 和 TNF-α 的降低,提高 IL-2 和 IL-2 受体的水平[246]。

关于硒蛋白对细胞因子的调节,最近的研究表明,硒蛋白 MsrB1 可刺激巨噬细胞抗炎细胞因子的表达。在 LPS 刺激下,MsrB1 的表达高度上调,MsrB1 基因敲除导致血浆抗炎细胞因子 IL-10 和 IL-1 受体拮抗剂的水平显著降低,而促炎细胞因子 TNF-α、IL-6 和 IL-1β 的水平未受影响[247],这与炎症反应增强和炎症严重程度增加有关。有趣的是,在刀豆蛋白诱导的 T 细胞介导的肝损伤模型中,GPX1 缺乏抑制了细胞因子 TNF-α、IL-2 和 IFN-γ 的表达(T 细胞活化),这归因于氧化还原平衡的改变[248]。然而,在变应原诱导的气道炎症中,GPX1 缺乏增加了 Th 细胞 IFN-γ 和 IL-2 的表达,而 Th2 型细胞因子 IL-4、IL-5 和 IL-13 的表达下降[249]。SelS 是一种内质网驻留的硒蛋白,可防止内质网应激,并参与错误折叠蛋白的降解。阻断 SelS 可导致佛波酯刺激的巨噬细胞 IL-6 和 TNF-α 水平的增加[250]。

另一种内质网硒蛋白 SelK 对细胞因子模式也有明显影响，SelK 基因敲除小鼠血清 TNF-α 水平升高，单核细胞趋化蛋白-1 水平降低[251]。有趣的是，SelK 基因敲除对细胞因子的影响与因作用时间的不同而呈现出不同的作用方式，基因沉默 24 h 后，原代成肌细胞 IL-1β、IL-6、IL-7、IL-8、IL-17 和 IFN-γ 水平升高，72h 后仅 IL-6、IL-7 和 IL-8 水平升高，之后 IL-1β、IL-17 和干扰素-γ 水平显著降低[252]。围产期奶牛乳腺氧化应激水平较高时，外周血单核细胞中 TrxR 活性降低[224]，添加硒显著提高了人和牛的内皮细胞 GPX 和 TrxR 的活性，降低了氧化诱导的促炎细胞因子的提高和细胞凋亡[253]。围产期奶牛乳腺组织中 GPX1、GPX4 和 TrxR 与促炎因子的基因表达具有强烈的正相关[223]，提示这些硒蛋白酶的抗氧化作用在调节免疫功能方面起关键作用。

综上所述，在免疫挑战条件下，硒降低了促炎细胞因子的产生，而在非刺激条件下，硒增加了抗炎细胞因子的水平。对于巨噬细胞，硒使细胞因子环境更倾向于抗炎 M2 细胞因子。然而，Th 细胞因子的表达在两种表型中都是平衡和增加的，这表明硒促进 Th 细胞反应依赖于所面对的病原体。

2. 硒对氧化脂质合成的调节

氧化脂质是由 PUFA 经酶和非酶氧化途径代谢生成的一类脂质代谢产物。目前为止，在人的血浆中已鉴定出至少 158 种 PUFA 代谢产物[254]。通常，n-6 PUFA 代谢生成的氧化脂质多数具有促炎作用，例如我们熟知的前列腺素（prostaglandin，PG）E_2 和白三烯（leukotriene，LT）B_4 就具有强有力的促炎作用，而 n-3 PUFA 代谢生成的氧化脂质多数具有抗炎或促进炎症消退的作用[255]。研究表明，硒及其合成的硒蛋白可在多个水平上对氧化脂质的生物合成进行调节，主要通过限制具有促炎作用的氧化脂质的生成和促进具有抗炎作用的氧化脂质的合成来维持二者的平衡[256]。就目前的研究进展而言，硒主要从抗氧化调节、氧化还原调节以及反馈调节三个方面对氧化脂质的生物合成进行调控。

(1)抗氧化调节

利用其抗氧化作用，硒及其合成的硒蛋白可通过控制环氧合酶（COX）和脂氧合酶（LOX）的活性来调节氧化脂质的生物合成。如图 5-3 所示，从细胞膜磷脂池中释放出来的花生四烯酸（ARA）在 COX 的催化下生成 PGH_2，PGH_2 在相应的 PG 合成酶和血栓素（TX）合成酶的作用下生成 PGD_2、PGI_2、PGF_2 和 PGE_2 以及 TXA_2 和 TXB_2。PGD_2 可进一步经两次脱水反应生成 15-dPGJ_2，而 PGE_2 可在前列腺素脱氢酶（PGDH）的作用下进一步代谢生成 15-酮基 PGE_2（15-keto PGE_2）。COX 存在持续表达的

COX-1 和诱导表达的 COX-2 两种亚型。COX-2 的活化需要其活性位点上的酪氨酸残基在活性氧的作用下形成自由基，该自由基进而催化 ARA 的去氢和随后的氧插入[257]。在氧化应激和炎症期间，COX-2 的活性显著提高，下游脂质代谢产物为 PGE_2 的生成显著增加。GPX1 可直接化学还原胞质中的 H_2O_2，从而抑制其对 COX-2 的激活[258]。GPX2 和 COX-2 在炎症区域的共同定位表明，GPX2 也可能通过直接还原 H_2O_2 来抑制 COX-2 的活性[236]。这些研究提示，硒可通过限制细胞内活性氧的生成降低 COX 的活性，进而减少促炎氧化脂质 PGE_2 的生成。然而一些研究表明，当细胞内活性氧水平太高时也会导致 COX 活性降低[259]，这种现象被解释为“自杀性灭活”[260]。发生“自杀性灭活”的原因还不清楚，可能是机体的一种保护性反应。5-LOX 具有双加氧酶和 LTA_4 合成酶两种催化功能。如图 5-3 所示，ARA 首先在双加氧酶的作用下生成 5-羟基过氧二十碳四烯酸（5-HPETE），不稳定的 5-HPETE 在 LTA_4 合成酶的作用下生成促炎氧化脂质 LTA_4，LTA_4 可在 LTA_4 水解酶（LTA_4H）的作用下生成 LTB_4，也可在 LTC_4 合成酶的作用下与 GSH 耦合生成 LTC_4。与 LTA_4 相比，LTB_4 和 LTC_4 的促炎作用更强。与 5-LOX 相似，12-LOX 和 15-LOX 可催化 ARA 生成 12/15-HPETE，12/15-HPETE 随后通过异构化作用生成具有促炎作用的羟基环氧二十碳三烯酸（Hxs）或者在 5-LOX 的催化下生成具有抗炎作用的脂氧素（LX）A_4 和 B_4。LOX 的激活依赖于其活性位点上的非血红素铁在活性氧的作用下形成自由基。研究表明，含硒化合物依布硒啉可通过限制 15-LOX 活性位点上非血红素铁的氧化而抑制该酶的活性[261]。用 GPX1 预培养中性粒细胞可抑制 5-LOX 活性，从而降低 5-HPETE 的生成[262]。用 GPX4 预培养内皮细胞可抑制 15-LOX 的活性因而降低 15-HPETE 的生成[263]。GPX4 基因敲除小鼠的胚胎成纤维细胞中 12/15-LOX 的活性明显提高，而且导致 AIF 诱导的细胞死亡[264]。利用其抗氧化作用，硒及其合成的硒蛋白也可通过直接还原 PUFA 氧化代谢网络中的中间代谢产物（脂质氢过氧化物）来调节氧化脂质的生物合成。ARA 和亚油酸（LA）经 LOX 途径分别代谢生成的 HPETE 和羟基过氧十八碳二烯酸（HPODE）就属于脂质氢过氧化物。这些过氧化物可在抗氧化酶的作用下分别还原为相应的羟基衍生物（HETE 和 HODE），并分别代谢为相应的酮基衍生物（oxoETE 和 oxoODE）。这些过氧化物的还原可有效降低其向其他氧化脂质的代谢。研究表明，GPX1 可将 5-HPETE 直接还原为 5-HETE[262]，GPX1 基因敲除，小鼠的血小板中 12-HETE 的生成显著降低[265]。用 GPX4 预培养内皮细胞显著降低了细胞内 15-HPETE 的水平[263]，GPX4 基因过表达显著促进了 5-HPETE 向 5-HETE 的还原，因而降低了 5-HPETE 向促炎氧脂 LTB_4 和 LTC_4 的生成[266]。除了 GPX，其他硒蛋白如硫氧还蛋白酶 1(TrxR1)和 SelP 也具有脂质氢过氧化物酶活性。在体外条件下，SelP 可直接化学还原 15-HPETE，而且可抑制 15-HPETE 诱导引起的细胞内活性氧自由基的生成[267]。在 NADPH 存在的情况下，TrxR1 也可直接将 15-HPETE 还原为 15-HETE[268]。

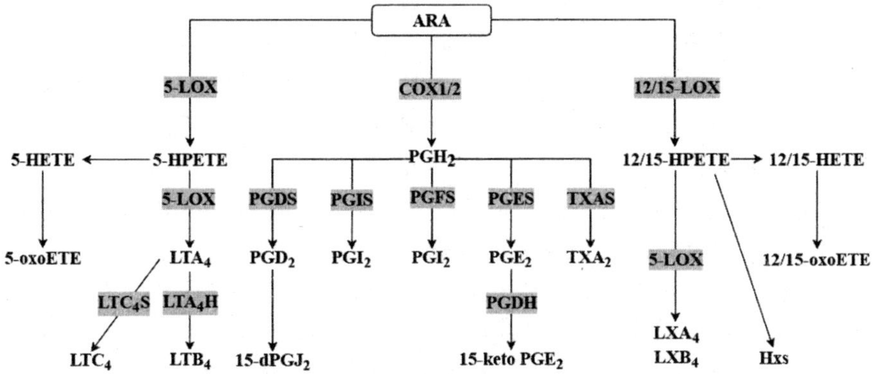

图 5-3　花生四烯酸氧化代谢途径

(2)氧化还原调节

除了对氧化脂质生物合成相关酶活性的调节外，硒及其合成的硒蛋白还可以调节氧化脂质生物合成相关酶的基因表达。研究表明，硒缺乏上调了 LPS 诱导的小鼠巨噬细胞 COX-2 的基因表达，提高了 PGE_2 的生成[269,270]。用生理水平的 SeMet(0.5 μM)预处理人软骨细胞 24 h，降低了 IL-1β 诱导的 COX-2 的基因表达以及 PGE_2 的生成[239]。在小鼠结肠癌模型中，两周的硒酸盐添加(30 μg/g 体重)下调了 COX-2 的基因表达，降低了肿瘤的尺寸[271]。此外，硒添加还下调了巨噬细胞 PGE 合成酶(PGES)和 TXA 合成酶(TXAS)的基因表达，因而降低了促炎氧化脂质 PGE_2 和 TXA_2 的生成；同时，硒添加还上调了 PGD 合成酶(PGDS)的基因表达，因而提高了抗炎氧化脂质 15-dPGJ_2 的生成[272]。Kaushal 等[273]研究表明，硒添加还上调了 PGDH 的基因表达，因而促进了促炎氧化脂质 PGE_2 向无活性的 15-keto PGE_2 的生成。在异丙肾上腺素诱导的大鼠心肌梗死模型中，亚硒酸钠添加(8 μg/g 体重)上调了 LTA_4H 的基因表达，因而降低了 LTA_4 向促炎作用更强的 LTB_4 的生成[274]。通过敲除编码 $tRNA^{Sec}$ 的基因使硒蛋白无法合成的研究发现，小鼠巨噬细胞 COX-2、5-LOX、15-LOX1、15-LOX2 以及肿瘤坏死因子(TNF-α)的基因表达明显提高，LXA_4 的生成显著降低[275]。在人结肠癌细胞(HT-29)中，GPX2 基因敲除明显提高了 COX-2 和 PGES 的基因表达[236]。氧化脂质生物合成过程中许多酶的基因表达受到 NF-κB、激活蛋白 1(AP-1)、缺氧诱导因子 1α(HIF-1α)、STATs、过氧化物酶体增值激活受体(PPARs)等转录因子的调控。由于含有以硫醇(-SH)形式存在的半胱氨酸残基,这些转录因子可以和过氧化物的 O—O 键发生亲核攻击反应形成 S—S 键,因而是氧化还原敏感的。硒及其合成的硒蛋白可以调节这些氧化还原敏感的信号通路的活化进而调控氧化脂质生物合成相关酶的基因表达[240]。目前研究较多的是硒及其合成的硒蛋白通过调控 NF-κB 信号通路进而调节氧化脂质生物合成相关酶的基因表达。小鼠巨噬细胞体外培养研究表明,硒添加明显抑制了 LPS 诱导的 NF-κB 的活化[276]。对人结肠癌细胞的研究发现,超营养的硒添加(250~500 μM)抑制了 LPS 诱导的 NF-κB 的活化[271]。Zamamiri-Davis 等[269]研究表明,硒缺乏导致的 COX-2 基因表达的提高依赖 NF-κB 信号通路的激活。用小白菊内酯(parthenolide)抑制巨噬细胞 NF-κB 活化的研究表明,亚硒酸钠以及甲基硒酸添加引起的 PGES 和 TXA 合成酶(TXAS)基因表达的降低以及 PGDS 基因表达的提高受 NF-κB 信号通路的调控[272]。硒蛋白 GPX4 基因过表达抑制了兔动脉平滑肌细胞[277]和人皮肤成纤维细胞内 NF-κB 的活化[278],降低了 LTB_4 和 LTC_4 的生成,而当 GPX4 基因过表达细胞中 GSH 缺乏时,NF-κB 的核转录活性明显提高[266]。细胞在静息状态下,NF-κB 抑制因子(IκB)通过与 NF-κB 结合掩蔽了 NF-κB 二聚体的核定位信号,同时也干扰了 NF-κB 与 DNA 的结合,从而阻抑了 NF-κB 的转录功能。在生长因子、细胞因子、淋巴因子、紫外线、氧化应激以及病原微生物等多种外来刺激的作用下,NF-κB 与某些配体以及细胞表面受体结合,招募接头蛋白继而招募 IκB 激酶(IKK)复合物并导致 IKK 复合物激活,IKK 复合物的活化导致 IκB 磷酸化,继而导致 IκB 的泛素化和降解,IκB 的降解使 NF-κB 二聚体的核定位信号暴露,因而易位到细胞核中与 DNA 结合诱导相关基因的表达。硒依赖的抗氧化酶 GPX 和 TrxR 可通过调节接头蛋白的氧化还原状况抑制 IKK 活性,进而抑制 NF-κB 信号通路的活化[256]。

(3)反馈调节

如前所述,在硒缺乏巨噬细胞中添加硒可提高抗炎氧化脂质 15d-PGJ_2 的生成[272]。现有文献表明,硒对氧化脂质生物合成的反馈调节主要是通过提高 15d-PGJ_2 的生成实现的。由于其亲电特性,15d-PGJ_2 可以和 IKKβ 活性位点上硫醇形式的半胱氨酸残基发生 Michael 加成反应,导致 IKKβ 活性抑制,这样,IκB 不能被磷酸化而降解,因而抑制了促炎信号通路 NF-κB 的活化[279]。除了抑制 NF-κB 信号通路外,15d-PGJ_2 还可以抑制促炎信号通路 AP-1 和 STAT,也可以作为内源配体激活 PPARγ。PPARγ

的活化起着使促炎的M1型巨噬细胞向抗炎的M2型巨噬细胞极化的关键作用[280]。研究表明，硒依赖的15d-PGJ_2的生成通过激活PPARγ下调了LPS诱导的M1巨噬细胞标志物TNF-α和白介素-1β(IL-1β)的基因表达，上调了IL-4处理后M2巨噬细胞标志物精氨酸酶-1(Arg-1)和Fizz1的基因表达[281,282]；下调了白介素-1β诱导的COX-2和PGES的基因表达，上调了PGDS的基因表达，因而降低了PGE_2的生成，维持了15d-PGJ_2的生成[272,283]。因此，硒及其合成的硒蛋白通过提高抗炎氧化脂质15d-PGJ_2的生成反馈性地抑制了促炎信号通路，激活了抗炎信号通路，因而在维护促炎反应与抗炎反应之间的平衡方面起着重要作用。此外，表观遗传学研究发现，组蛋白的乙酰化与一些促炎基因的表达提高有关，依赖硒的15d-PGJ_2的生成可通过抑制组蛋白乙酰转移酶p300的活性，进而抑制LPS诱导的COX-2和TNF-α的基因表达[284]。如前所述，ARA经LOX催化生成HPETE，HPETE作为脂质氢过氧化物也可激活LOX。因此，硒蛋白通过将HPETE还原为HETE切断了HPETE对LOX活性的反馈促进作用。

参考文献

[1] National Research Council (NRC). Nutrient requirements of dairy cattle [M]. Washington DC: National Academies Press, 2001.

[2] Guyot H, Rollln F. The diagnosis of selenium and iodine deficiencies in cattle[J]. Annales de Medecine Veterinaire, 2007, 151: 166-191.

[3] Surai P F, Kochish I I, Fisinin V I, et al. Revisiting Oxidative Stress and the Use of Organic Selenium in Dairy Cow Nutrition[J]. Animals, 2019, 9(7): 462.

[4] Gong J, Xiao M. Effect of organic selenium supplementation on selenium status, oxidative stress, and antioxidant status in selenium-adequate dairy cows during the periparturient period[J]. Biological Trace Element Research, 2018, 186(2): 430-440.

[5] National Research Council (NRC). Nutrient requirements of beef cattle [M]. Washington, DC: National Academy Press, 1996.

[6] Spears J W, Weiss W P. Role of antioxidants and trace elements in health and immunity of transition dairy cows[J]. Veterinary Journal, 2008, 176(1): 70-76.

[7] Harrison J H, Conrad H R. Effect of Dietary Calcium on Seleni-

um Absorption by the Nonlactating Dairy Cow[J]. Journal of Dairy Science，1984，67(8)：1860-1864.

[8] Garciavaquero M，Miranda M，Benedito J L，et al. Effect of type of muscle and Cu supplementation on trace element concentrations in cattle meat[J]. Food and Chemical Toxicology，2011，49(6)：1443-1449.

[9] Villar D，Arthur J R，Gonzalez J M，et al. Selenium status in cattle：interpretation of laboratory results[J]. Bovine Practitioner，2002，36：73-80.

[10] Pavlata L，Illek J，Pechova A，et al. Selenium Status of Cattle in the Czech Republic[J]. Acta Veterinaria Brno，2002，71(1)：3-8.

[11] Zust J，Hrovatin B，Simundic B，et al. Selenium and vitamin E status of cattle in Slovenia[J]. Sodobno Kmetijstvo，1995，20：570-573.

[12] Velasquezpereira J，Dowell L M，Conrad J，et al. Mineral status of soils，forages and cattle in Nicaragua. I. Microminerals[J]. Revista De La Facultad De Agronomia De La Universidad Del Zulia，1997，14(1)：73-89.

[13] Lazarus M，Orct T，Jurasoviae J，et al. The effect of dietary selenium supplementation on cadmium absorption and retention in suckling rats[J]. BioMetals，2009，22(6)：973-983.

[14] Cortese V. Selenium and reproductive performance in dairy cattle[J]. Agri-Practice，1988，9：5-7.

[15] Cohen R D H，King B D，Guenther C，et al. Effect of pre-partum parenteral supplementation of pregnant beef cows with selenium/vitamin E on cow and calf plasma selenium and productivity[J]. Canadian Veterinary Journal-revue Veterinaire Canadienne，1991，32(2)：113-115.

[16] Meyer C L，Berger P J，Koehler K J. Interactions among factors affecting stillbirths in Holstein cattle in the United States[J]. Journal of Dairy Science，2000，83(11)：2657-2663.

[17] 惠雪，吴中红，王美芝.规模化奶牛场牛群结构的算法 [J].黑龙江畜牧兽医，2008(8)：30-32.

[18] Kolb E，Seehawer J. The importance of selenium availability and deficiency in cattle-the prevention of deficiency[J]. Tierarztlich Umschau，2001，56：263-269.

[19] Kellogg D W，Pennington J A，Johnson Z B，et al. Survey of Management Practices Used for the Highest Producing DHI Herds in the United States[J]. Journal of Dairy Science，2001，84：120-127.

[20] Kimura K, Goff J P, Kehrli M E, et al. Decreased Neutrophil Function as a Cause of Retained Placenta in Dairy Cattle[J]. Journal of Dairy Science, 2002, 85(3): 544-550.

[21] Lucey S, Rowlands G J, Russell A M. Short-term associations between disease and milk yield of dairy cows[J]. Journal of Dairy Research, 1986, 53(1): 7-15.

[22] Emanuelson U, Oltenacu P A, Grohn Y T. Nonlinear Mixed Model Analyses of Five Production Disorders of Dairy Cattle[J]. Journal of Dairy Science, 1993, 76(9): 2765-2772.

[23] Sordillo L M, Shaferweaver K A, Derosa D. Immunobiology of the mammary gland[J]. Journal of Dairy Science, 1997, 80(8): 1851-1865.

[24] Spears, Jerry W. Micronutrients and immune function in cattle [J]. Proceedings of the Nutrition Society, 2000, 59(4): 587-594.

[25] Ibeagha A E, Ibeagha-Awemu E M, Mehrzad J, et al. Selenium, immune functions and health of dairy cattle. in proceedings of the 23rd Annual Symposium of Nutritional Biotechnology in the Feed and Food Industries [C]. Nottingham, UK: Nottingham University Press, 2007.

[26] Weiss W P, Todhunter D A, Hogan J S, et al. Effect of duration of supplementation of selenium and vitamin E on preparturient dairy cows[J]. Journal of Dairy Science, 1990, 73(11): 3187-3194.

[27] Bogush A A, Ivanov V E, Pankovets E A, et al. The effectiveness of using a complex mineral preparation in the treatment and prevention of mastitis in cows, and a possibility of using the atom emission spectroscopy in the control of the maximum permitted concentrations of selenium in milk[J]. Vesti Akedemii Agrarnykh Navuk Respubliki Belarus, 2000, 4: 79-82.

[28] Harrison J H, Hancock D D. The role of selenium and vitamin E deficiency in post partum reproductive disease in bovine. in proceedings of the Alvin Lloyd Moxon honorary lectures on selenium and vitamin E [C]. Ohio State University, Ohio, USA, 1999.

[29] Malbe M, Attila M, Atroshi F. Possible involvement of selenium in Staphylococcus aureus inhibition in cow's whey[J]. Journal of Animal Physiology and Animal Nutrition, 2006, 90: 159-164.

[30] Bruzelius K, Hoac T, Sundler R, et al. Occurrence of Selenoprotein Enzyme Activities and mRNA in Bovine Mammary Tissue[J].

Journal of Dairy Science, 2007, 90(2): 918-927.

[31] Abutarbush S M, Radostits O M. Congenital nutritional muscular dystrophy in a beef calf[J]. Canadian Veterinary Journal-revue Veterinaire Canadienne, 2003, 44(9): 738-739.

[32] Ortman K. Organic vs. Inorganic selenium in farm animal nutrition with special reference to supplementation of cattle[D]. Uppsala, Sweden: Swedish University of Agricultural Sciences, 1999.

[33] Surai P F. Selenium in Nutrition and Health [M]. Nottingham. : Nottingham University Press, 2006.

[34] Youssef R H, Mohamed O M, Hassanin H A, et al. Growth performance and hormonal profile of newborn calves after pre-partum selenium injection of their dams[J]. Egyptian Journal of Agricultural Research, 2000, 78(4): 1749-1761.

[35] Bonomi A, Bonomi B M, Quarantelli A, et al. Selenium in weaning calf feeding[J]. Rivista Di Scienza Dellalimentazione, 2001, 30: 187-196.

[36] Schone F, Steinhofel O, Weigel K, et al. Selenium in feedstuffs and rations for dairy cows including a view of the food chain up to the consumer[J]. Journal fur Verbraucherschutz und Lebensmittelsicherheit-Journal of Consumer, 2013, 8(4): 271-280.

[37] Gerloff B J. Effect of selenium supplementation on dairy cattle [J]. Journal of Animal Science, 1992, 70(12): 3934-3940.

[38] Ahsan U, Kamran Z, Raza I, et al. Role of selenium in male reproduction—A review[J]. Animal Reproduction Science, 2014, 146(1): 55-62.

[39] Fairweathertait S J, Collings R, Hurst R. Selenium bioavailability: current knowledge and future research requirements[J]. The American Journal of Clinical Nutrition, 2010, 91(5): 1484-1491.

[40] Mehdi Y, Clinquart A, Hornick J, et al. Meat composition and quality of young growing Belgian Blue bulls offered a fattening diet with selenium enriched cereals[J]. Canadian Journal of Animal Science, 2015, 95(3): 465-473.

[41] Juniper D T, Phipps R H, Ramosmorales E, et al. Effect of dietary supplementation with selenium-enriched yeast or sodium selenite on selenium tissue distribution and meat quality in beef cattle[J]. Journal of Animal Science, 2008, 86(11): 3100-3109.

[42] Lawler T L, Taylor J B, Finley J W, et al. Effect of supranutritional and organically bound selenium on performance, carcass characteristics, and selenium distribution in finishing beef steers[J]. Journal of Animal Science, 2004, 82(5): 1488-1493.

[43] Maiorino M, Flohe L, Roveri A, et al. Selenium and reproduction[J]. BioFactors, 1999, 10: 251-256.

[44] Ceko M J, Hummitzsch K, Hatzirodos N, et al. X-Ray fluorescence imaging and other analyses identify selenium and GPX1 as important in female reproductive function[J]. Metallomics, 2015, 7(1): 71-82.

[45] Wu X, Yao J, Yang Z, et al. Improved fetal hair follicle development by maternal supplement of selenium at nano size (Nano-Se)[J]. Livestock Science, 2011, 142(1): 270-275.

[46] Wilde D. Influence of macro and micro minerals in the peri-parturient period on fertility in dairy cattle[J]. Animal Reproduction Science, 2006, 96: 240-249.

[47] Kommisrud E, Osteras O, Vatn T. Blood selenium associated with health and fertility in Norwegian dairy herds[J]. Acta Veterinaria Scandinavica, 2005, 46(4): 229-240.

[48] Gunter S A, Beck P A, Phillips J M. Effects of supplementary selenium source on the performance and blood measurements in beef cows and their calves[J]. Journal of Animal Science, 2003, 81(4): 856-864.

[49] Żarczyńska K, Sobiech P, Radwińska J, et al. Effects of selenium on animal health[J]. Journal of Elementology, 2012, 18(2): 329-340.

[50] Wichtel J J, Keefe G P, Van Leeuwen J A, et al. The selenium status of dairy herds in Prince Edward Island[J]. Canadian Veterinary Journal-revue Veterinaire Canadienne, 2004, 45(2): 124-132.

[51] Muniznaveiro O, Dominguezgonzalez R, Bermejobarrera A, et al. Determination of total selenium and selenium distribution in the milk phases in commercial cow's milk by HG-AAS[J]. Analytical and Bioanalytical Chemistry, 2005, 381(6): 1145-1151.

[52] Van Dael P, Vlaemynck G, Van Renterghem R, et al. Selenium content of cow's milk and its distribution in protein fractions[J]. Zeitschrift Für Lebensmittel Untersuchung Und Forschung, 1991, 192: 422-426.

[53] Muniznaveiro O, Dominguezgonzalez R, Bermejobarrera A, et

al. Selenium Content and Distribution in Cow's Milk Supplemented with Two Dietary Selenium Sources[J]. Journal of Agricultural and Food Chemistry, 2005, 53(25): 9817-9822.

[54] Ivancic J J, Weiss W P. Effect of Dietary Sulfur and Selenium Concentrations on Selenium Balance of Lactating Holstein Cows[J]. Journal of Dairy Science, 2001, 84(1): 225-232.

[55] Rowntree J E, Hill G M, Hawkins D R, et al. Effect of Se on selenoprotein activity and thyroid hormone metabolism in beef and dairy cows and calves[J]. Journal of Animal Science, 2004, 82(10): 2995-3005.

[56] Ceballosmarquez A, Barkema H W, Stryhn H, et al. Milk selenium concentration and its association with udder health in Atlantic Canadian dairy herds[J]. Journal of Dairy Science, 2010, 93(10): 4700-4709.

[57] Waegeneers N, Thiry C, De Temmerman L, et al. Predicted dietary intake of selenium by the general adult population in Belgium[J]. Food Additives and Contaminants Part A-chemistry Analysis Control Exposure & Risk Assessment, 2013, 30(2): 278-285.

[58] Choi Y, Kim J, Lee H, et al. Selenium content in representative Korean foods[J]. Journal of Food Composition and Analysis, 2009, 22(2): 117-122.

[59] Pappa E C, Pappas A C, Surai P F. Selenium content in selected foods from the Greek market and estimation of the daily intake[J]. Science of the Total Environment, 2006, 372(1): 100-108.

[60] Tinggi U, Patterson C, Reilly C. Selenium levels in cow's milk from different regions of Australia[J]. International Journal of Food Sciences and Nutrition, 2001, 52(1): 43-51.

[61] Ceballos A, Sanchez J, Stryhn H, et al. Meta-analysis of the effect of oral selenium supplementation on milk selenium concentration in cattle[J]. Journal of Dairy Science, 2009, 92(1): 324-342.

[62] Horky P. Effect of selenium on its content in milk and performance of dairy cows in ecological farming[J]. Potravinarstvo, 2015, 9(1): 324-329.

[63] Ran L, Wu X, Shen X, et al. Effects of selenium form on blood and milk selenium concentrations, milk component and milk fatty acid composition in dairy cows[J]. Journal of the Science of Food and Agriculture, 2010, 90(13): 2214-2219.

[64] Ortman K, Pehrson B. Effect of selenate as a feed supplement to dairy cows in comparison to selenite and selenium yeast[J]. Journal of Animal Science, 1999, 77(12): 3365-3370.

[65] Salman S, Dinse D, Kholparisini A, et al. Colostrum and milk selenium, antioxidative capacity and immune status of dairy cows fed sodium selenite or selenium yeast[J]. Archives of Animal Nutrition, 2013, 67(1): 48-61.

[66] Gong J, Ni L, Wang D, et al. Effect of dietary organic selenium on milk selenium concentration and antioxidant and immune status in midlactation dairy cows[J]. Livestock Science, 2014, 170: 84-90.

[67] Cook J G, Green M J. Milk production in early lactation in a dairy herd following supplementation with iodine, selenium and cobalt [J]. Veterinary Record, 2010, 167(20): 788-789.

[68] Juan G L, Jor S O, Hu C V, et al. Effects of the Selenium and Vitamin E in the Production, Physicochemical Composition and Somatic Cell Count in Milk of Ayrshire Cows[J]. Journal of Animal and Veterinary Advances, 2012, 11(5): 687-691.

[69] Machado V S, Bicalho M L S, Pereira R V V, et al. Effect of an injectable trace mineral supplement containing selenium, copper, zinc, and manganese on the health and production of lactating Holstein cows [J]. Veterinary Journal, 2013, 197(2): 451-456.

[70] Enjalbert F, Lebreton P, Salat O, et al. Effects of pre-or postpartum selenium supplementation on selenium status in beef cows and their calves[J]. Journal of Animal Science, 1999, 77(1): 223-229.

[71] Hefnawy A E G, Tórtora-Pérez J L. The importance of selenium and the effects of its deficiency in animal health[J]. Small Ruminant Research, 2010, 89: 185-192.

[72] House W A, Bell A W. Sulfur and Selenium Accretion in the Gravid Uterus During Late Gestation In Holstein Cows[J]. Journal of Dairy Science, 1994, 77(7): 1860-1869.

[73] Mcdowell L R, Valle G, Cristaldi L, et al. Selenium availability and methods of selenium supplementation for grazing ruminants. in proceedings of the 13th Annual Florida Ruminant Nutrition Symposium [C]. 2002.

[74] Castellan D M, Maas J, Gardner I A, et al. Growth of suckling

beef calves in response to parenteral administration of selenium and the effect of dietary protein provided to their dams[J]. Javma-journal of The American Veterinary Medical Association, 1999, 214(6): 816-821.

[75] Salles M S V, Zanetti M A, Junior L C R, et al. Performance and immune response of suckling calves fed organic selenium[J]. Animal Feed Science and Technology, 2014, 188: 28-35.

[76] Guyot H, Spring P, Andrieu S, et al. Comparative responses to sodium selenite and organic selenium supplements in Belgian Blue cows and calves[J]. Livestock Science, 2007, 111(3): 259-263.

[77] Schrama J W, Arieli A, Brandsma H A, et al. Thermal requirements of young calves during standing and lying[J]. Journal of Animal Science, 1993, 71(12): 3285-3292.

[78] Dargatz D A, Ross P F. Blood selenium concentrations in cows and heifers on 253 cow-calf operations in 18 states[J]. Journal of Animal Science, 1996, 74(12): 2891-2895.

[79] Longnecker M P, Stram D O, Taylor P R, et al. Use of Selenium Concentration in Whole Blood, Serum, Toenails, or Urine as a Surrogate Measure of Selenium Intake[J]. Epidemiology, 1996, 7(4): 384-390.

[80] Hall J A, Bobe G, Hunter J K, et al. Effect of Feeding Selenium-Fertilized Alfalfa Hay on Performance of Weaned Beef Calves[J]. PloS One, 2013, 8(3): 96-110.

[81] Chorfi Y, Girard V, Fournier A, et al. Effect of subcutaneous selenium injection and supplementary selenium source on blood selenium and glutathione peroxidase in feedlot heifers[J]. Canadian Veterinary Journal-revue Veterinaire Canadienne, 2011, 52(10): 1089-1094.

[82] Weiss W P, Hogan J S. Effect of Selenium Source on Selenium Status, Neutrophil Function, and Response to Intramammary Endotoxin Challenge of Dairy Cows * [J]. Journal of Dairy Science, 2005, 88(12): 4366-4374.

[83] Hall J A, Bobe G, Vorachek W R, et al. Effect of supranutritional organic selenium supplementation on postpartum blood micronutrients, antioxidants, metabolites, and inflammation biomarkers in selenium-replete dairy cows[J]. Biological Trace Element Research, 2014, 161(3): 272-287.

[84] Hidiroglou M, Jenkins K J. Effect of selenium and vitamin E,

and copper administrations on weight gains of beef cattle raised in a selenium-deficient area[J]. Canadian Journal of Animal Science, 1975, 55(3): 307-313.

[85] Awadeh F T, Kincaid R L, Johnson K. Effect of level and source of dietary selenium on concentrations of thyroid hormones and immunoglobulins in beef cows and calves[J]. Journal of Animal Science, 1998, 76(4): 1204-1215.

[86] Swecker W S, Eversole D E, Thatcher C D, et al. Influence of supplemental selenium on humoral immune responses in weaned beef calves [J]. American Journal of Veterinary Research, 1989, 50(10): 1760-1763.

[87] Netto A S, Zanetti M A, Claro G R D, et al. Copper and selenium supplementation in the diet of Brangus steers on the nutritional characteristics of meat[J]. Revista Brasileira de Zootecnia, 2013, 42(1): 70-75.

[88] Wichtel J J, Craigie A L, Freeman D A, et al. Effect of Selenium and Iodine Supplementation on Growth Rate and on Thyroid and Somatotropic Function in Dairy Calves at Pasture[J]. Journal of Dairy Science, 1996, 79(10): 1865-1872.

[89] Gleed P T, Allen W M, Mallinson C B, et al. Effects of selenium and copper supplementation on the growth of beef steers[J]. Veterinary Record, 1983, 113(17): 388-392.

[90] Galbraith M L, Vorachek W R, Estill C T, et al. Rumen Microorganisms Decrease Bioavailability of Inorganic Selenium Supplements[J]. Biological Trace Element Research, 2016, 171(2): 338-343.

[91] Behne D, Kyriakopoulos A, Scheid S, et al. Effects of Chemical Form and Dosage on the Incorporation of Selenium into Tissue Proteins in Rats[J]. Journal of Nutrition, 1991, 121(6): 806-814.

[92] Combs F, Gerald. Biomarkers of selenium status[J]. Nutrients, 2015, 7(4): 2209-2236.

[93] Kieliszek M, Blazejak S, Gientka I, et al. Accumulation and metabolism of selenium by yeast cells[J]. Applied Microbiology and Biotechnology, 2015, 99(13): 5373-5382.

[94] Bermingham E N, Hesketh J E, Sinclair B R, et al. Selenium-Enriched Foods Are More Effective at Increasing Glutathione Peroxidase (GPx) Activity Compared with Selenomethionine: A Meta-Analysis[J]. Nutrients, 2014, 6(10): 4002-4031.

[95] Dermauw V, Alonso M L, Duchateau L, et al. Trace Element Distribution in Selected Edible Tissues of Zebu (Bos indicus) Cattle Slaughtered at Jimma, SW Ethiopia[J]. PloS One, 2014, 9(1): e85300.

[96] Herdt T H, Rumbeiha W K, Braselton W E. The Use of Blood Analyses to Evaluate Mineral Status in Livestock[J]. Veterinary Clinics of North America-food Animal Practice, 2000, 16(3): 423-444.

[97] Archer J A, Judson G J. Selenium concentrations in tissues of sheep given a subcutaneous injection of barium selenate or sodium selenate[J]. Australian Journal of Experimental Agriculture, 1994, 34(5): 581-588.

[98] Joksimovictodorovic M, Davidovic V, Sretenovic L. The effect of diet selenium supplement on meat quality[J]. Biotechnology in Animal Husbandry, 2012, 28(3): 553-561.

[99] Khan M I, Min J, Lee S, et al. Cooking, storage, and reheating effect on the formation of cholesterol oxidation products in processed meat products[J]. Lipids in Health and Disease, 2015, 14(1): 89-89.

[100] Taylor J B, Marchello M J, Finley J W, et al. Nutritive value and display-life attributes of selenium-enriched beef-muscle foods[J]. Journal of Food Composition and Analysis, 2008, 21(2): 183-186.

[101] Cozzi G, Prevedello P, Stefani A, et al. Effect of dietary supplementation with different sources of selenium on growth response, selenium blood levels and meat quality of intensively finished Charolais young bulls[J]. Animal, 2011, 5(10): 1531-1538.

[102] Dokoupilova A, Marounek M, Skřivanova V, et al. Selenium content in tissues and meat quality in rabbits fed selenium yeast[J]. Czech Journal of Animal Science, 2018, 52(6): 165-169.

[103] Zhan X, Wang M, Zhao R, et al. Effects of different selenium source on selenium distribution, loin quality and antioxidant status in finishing pigs[J]. Animal Feed Science and Technology, 2007, 132(3): 202-211.

[104] Svoboda M, Salakova A, Fajt Z, et al. Selenium from Se-enriched lactic acid bacteria as a new Se source for growing-finishing pigs[J]. Polish Journal of Veterinary Sciences, 2009, 12(3): 355-361.

[105] Pereira A S C, Santos M V D, Aferri G, et al. Lipid and selenium sources on fatty acid composition of intramuscular fat and muscle selenium concentration of Nellore steers[J]. Revista Brasileira de Zootecnia, 2012, 41(11): 2357-2363.

[106] Netto A S, Zanetti M A, Claro G R D, et al. Effects of copper and selenium supplementation on performance and lipid metabolism in confined brangus bulls [J]. Asian-Australasian Journal of Animal Sciences, 2014, 27(4): 488-494.

[107] Hogan J S, Smith K L, Weiss W P, et al. Relationships among vitamin E, selenium, and bovine blood neutrophils[J]. Journal of Dairy Science, 1990, 73(9): 2372-2378.

[108] Arthur J R, Mckenzie R C, Beckett G J. Selenium in the Immune System[J]. Journal of Nutrition, 2003, 133(5): 1457-1459.

[109] Ndiweni N, Finch J M. Effects of in vitro supplementation of bovine mammary gland macrophages and peripheral blood lymphocytes with α-tocopherol and sodium selenite: implications for udder defences [J]. Veterinary Immunology and Immunopathology, 1995, 47: 111-121.

[110] Cebra C K, Heidel J R, Crisman R O, et al. The Relationship between Endogenous Cortisol, Blood Micronutrients, and Neutrophil Function in Postparturient Holstein Cows[J]. Journal of Veterinary Internal Medicine, 2003, 17(6): 902-907.

[111] Finch J M, Turner R J. Effects of selenium and vitamin E on the immune responses of domestic animals[J]. Research in Veterinary Science, 1996, 60(2): 97-106.

[112] Eulogio G L J, Hugo C V, Antonio C N, et al. Effects of the selenium and vitamin E in the production, physicochemical composition and somatic cell count in milk of Ayrshire cows[J]. J Anim Vet Adv, 2012, 11: 687-691.

[113] Latorre A O, Greghi G F, Netto A S, et al. Selenium and vitamin E enriched diet increases NK cell cytotoxicity [J]. Pesquisa Veterinária Brasileira Brazilian Journal of Veterinary Research, 2014, 34: 1141-1145.

[114] Cao Y, Maddox J F, Mastro A M, et al. Selenium Deficiency Alters the Lipoxygenase Pathway and Mitogenic Response in Bovine Lymphocytes[J]. Journal of Nutrition, 1992, 122(11): 2121-2127.

[115] Larsen H J, Moksnes K, Overnes G. Influence of selenium on antibody production in sheep[J]. Research in Veterinary Science, 1988, 45(1): 4-10.

[116] Kiremidjianschumacher L, Roy M, Wishe H I, et al. Supple-

meentation with selenium and human immune cell functions. II: effect on cutotoxic lymphocytes and natural killer cells[J]. Biological Trace Element Research, 1994, 41(1): 115-127.

[117] Pavlata L, Prasek J, Filipek J, et al. Influence of parenteral administration of selenium and vitamin E during pregnancy on selected metabolic parameters and colostrum quality in dairy cows at parturition [J]. Veterinarni Medicina, 2018, 49(5): 149-155.

[118] Kamada H, Nonaka I, Ueda Y, et al. Selenium Addition to Colostrum Increases Immunoglobulin G Absorption by Newborn Calves [J]. Journal of Dairy Science, 2007, 90(12): 5665-5670.

[119] Sordillo L M. Selenium-Dependent Regulation of Oxidative Stress and Immunity in Periparturient Dairy Cattle[J]. Veterinary Medicine International, 2013, 2013: 154045-154045.

[120] Abuelo T, Pérez-Santos M, Hernández J, et al. Effect of colostrum redox balance on the oxidative status of calves during the first 3months of life and the relationship with passive immune acquisition[J]. Veterinary Journal, 2014, 199(2): 295-299.

[121] Trevisi E, Minuti A. Assessment of the innate immune response in the periparturient cow[J]. Research in Veterinary Science, 2018, 116: 47-54.

[122] Skibiel A L, Zachut M, Amaral B C D, et al. Liver proteomic analysis of postpartum Holstein cows exposed to heat stress or cooling conditions during the dry period[J]. Journal of Dairy Science, 2018, 101 (1): 705-716.

[123] Laubenthal L, Ruda L, Sultana N, et al. Effect of increasing body condition on oxidative stress and mitochondrial biogenesis in subcutaneous adipose tissue depot of nonlactating dairy cows[J]. Journal of Dairy Science, 2017, 100: 4976-4986.

[124] Pedernera M, Celi P, García S C, et al. Effect of diet, energy balance and milk production on oxidative stress in early-lactating dairy cows grazing pasture[J]. Veterinary Journal, 2010, 186(3): 352-357.

[125] Kargar S, Ghorbani G R, Fievez V, et al. Performance, bioenergetic status, and indicators of oxidative stress of environmentally heat-loaded Holstein cows in response to diets inducing milk fat depression[J]. Journal of Dairy Science, 2015, 98(7): 4772-4784.

[126] Kuhn M J, Mavangira V, Gandy J C, et al. Production of 15-F(2t)-isoprostane as an assessment of oxidative stress in dairy cows at different stages of lactation[J]. Journal of Dairy Science, 2018, 101(10): 9287-9295.

[127] Hanschke N, Kankofer M, Ruda L, et al. The effect of conjugated linoleic acid supplements on oxidative and antioxidative status of dairy cows[J]. Journal of Dairy Science, 2016, 99(10): 8090-8102.

[128] Da Silva A D, Da Silva A S, Baldissera M D, et al. Oxidative stress in dairy cows naturally infected with the lungworm Dictyocaulus viviparus (Nematoda: Trichostrongyloidea)[J]. Journal of Helminthology, 2017, 91(4): 462-469.

[129] Fidan A F, Cingi C C, Karafakioglu Y S, et al. The levels of antioxidant activity, malondialdehyde and nitric oxide in cows naturally infected with Neospora caninum[J]. Journal of Animal and Veterinary Advances, 2010, 9(12): 1707-1711.

[130] Li Y, Ding H, Wang X, et al. An association between the level of oxidative stress and the concentrations of NEFA and BHBA in the plasma of ketotic dairy cows[J]. Journal of Animal Physiology and Animal Nutrition, 2016, 100(5): 844-851.

[131] Gabai G, De Luca E, Miotto G, et al. Relationship between Protein Oxidation Biomarkers and Uterine Health in Dairy Cows during the Postpartum Period[J]. Antioxidants, 2019, 8(1): 21.

[132] Celi P. The role of oxidative stress in small ruminants' health and production[J]. Revista Brasileira de Zootecnia, 2010, 39: 348-363.

[133] Sordillo L M, Mavangira V. The nexus between nutrient metabolism, oxidative stress and inflammation in transition cows[J]. Animal Production Science, 2014, 54(9): 1204-1214.

[134] Agarwal A, Gupta S, Sekhon L, et al. Redox Considerations in Female Reproductive Function and Assisted Reproduction: From Molecular Mechanisms to Health Implications[J]. Antioxidants & Redox Signaling, 2008, 10(8): 1375-1404.

[135] Talukder S, Kerrisk K L, Gabai G, et al. Role of oxidant-antioxidant balance in reproduction of domestic animals[J]. Animal Production Science, 2017, 57(8): 1588-1597.

[136] Gilbert R O. Symposium review: Mechanisms of disruption of

fertility by infectious diseases of the reproductive tract[J]. Journal of Dairy Science, 2019, 102(4): 3754-3765.

[137] Celi P, Merlo M, Dalt L D, et al. Relationship between late embryonic mortality and the increase in plasma advanced oxidised protein products (AOPP) in dairy cows[J]. Reproduction, Fertility and Development, 2011, 23(4): 527-533.

[138] Antoncicsvetina M, Turk R, Svetina A, et al. Lipid status, paraoxonase-1 activity and metabolic parameters in serum of heifers and lactating cows related to oxidative stress[J]. Research in Veterinary Science, 2011, 90(2): 298-300.

[139] Ling T, Hernandezjover M, Sordillo L M, et al. Maternal late-gestation metabolic stress is associated with changes in immune and metabolic responses of dairy calves[J]. Journal of Dairy Science, 2018, 101(7): 6568-6580.

[140] Han L Q, Zhou Z, Ma Y F, et al. Phosphorylation of nuclear factor erythroid 2-like 2 (NFE2L2) in mammary tissue of Holstein cows during the periparturient period is associated with mRNA abundance of antioxidant gene networks[J]. Journal of Dairy Science, 2018, 101(7): 6511-6522.

[141] Sun P, Wang J, Liu W, et al. Hydroxy-selenomethionine: A novel organic selenium source that improves antioxidant status and selenium concentrations in milk and plasma of mid-lactation dairy cows[J]. Journal of Dairy Science, 2017, 100(12): 9602-9610.

[142] Sun L L, Gao S T, Wang K, et al. E? ects of source on bioavailability of selenium, antioxidant status, and performance in lactating dairy cows during oxidative stress-inducing conditions[J]. Journal of Dairy Science, 2019, 102: 311-319.

[143] Roche J R, Bell A W, Overton T R, et al. Nutritional management of the transition cow in the 21st century-a paradigm shift in thinking [J]. Animal Production Science, 2013, 53(9): 1000-1023.

[144] Baumgard L H, Collier R J, Bauman D E. A 100-Year Review: Regulation of nutrient partitioning to support lactation[J]. Journal of Dairy Science, 2017, 100(12): 10353-10366.

[145] Holtenius K, Agenas S, Delavaud C, et al. Effects of Feeding Intensity During the Dry Period. 2. Metabolic and Hormonal Responses

[J]. Journal of Dairy Science, 2003, 86(3): 883-891.

[146] Bossaert P, Leroy J L M R, De Campeneere S, et al. Differences in the glucose-induced insulin response and the peripheral insulin responsiveness between neonatal calves of the Belgian Blue, Holstein-Friesian, and East Flemish breeds[J]. Journal of Dairy Science, 2009, 92(9): 4404-4411.

[147] Lomax M A, Baird G D, Mallinson C B, et al. Differences between lactating and non-lactating dairy cows in concentration and secretion rate of insulin[J]. Biochemical Journal, 1979, 180(2): 281-289.

[148] Komatsu T, Itoh F, Kushibiki S, et al. Changes in gene expression of glucose transporters in lactating and nonlactating cows[J]. Journal of Animal Science, 2005, 83(3): 557-564.

[149] Sano H, Narahara S, Kondo T, et al. Insulin responsiveness to glucose and tissue responsiveness to insulin during lactation in dairy cows [J]. Domestic Animal Endocrinology, 1993, 10(3): 191-197.

[150] Douglas G N, Rehage J, Beaulieu A D, et al. Prepartum nutrition alters fatty acid composition in plasma, adipose tissue, and liver lipids of periparturient dairy cows[J]. Journal of Dairy Science, 2007, 90(6): 2941-2959.

[151] Rukkwamsuk T, Geelen M J H, Kruip T a M, et al. Interrelation of Fatty Acid Composition in Adipose Tissue, Serum, and Liver of Dairy Cows During the Development of Fatty Liver Postpartum[J]. Journal of Dairy Science, 2000, 83(1): 52-59.

[152] Maedler K, Oberholzer J, Bucher P a R, et al. Monounsaturated Fatty Acids Prevent the Deleterious Effects of Palmitate and High Glucose on Human Pancreatic β-Cell Turnover and Function[J]. Diabetes, 2003, 52(3): 726-733.

[153] De Souza J, Preseault C L, Lock A L. Altering the ratio of dietary palmitic, stearic, and oleic acids in diets with or without whole cottonseed affects nutrient digestibility, energy partitioning, and production responses of dairy cows[J]. Journal of Dairy Science, 2018, 101(1): 172-185.

[154] Boden G, Shulman G I. Free fatty acids in obesity and type 2 diabetes: defining their role in the development of insulin resistance and beta-cell dysfunction[J]. European Journal of Clinical Investigation, 2002, 32: 14-23.

[155] Lepretti M, Martucciello S, Aceves M a B, et al. Omega-3 Fatty Acids and Insulin Resistance: Focus on the Regulation of Mitochondria and Endoplasmic Reticulum Stress[J]. Nutrients, 2018, 10(3): 350.

[156] Pires J a A, Souza A H, Grummer R R. Induction of Hyperlipidemia by Intravenous Infusion of Tallow Emulsion Causes Insulin Resistance in Holstein Cows[J]. Journal of Dairy Science, 2007, 90(6): 2735-2744.

[157] Pires J A, Pescara J B, Brickner A E, et al. Effects of Abomasal Infusion of Linseed Oil on Responses to Glucose and Insulin in Holstein Cows[J]. Journal of Dairy Science, 2008, 91(4): 1378-1390.

[158] Cartiff S E, Fellner V, Eisemann J H. Eicosapentanoic and docosahexanoic acids increase insulin sensitivity in growing steers[J]. Journal of Animal Science, 2013, 91(5): 2332-2342.

[159] Schooneman M G, Vaz F M, Houten S M, et al. Acylcarnitines: Reflecting or Inflicting Insulin Resistance? [J]. Diabetes, 2013, 62(1): 1-8.

[160] Aguer C, Mccoin C S, Knotts T A, et al. Acylcarnitines: potential implications for skeletal muscle insulin resistance[J]. The FASEB Journal, 2015, 29(1): 336-345.

[161] Yang Y, Sadri H, Prehn C, et al. Acylcarnitine profiles in serum and muscle of dairy cows receiving conjugated linoleic acids or a control fat supplement during early lactation[J]. Journal of Dairy Science, 2019, 102: 754-767.

[162] Rico J E, Zang Y, Haughey N J, et al. Short communication: Circulating fatty acylcarnitines are elevated in overweight periparturient dairy cows in association with sphingolipid biomarkers of insulin resistance[J]. Journal of Dairy Science, 2018, 101(1): 812-819.

[163] Boon J Y C, Hoy A J, Stark R, et al. Ceramides Contained in LDL Are Elevated in Type 2 Diabetes and Promote Inflammation and Skeletal Muscle Insulin Resistance[J]. Diabetes, 2013, 62(2): 401-410.

[164] Rico J E, Samii S S, Mathews A T, et al. Temporal changes in sphingolipids and systemic insulin sensitivity during the transition from gestation to lactation[J]. PloS One, 2015, 12(5): 554.

[165] Davis A N, Rico J E, Myers W A, et al. Circulating low-density lipoprotein ceramide concentrations increase in Holstein dairy cows

transitioning from gestation to lactation[J]. Journal of Dairy Science, 2019, 102(6): 5634-5646.

[166] Davis A N, Clegg J L, Perry C A, et al. Nutrient Restriction Increases Circulating and Hepatic Ceramide in Dairy Cows Displaying Impaired Insulin Tolerance[J]. Lipids, 2017, 52(9): 771-780.

[167] Rico J E, Giesy S L, Haughey N J, et al. Intravenous Triacylglycerol Infusion Promotes Ceramide Accumulation and Hepatic Steatosis in Dairy Cows[J]. Journal of Nutrition, 2018, 148(10): 1529-1535.

[168] Rico J E, Mathews A T, Lovett J, et al. Palmitic acid feeding increases ceramide supply in association with increased milk yield, circulating nonesterified fatty acids, and adipose tissue responsiveness to a glucose challenge[J]. Journal of Dairy Science, 2016, 99(11): 8817-8830.

[169] Mcfadden J W, Rico J E. Invited review: Sphingolipid biology in the dairy cow: The emerging role of ceramide[J]. Journal of Dairy Science, 2019, 102(9): 7619-7639.

[170] Qin N, Kokkonen T, Salin S, et al. Prepartal overfeeding alters the lipidomic profiles in the liver and the adipose tissue of transition dairy cows[J]. Metabolomics, 2017, 13(2): 21.

[171] Zachut M, Honig H, Striem S, et al. Periparturient dairy cows do not exhibit hepatic insulin resistance, yet adipose-specific insulin resistance occurs in cows prone to high weight loss[J]. Journal of Dairy Science, 2013, 96(9): 5656-5669.

[172] Kokkonen T, Taponen J, Anttila T, et al. Effect of Body Fatness and Glucogenic Supplement on Lipid and Protein Mobilization and Plasma Leptin in Dairy Cows[J]. Journal of Dairy Science, 2005, 88(3): 1127-1141.

[173] Block S S, Butler W R, Ehrhardt R A, et al. Decreased concentration of plasma leptin in periparturient dairy cows is caused by negative energy balance[J]. Journal of Endocrinology, 2001, 171(2): 339-348.

[174] Chelikani P K, Keisler D H, Kennelly J J. Response of Plasma Leptin Concentration to Jugular Infusion of Glucose or Lipid Is Dependent on the Stage of Lactation of Holstein Cows[J]. Journal of Nutrition, 2003, 133(12): 4163-4171.

[175] Ehrhardt R A, Foskolos A, Giesy S L, et al. Increased plasma leptin attenuates adaptive metabolism in early lactating dairy cows[J].

Journal of Endocrinology, 2016, 229(2): 145-157.

[176] Giesy S L, Yoon B, Currie W B, et al. Adiponectin deficit during the precarious glucose economy of early lactation in dairy cows[J]. Endocrinology, 2012, 153(12): 5834-5844.

[177] Urh C, Denisen J, Harder I, et al. Circulating adiponectin concentrations during the transition from pregnancy to lactation in high yielding dairy cows: testing the effects of farm, parity, and dietary energy level in large animal numbers[J]. Domestic Animal Endocrinology, 2019, 69: 1-12.

[178] Saremi B, Winand S, Friedrichs P, et al. Longitudinal Profiling of the Tissue-Specific Expression of Genes Related with Insulin Sensitivity in Dairy Cows during Lactation Focusing on Different Fat Depots[J]. PloS One, 2014, 9(1): e86211.

[179] Reynolds C K, Durst B, Humphries D J, et al. Visceral tissue mass in transition dairy cows[J]. Journal of Dairy Science, 2000, 83(1): 257.

[180] Reynolds C K, Aikman P C, Humphries D J, et al. Splanchnic metabolism in transition dairy cows[J]. Journal of Dairy Science, 2000, 83(1): 257.

[181] Veenhuizen J J, Russell R W, Young J W. Kinetics of metabolism of glucose, propionate and CO_2 in steers as affected by injecting phlorizin and feeding propionate[J]. Journal of Nutrition, 1988, 118(11): 1366-1375.

[182] Aiello R J, Kenna T M, Herbein J H. Hepatic Gluconeogenic and Ketogenic Interrelationships in the Lactating Cow[J]. Journal of Dairy Science, 1984, 67(8): 1707-1715.

[183] Lomax M A, Donaldson I A, Pogson C I. The effect of fatty acids and starvation on the metabolism of gluconeogenic precursors by isolated sheep liver cells[J]. Biochemical Journal, 1986, 240(1): 277-280.

[184] Armentano L E, Grummer R R, Bertics S J, et al. Effects of Energy Balance on Hepatic Capacity for Oleate and Propionate Metabolism and Triglyceride Secretion[J]. Journal of Dairy Science, 1991, 74(1): 132-139.

[185] Overton T R, Drackley J K, Douglas G N, et al. Hepatic gluconeogenesis and whole-body protein metabolism of periparturient dairy cows as affected by source of energy and intake of the prepartum diet[J].

Journal of Dairy Science, 1998, 81(1): 295.

[186] Aiello R J, Armentano L E. Gluconeogenesis in goat hepatocytes is affected by calcium, ammonia and other key metabolites but not primarily through cytosolic redox state[J]. Comparative Biochemistry and Physiology B, 1987, 88(1): 193-201.

[187] Bergman E N. Splanchnic and peripheral uptake of amino acids in relation to gut[J]. Federation Proceedings, 1986, 45(8): 2277-2282.

[188] Greenfield R B, Cecava M J, Donkin S S. Changes in mRNA expression for gluconeogenic enzymes in liver of dairy cattle during the transition to lactation[J]. Journal of Dairy Science, 2000, 83(6): 1228-1236.

[189] Bell A W, Burhans W S, Overton T R. Protein nutrition in late pregnancy, maternal protein reserves and lactation performance in dairy cows[J]. Proceedings of the Nutrition Society, 2000, 59(1): 119-126.

[190] Krehbiel C R, Harmon D L, Schneider J E. Effect of increasing ruminal butyrate on portal and hepatic nutrient flux in steers[J]. Journal of Animal Science, 1992, 70: 904-914.

[191] Bell A W. Regulation of organic nutrient metabolism during transition from late pregnancy to early lactation[J]. Journal of Animal Science, 1995, 73(9): 2804-2819.

[192] Grum D E, Drackley J K, Younker R S, et al. Nutrition During the Dry Period and Hepatic Lipid Metabolism of Periparturient Dairy Cows[J]. Journal of Dairy Science, 1996, 79(10): 1850-1864.

[193] Kunz P L, Blum J W, Hart I C, et al. Effects of different energy intakes before and after calving on food intake, performance and blood hormones and metabolites in dairy cows[J]. Animal Production, 1985, 40(02): 219-231.

[194] De Boer G, Trenkle A, Young J W. Secretion and clearance rates of glucagon in dairy cows[J]. Journal of Dairy Science, 1986, 69(3): 721-733.

[195] Patel O V, Takahashi T, Takenouchi N, et al. Peripheral cortisol levels throughoutgestation in the cow: Effect of stage of gestation and foetal number[J]. British Veterinary Journal, 1996, 152: 425-432.

[196] Tiedge M, Lortz S, Drinkgern J, et al. Relation Between Antioxidant Enzyme Gene Expression and Antioxidative Defense Status of Insulin-Producing Cells[J]. Diabetes, 1997, 46(11): 1733-1742.

[197] Maechler P, Jornot L, Wollheim C B. Hydrogen Peroxide Alters Mitochondrial Activation and Insulin Secretion in Pancreatic Beta Cells[J]. Journal of Biological Chemistry, 1999, 274(39): 27905-27913.

[198] Lei X G, Vatamaniuk M Z. Two Tales of Antioxidant Enzymes on β Cells and Diabetes[J]. Antioxidants & Redox Signaling, 2011, 14(3): 489-503.

[199] Wolf G, Aumann N, Michalska M, et al. Peroxiredoxin III protects pancreatic β cells from apoptosis[J]. Journal of Endocrinology, 2010, 207(2): 163-175.

[200] Li X, Chen H, Epstein P N. Metallothionein and Catalase Sensitize to Diabetes in Nonobese Diabetic Mice: Reactive Oxygen Species May Have a Protective Role in Pancreatic β-Cells[J]. Diabetes, 2006, 55(6): 1592-1604.

[201] Mailloux R J, Fu A, Robsondoucette C A, et al. Glutathionylation state of uncoupling protein-2 and the control of glucose-stimulated insulin secretion[J]. Journal of Biological Chemistry, 2012, 287(47): 39673-39685.

[202] Steinbrenner H. Interference of selenium and selenoproteins with the insulin-regulated carbohydrate and lipid metabolism[J]. Free Radical Biology and Medicine, 2013, 65: 1538-1547.

[203] Daha M R. Grand Challenges in Molecular Innate Immunity [J]. Frontiers in Immunology, 2011, 2: 16-16.

[204] Burton J L, Kehrli M E, Kapil S, et al. Regulation of L-selectin and CD18 on bovine neutrophils by glucocorticoids: effects of cortisol and dexamethasone[J]. Journal of Leukocyte Biology, 1995, 57(2): 317-325.

[205] Taub D D. Neuroendocrine Interactions in the Immune System [J]. Cellular Immunology, 2008, 252(12): 1-6.

[206] De Vries M A, Alipour A, Klop B, et al. Glucose-dependent leukocyte activation in patients with type 2 diabetes mellitus, familial combined hyperlipidemia and healthy controls[J]. Metabolism-clinical and Experimental, 2015, 64(2): 213-217.

[207] Lacetera N, Scalia D, Bernabucci U, et al. Lymphocyte functions in overconditioned cows around parturition[J]. Journal of Dairy Science, 2005, 88(6): 2010-2016.

[208] Mallard B A, Dekkers J C, Ireland M J, et al. Alteration in

immune responsiveness during the peripartum period and its ramification on dairy cow and calf health[J]. Journal of Dairy Science, 1998, 81(2): 585-595.

[209] Trevisi E, Zecconi A, Bertoni G, et al. Blood and milk immune and in? ammatory responses in periparturient dairy cows showing a di? erent liver activity index[J]. Journal of Dairy Research, 2010, 77: 310-317.

[210] Trevisi E, Amadori M, Cogrossi S, et al. Metabolic stress and inflammatory response in high-yielding, periparturient dairy cows[J]. Research in Veterinary Science, 2012, 93(2): 695-704.

[211] Hammon D S, Evjen I M, Dhiman T R, et al. Neutrophil function and energy status in Holstein cows with uterine health disorders[J]. Veterinary Immunology and Immunopathology, 2006, 113(1): 21-29.

[212] Bertoni G, Trevisi E, Lombardelli R. Some new aspects of nutrition, health conditions and fertility of intensively reared dairy cows[J]. Italian Journal of Animal Science, 2009, 8(4): 491-518.

[213] Kehrli M E, Nonnecke B J, Roth J A. Alterations in bovine neutrophil function during the periparturient period[J]. American Journal of Veterinary Research, 1989, 50(2): 207-214.

[214] Goff J P, Horst R L. Physiological Changes at Parturition and Their Relationship to Metabolic Disorders[J]. Journal of Dairy Science, 1997, 80(7): 1260-1268.

[215] Jahan N, Minuti A, Trevisi E. Assessment of immune response in periparturient dairy cows using ex vivo whole blood stimulation assay with lipopolysaccharides and carrageenan skin test[J]. Veterinary Immunology and Immunopathology, 2015, 165(3): 119-126.

[216] Bertoni G, Trevisi E. Use of the Liver Activity Index and Other Metabolic Variables in the Assessment of Metabolic Health in Dairy Herds[J]. Veterinary Clinics of North America-food Animal Practice, 2013, 29(2): 413-431.

[217] Bionaz M, Trevisi E, Calamari L, et al. Plasma paraoxonase, health, inflammatory conditions, and liver function in transition dairy cows[J]. Journal of Dairy Science, 2007, 90(4): 1740-1750.

[218] Trevisi E, Amadori M, Bakudila A, et al. Metabolic changes in dairy cows induced by oral, low-dose interferon-alpha treatment[J]. Journal of Animal Science, 2009, 87(9): 3020-3029.

[219] Trevisi E, Jahan N, Bertoni G, et al. Pro-inflammatory cytokine profile in dairy cows: consequences for new lactation[J]. Italian Journal of Animal Science, 2015, 14(3): 3862.

[220] Minuti A, Jahan N, Picciolicapelli F, et al. 1072 Evaluation of immune function of circulating leukocytes during the transition period in dairy cows[J]. Journal of Animal Science, 2016, 94: 513-514.

[221] Sordillo L M, Raphael W. Significance of Metabolic Stress, Lipid Mobilization, and Inflammation on Transition Cow Disorders[J]. Veterinary Clinics of North America Food Animal Practice, 2013, 29(2): 267-278.

[222] Sch? nfeld P, Wojtczak L. Fatty acids as modulators of the cellular production of reactive oxygen species[J]. Free Radical Biology and Medicine, 2008, 45(3): 231-241.

[223] Aitken S, Karcher E, Rezamand P, et al. Evaluation of antioxidant and proinflammatory gene expression in bovine mammary tissue during the periparturient period[J]. Journal of Dairy Science, 2009, 92(2): 589-598.

[224] Sordillo L M, Oboyle N J, Gandy J C, et al. Shifts in Thioredoxin Reductase Activity and Oxidant Status in Mononuclear Cells Obtained from Transition Dairy Cattle[J]. Journal of Dairy Science, 2007, 90(3): 1186-1192.

[225] Martinez N, Risco C A, Lima F S, et al. Evaluation of peripartal calcium status, energetic profile, and neutrophil function in dairy cows at low or high risk of developing uterine disease[J]. Journal of Dairy Science, 2012, 95(12): 7158-7172.

[226] Waldron R M, Nonnecke B J, Nishida T, et al. E? ect of lipopolysaccharide infusion on serum macromineral and vitamin D concentrations in dairy cows[J]. Journal of Dairy Science, 2003, 86(11): 3440-3446.

[227] Minuti A, Ahmed S, Trevisi E, et al. Experimental acute rumen acidosis in sheep: Consequences on clinical, rumen, and gastrointestinal permeability conditions and blood chemistry[J]. Journal of Animal Science, 2014, 92(9): 3966-3977.

[228] Eckel E F, Ametaj B N. Invited review: Role of bacterial endotoxins in the etiopathogenesis of periparturient diseases of transition dairy

cows[J]. Journal of Dairy Science, 2016, 99(8): 5967-5990.

[229] Griffiths H R, Gao D, Pararasa C. Redox regulation in metabolic programming and inflammation[J]. Redox biology, 2017, 12: 50-57.

[230] Boosalis M G. The role of selenium in chronic disease[J]. Nutrition in Clinical Practice, 2008, 23(2): 152-160.

[231] Sordillo L M, Contreras G A, Aitken S L. Metabolic factors affecting the inflammatory response of periparturient dairy cows[J]. Animal Health Research Reviews, 2009, 10(1): 53-63.

[232] Mozaffarian D, Benjamin E J, Go A S, et al. Heart disease and stroke statistics-2016 update a report from the American Heart Association[J]. Circulation, 2016, 133(4): e38-e48.

[233] Zhou Z, Loor J J, Piccioli-Cappelli F, et al. Circulating amino acids during the peripartal period in cows with di? erent liver functionality index[J]. Journal of Dairy Science, 2016, 99: 1-11.

[234] Soriani N, Trevisi E, Calamari L. Relationships between rumination time, metabolic conditions and health status in dairy cows during the transition period. [J]. Journal of Animal Science, 2012, 90: 4544-4554.

[235] Trevisi E, Moscati L, Amadori M. Disease-Predicting and prognostic potential of innate immune responses to noninfectious stressors: human and animal models[A]. Amadori M. The Innate Immune Response to Noninfectious Stressors[C]. The Netherland: Elsevier Inc. 2016: 209-135.

[236] Banning A, Florian S, Deubel S, et al. GPx2 counteracts PGE2 production by dampening COX-2 and mPGES-1 expression in human colon cancer cells[J]. Antioxidants & Redox Signaling, 2008, 10(9): 1491-1500.

[237] Flores-Mateo G, Navas-Acien A, Pastor-Barriuso R, et al. Selenium and coronary heart disease: a meta-analysis[J]. The American Journal of Clinical Nutrition, 2006, 84(4): 762-773.

[238] Aitken S L, Karcher E L, Rezamand P, et al. Evaluation of antioxidant and proinflammatory gene expression in bovine mammary tissue during the periparturient period[J]. Journal of Dairy Science, 2009, 92(2): 589-598.

[239] Cheng A W M, Stabler T V, Bolognesi M, et al. Selenomethionine inhibits IL-1β inducible nitric oxide synthase (iNOS) and cyclooxy-

genase 2 (COX2) expression in primary human chondrocytes[J]. Osteoarthritis and Cartilage, 2011, 19(1): 118-125.

[240] Lei Y, Wang K, Deng L, et al. Redox Regulation of Inflammation: Old Elements, a New Story[J]. Medicinal Research Reviews, 2015, 35(2): 306-340.

[241] Bi C, Wang H, Wang Y, et al. Selenium inhibits Staphylococcus aureus-induced inflammation by suppressing the activation of the NF-κB and MAPK signalling pathways in RAW264. 7 macrophages[J]. European Journal of Pharmacology, 2016, 780: 159-165.

[242] Zhu C, Ling Q, Cai Z, et al. Selenium-Containing Phycocyanin from Se-Enriched Spirulina platensis Reduces Inflammation in Dextran Sulfate Sodium-Induced Colitis by Inhibiting NF-κB Activation[J]. Journal of Agricultural and Food Chemistry, 2016, 64(24): 5060-5070.

[243] Stoedter M, Renko K, Hog A, et al. Selenium controls the sex-specific immune response and selenoprotein expression during the acute-phase response in mice[J]. Biochemical Journal, 2010, 429(1): 43-51.

[244] Khoso P A, Yang Z, Liu C, et al. Selenoproteins and heat shock proteins play important roles in immunosuppression in the bursa of Fabricius of chickens with selenium deficiency[J]. Cell Stress & Chaperones, 2015, 20(6): 967-978.

[245] Tsuji P A, Carlson B A, Anderson C B, et al. Dietary Selenium Levels Affect Selenoprotein Expression and Support the Interferon-γ and IL-6 Immune Response Pathways in Mice[J]. Nutrients, 2015, 7(8): 6529-6549.

[246] Hoffmann F W, Hashimoto A C, Shafer L A, et al. Dietary Selenium Modulates Activation and Differentiation of CD4+ T Cells in Mice through a Mechanism Involving Cellular Free Thiols[J]. Journal of Nutrition, 2010, 140(6): 1155-1161.

[247] Lee B C, Lee S G, Choo M, et al. Selenoprotein MsrB1 promotes anti-inflammatory cytokine gene expression in macrophages and controls immune response in vivo[J]. Scientific Reports, 2017, 7(1): 5119.

[248] Lee D H, Son D J, Park M H, et al. Glutathione peroxidase 1 deficiency attenuates concanavalin A-induced hepatic injury by modulation of T-cell activation[J]. Cell Death and Disease, 2016, 7(4): 0.

[249] Won H Y, Sohn J H, Min H J, et al. Glutathione Peroxidase 1

Deficiency Attenuates Allergen-Induced Airway Inflammation by Suppressing Th2 and Th17 Cell Development[J]. Antioxidants & Redox Signaling, 2010, 13(5): 575-587.

[250] Curran J E, Jowett J B M, Elliott K S, et al. Genetic variation in selenoprotein S influences inflammatory response[J]. Nature Genetics, 2005, 37(11): 1234-1241.

[251] Verma S, Hoffmann F W, Kumar M, et al. Selenoprotein K Knockout Mice Exhibit Deficient Calcium Flux in Immune Cells and Impaired Immune Responses[J]. Journal of Immunology, 2011, 186(4): 2127-2137.

[252] Fan R, Yao H, Cao C, et al. Gene Silencing of Selenoprotein K Induces Inflammatory Response and Activates Heat Shock Proteins Expression in Chicken Myoblasts[J]. Biological Trace Element Research, 2017, 180(1): 135-145.

[253] Sordillo L M, Streicher K L, Mullarky I K, et al. Selenium inhibits 15-hydroperoxyoctadecadienoic acid-induced intracellular adhesion molecule expression in aortic endothelial cells[J]. Free Radical Biology and Medicine, 2008, 44(1): 34-43.

[254] Wang Y, Armando A M, Quehenberger O, et al. Comprehensive ultra-performance liquid chromatographic separation and mass spectrometric analysis of eicosanoid metabolites in human samples[J]. Journal of Chromatography A, 2014, 1359: 60-69.

[255] Gil A. Polyunsaturated fatty acids and inflammatory diseases [J]. Biomedicine and Pharmacotherapy, 2002, 56(8): 388-396.

[256] Mattmiller S, Carlson B A, Sordillo L. Regulation of inflammation by selenium and selenoproteins: impact on eicosanoid biosynthesis [J]. Journal of Nutritional Science, 2013, 2: 1-13.

[257] Marnett L J, Rowlinson S W, Goodwin D C, et al. Arachidonic acid oxygenation by COX-1 and COX-2 Mechanisms of catalysis and inhibition[J]. Journal of Biological Chemistry, 1999, 274(33): 22903-22906.

[258] Cook H W, Lands W E. Mechanism for suppression of cellular biosynthesis of prostaglandins[J]. Nature, 1976, 260(5552): 630-632.

[259] Hampel G, Watanabe K, Weksler B B, et al. Selenium deficiency inhibits prostacyclin release and enhances production of platelet activating factor by human endothelial cells[J]. Biochimica et Biophysica Ac-

ta (BBA)-Lipids and Lipid Metabolism, 1989, 1006(2): 151-158.

[260] Smith W L, Lands W E. Oxygenation of polyunsaturated fatty acids during prostaglandin biosynthesis by sheep vesicular glands[J]. Biochemistry, 1972, 11(17): 3276-3285.

[261] Walther M, Holzhütter H G, Kuban R J, et al. The inhibition of mammalian 15-lipoxygenases by the anti-inflammatory drug ebselen: dual-type mechanism involving covalent linkage and alteration of the iron ligand sphere[J]. Molecular Pharmacology, 1999, 56(1): 196-203.

[262] Burkert E, Arnold C, Hammarberg T, et al. The C2-like β-barrel domain mediates the Ca^{2+}-dependent resistance of 5-lipoxygenase activity against inhibition by glutathione peroxidase-1[J]. Journal of Biological Chemistry, 2003, 278(44): 42846-42853.

[263] Schnurr K, Belkner J, Ursini F, et al. The selenoenzyme phospholipid hydroperoxide glutathione peroxidase controls the activity of the 15-lipoxygenase with complex substrates and preserves the specificity of the oxygenation products[J]. Journal of Biological Chemistry, 1996, 271(9): 4653-4658.

[264] Seiler A, Schneider M, Förster H, et al. Glutathione peroxidase 4 senses and translates oxidative stress into 12/15-lipoxygenase dependent-and AIF-mediated cell death[J]. Cell Metabolism, 2008, 8(3): 237-248.

[265] Ho Y S, Magnenat J L, Bronson R T, et al. Mice deficient in cellular glutathione peroxidase develop normally and show no increased sensitivity to hyperoxia[J]. Journal of Biological Chemistry, 1997, 272(26): 16644-16651.

[266] Imai H, Narashima K, Arai M, et al. Suppression of leukotriene formation in RBL-2H3 cells that overexpressed phospholipid hydroperoxide glutathione peroxidase[J]. Journal of Biological Chemistry, 1998, 273(4): 1990-1997.

[267] Rock C, Moos P J. Selenoprotein P protects cells from lipid hydroperoxides generated by 15-LOX-1[J]. Prostaglandins, Leukotrienes and Essential Fatty Acids (PLEFA), 2010, 83(4-6): 203-210.

[268] Bjornstedt M, Hamberg M, Kumar S, et al. Human thioredoxin reductase directly reduces lipid hydroperoxides by NADPH and selenocystine strongly stimulates the reaction via catalytically generated sele-

nols[J]. Journal of Biological Chemistry, 1995, 270(20): 11761-11764.

[269] Zamamiri-Davis F, Lu Y, Thompson J T, et al. Nuclear factor-κB mediates over-expression of cyclooxygenase-2 during activation of RAW 264.7 macrophages in selenium deficiency[J]. Free Radical Biology and Medicine, 2002, 32(9): 890-897.

[270] Vunta H, Davis F, Palempalli U D, et al. The anti-inflammatory effects of selenium are mediated through 15-deoxy-Δ12, 14-prostaglandin J2 in macrophages[J]. Journal of Biological Chemistry, 2007, 282(25): 17964-17973.

[271] Hwang J T, Kim Y M, Surh Y J, et al. Selenium regulates cyclooxygenase-2 and extracellular signal-regulated kinase signaling pathways by activating AMP-activated protein kinase in colon cancer cells[J]. Cancer Research, 2006, 66(20): 10057-10063.

[272] Gandhi U H, Kaushal N, Ravindra K C, et al. Selenoprotein-dependent up-regulation of hematopoietic prostaglandin D2 synthase in macrophages is mediated through the activation of peroxisome proliferator-activated receptor (PPAR) γ[J]. Journal of Biological Chemistry, 2011, 286(31): 27471-27482.

[273] Kaushal N, Kudva A K, Patterson A D, et al. Crucial role of macrophage selenoproteins in experimental colitis[J]. The Journal of Immunology, 2014, 193(7): 3683-3692.

[274] Panicker S, Swathy S S, John F, et al. Impact of Selenium on the Leukotriene B 4 Synthesis Pathway during Isoproterenol-Induced Myocardial Infarction in Experimental Rats[J]. Inflammation, 2012, 35(1): 74-80.

[275] Mattmiller S A, Carlson B A, Gandy J C, et al. Reduced macrophage selenoprotein expression alters oxidized lipid metabolite biosynthesis from arachidonic and linoleic acid[J]. The Journal of nutritional biochemistry, 2014, 25(6): 647-654.

[276] Kim S H, Johnson V J, Shin T Y, et al. Selenium attenuates lipopolysaccharide-induced oxidative stress responses through modulation of p38 MAPK and NF-κB signaling pathways[J]. Experimental Biology and Medicine, 2004, 229(2): 203-213.

[277] Brigelius-Flohe R, Maurer S, Lötzer K, et al. Overexpression of PHGPx inhibits hydroperoxide-induced oxidation, NFκB activation and

apoptosis and affects oxLDL-mediated proliferation of rabbit aortic smooth muscle cells[J]. Atherosclerosis, 2000, 152(2): 307-316.

[278] Wenk J, Schüller J, Hinrichs C, et al. Overexpression of phospholipid-hydroperoxide glutathione peroxidase in human dermal fibroblasts abrogates UVA irradiation-induced expression of interstitial collagenase/matrix metalloproteinase-1 by suppression of phosphatidylcholine hydroperoxide-mediated NF-κB activation and interleukin-6 release[J]. Journal of Biological Chemistry, 2004, 279(44): 45634-45642.

[279] Brigelius-Flohe R, Flohe L. Selenium and redox signaling[J]. Archives of Biochemistry and Biophysics, 2017, 617: 48-59.

[280] Ricote M, Li A C, Willson T M, et al. The peroxisome proliferator-activated receptor-γ is a negative regulator of macrophage activation [J]. Nature, 1998, 391(6662): 79.

[281] Nelson S M, Lei X, Prabhu K S. Selenium levels affect the IL-4-induced expression of alternative activation markers in murine macrophages[J]. The Journal of nutrition, 2011, 141(9): 1754-1761.

[282] Nelson S M, Shay A E, James J L, et al. Selenoprotein expression in macrophages is critical for optimal clearance of parasitic helminth Nippostrongylus brasiliensis[J]. Journal of Biological Chemistry, 2016, 291(6): 2787-2798.

[283] Mendez M, Lapointe M C. PPARγ inhibition of cyclooxygenase-2, PGE2 synthase, and inducible nitric oxide synthase in cardiac myocytes[J]. Hypertension, 2003, 42(4): 844-850.

[284] Narayan V, Ravindra K C, Liao C, et al. Epigenetic regulation of inflammatory gene expression in macrophages by selenium[J]. The Journal of nutritional biochemistry, 2015, 26(2): 138-145.